T0138293

The Ethnobotany of Eden

The Ethnobotany of Eden

Rethinking the Jungle Medicine Narrative

ROBERT A. VOEKS

The University of Chicago Press   Chicago and London

The University of Chicago Press, Chicago 60637
The University of Chicago Press, Ltd., London
© 2018 by The University of Chicago
Published 2018
Printed in the United States of America

27 26 25 24 23 22   2 3 4 5

ISBN-13: 978-0-226-54771-8 (cloth)
ISBN-13: 978-0-226-54785-5 (e-book)
DOI: https://doi.org/10.7208/chicago/9780226547855.001.0001

Library of Congress Cataloging-in-Publication Data

Names: Voeks, Robert A., 1950– author.
Title: The ethnobotany of Eden : rethinking the jungle medicine
    narrative / Robert A. Voeks.
Description: Chicago ; London : The University of Chicago Press, 2018. |
    Includes bibliographical references and index.
Identifiers: LCCN 2017042871 | ISBN 9780226547718 (cloth : alk. paper) |
    ISBN 9780226547855 (e-book)
Subjects: LCSH: Ethnobotany—Tropics. | Traditional medicine—Tropics.
Classification: LCC GN476.73 .V64 2018 | DDC 581.6/34—dc23
LC record available at https://lccn.loc.gov/2017042871

♾ This paper meets the requirements of ANSI/NISO Z39.48–1992
(Permanence of Paper).

For Kai, Bobby, Owen, and Leif, my favorite forest companions

Contents

Preface

The lands and peoples of the humid tropics have long constituted more myth and metaphor than geographic reality. In the imaginations of the ancient Greeks, the burning rays of the equatorial Sun spawned boiling seas, scorched landscapes, and monstrous races of semi-humans. Following the Columbian encounter, such armchair scholarship yielded to direct observations by settlers, scientists, and men of the cloth. But colonial empiricism in almost every instance passed through a filter of European prejudice and preconception. As a consequence, tropical landscapes and their indigenous forest-dwellers were saddled from the beginning with a collection of culturally constructed and often romanticized images, pivoting according to the needs of the narrator from virgin to defiled, sublime to horrific, and salubrious to disease-ridden. Among these was the idea that the biblical Garden of Eden, God's sacred oasis of perpetual spring, healing leaves, and life everlasting, was hidden deep in the primordial rainforest. Although the belief in a material Paradise proved illusory, the notion that tropical forests and fields were brimming with nature's mysterious medicinal plants has gained almost mythical standing over the centuries.

In the late twentieth century, as the Western world became aware of the ongoing destruction of the world's tropical forests, and as fear of the effects of HIV/AIDS and cancer peaked, a compelling environmental narrative appeared. The "jungle medicine narrative" coupled the perceived medicinal value of tropical forests with an array of complementary cultural, biological, and economic components. The narrative drew its inspiration from long-held myths about the nature of tropical lands and people, as well as the legitimacy of Western science and medicine. This book explores the perennating pieces of this evolving narrative—pristine nature, noble natives, mysterious shamans, miracle drug plants, and biopirates—in the context of current theory and practice in the field of ethnobotany. Taken together, the jungle medicine narrative resonated on a whole series of emotional and intellectual levels with its intended audience. Like other environmental narratives, it sought to translate complex science and social science into a comprehensible story. And it was a good story. But like most good stories, the jungle medicine narrative didn't let pesky facts and contrary evidence stand in the way of its big picture message. And like other environmental narratives crafted by outsiders, this one was always about us and our worries, not the monumental challenges facing indigenous forest dwellers.

The ideas presented here gestated slowly over many years. Some were gleaned from archival sources, others from the current literature, but most were stumbled upon through first-hand experience in the forests and fields of Brazil, northern Borneo, and most recently Mozambique. I owe a huge debt to my mentors, colleagues, and students, who helped me along this path of discovery. I thank my early Brazilian colleagues, especially Andre Mauricio de Carvalho, Sergio da Vinha, Louis Alberto Mattos e Silva, and Talmon Santos. It was from these botanists that I began to appreciate that much of what outsiders thought about tropical nature was a product of their fertile imaginations. I am likewise indebted to the wise counsel provided by Pierre Verger and Rui Póvoas on all matters dealing with the healing traditions and flora of the African Diaspora in Brazil. I benefitted greatly during those early years from the tolerance of my field collaborators, in particular the Candomblé community in Salvador, Ilhéus and Itabuna, Bahia. I thank Angela Leony for introducing me to the spectacular landscapes of the Chapada Diamantina National Park. Together we explored the question of ethnobotanical erosion under the mentorship of local elders, including the late Dona Belinha and Dona Senhorinha. Research in the park and nearby environs benefitted considerably from the knowledge and friendship provided by Roy and Ligia Funch, who continue to share my enthusiasm for Bahia's natural and cultural landscapes. Finally, I thank Ulysses Albuquerque and Rômulo Alves, both of whom brought important insights into how I perceive tropical people-plant relations, and who are actively blazing the frontier of Brazilian ethnobiology.

I had the opportunity to work with several Borneo specialists and indigenous elders while living and working in Brunei Darussalam. I am especially indebted to linguist Peter Sercombe, who was an outstanding collaborator and mentor on all things dealing with the Penan and Iban culture, and who was a fine backcountry companion. I thank also Samhan bin Nyawa, who introduced me to the Brunei Dusun, and who championed the cause of salvaging the Dusun's ethnobotany, as well as my colleague Francis Jana Lian, who introduced me to Kenyah culture. I thank all of my Penan, Kenyah, and Dusun research participants, as well as the students in my geography classes at Universiti Brunei Darussalam, who taught me at least as much as I taught them.

The ideas presented here were fine-tuned over the years through informal conversations with friends and colleagues. Among these, I wish to acknowledge Tinde van Andel, Herbert Baker, Mike Balick, Peter Becker, Curt Blondell, Rainer Bussmann, Judith Carney, John Carroll, Maria Fadiman, Susan Flaming, Fabiana Fonseca, Clarence Glacken, Aline Gregorio, Bruce Hoffman, Mark Merlin, Dan Moerman, James Parsons, Jeanine Pfeiffer, Tim Plowman, Erica Moret, Naveen Qureshi, Morteza Rahmatian, John Rashford, Anna Ribeiro, Bruna Santana, Annae Senkoro, James Sera, Leaa Short, Denise Stanley, Alexis Stavropoulos, Rick Stepp, Hilgard Sternberg, Scott Stine, Ina Vandebroek, Peter Vorster, and Case Watkins.

I thank the Society for Economic Botany for entrusting to me the editorship of their flagship journal, *Economic Botany*, these past eight years. I have learned so much from the variety of scholarship published, and from the unsung heroes of scholarly journals—the associate editors and peer reviewers.

The staff at several public and private archives provided invaluable assistance on this project. In Brazil, I thank the staff at the Archivo Público do Estado da Bahia in Salvador, the Biblioteca Nacional do Brasil in Rio de Janeiro, and the Instituto Histórico e Geográfico do Brasil in Rio. I am hugely indebted to Charlotte Greene, who did the archival research and French translations at the Muséum National d'Historie Naturelle in Paris. I also thank the staff at the Huntington Gardens and Library in Pasadena for access to their amazing resources. I gratefully acknowledge Stan Alpern for sharing a number of hard to find references and translations provided by early visitors to West Africa. I thank Susan Flaming and Ina Vandebroek, as well as three anonymous reviewers, for their insightful comments. And special appreciation goes to Kelly Donovan, who prepared nearly all of the figures and maps in this book.

Funding for the field and archive research was provided by grants from the National Institutes of Health—National Library of Medicine, the National Science Foundation, the National Geographic Society, the National Endowment for the Humanities, the Huntington Library (San Andreas Fellowship), and the Fulbright Scholar Program to Brazil and Mozambique. I also benefited considerably from the various seed grants supplied by California State University, Fullerton over the years, particularly from their Junior-Senior Grants Program.

ONE

God's Medicine Chest

The world "is not round as they describe it," proclaimed Christopher Columbus in his 1498 letter to his sovereign Queen Isabella, but rather "the shape of a pear which is everywhere very round except where the stalk is . . . like a woman's nipple . . . [and it] is the highest and nearest to the sky . . . beneath the equinoctial line" (Columbus 1932 [1498], 30). Still confusing his encounter with the Americas as a shortcut to fabled India, Admiral Columbus was now ready to impart his latest geographical delusion—that the long-lost site of the biblical Garden of Eden lay perched "on the very top" of this earthly protuberance deep in the rainforests of Venezuela's terra incognita. For Christians, Jews, and Muslims, Eden was imagined as an emerald paradise, blessed with a bounty of food and water, where the climate was forever balmy and spring-like. Flowers and fruit grew in abundance, including roses and violets associated with the Virgin Mary, and cherries symbolic of the Passion of Christ (Delumeau 1995, 123–127). It was also, as early biblical scholars and poets argued, brimming with healing plants—a vast and sacred pharmacopoeia of medicinal leaves and spices. The Garden of Eden was God's medicine chest, and Columbus had located it squarely in the bosom of tropical America.

Columbus was certainly not the first to divine the physical location of Paradise, but he may have been the first to situate it in the equatorial latitudes. Indeed, before his first voyage of discovery, it had been assumed since the time of the ancient Greeks that the lands bordering the Equator were too scorched by the Sun to be inhabited by humans. Mesopotamia seemed a better choice, or perhaps Persia (Friedman 1981, 9–21). But Columbus thought otherwise. For evidence of his nipple hypothesis, he described how erosion driven by the fabled four rivers of Paradise had shaped the archipelago of islands off the northern coast of South America. He also drew on long-held ethnographic stereotypes of tropical lands and peoples, reasoning that at this low latitude in distant West Africa the people were black and the climate too hot for Paradise. In his favored South American Eden, however, the natives were lighter skinned, and "shrewder and have greater intelligence and are not cowards" (Morison 1963, 277–279; Columbus 1932 [1498], 32), sure signals in his imagination that the mythical Garden must be near at hand.

The quest for Eden in the tropical forest did not end with Columbus, and in so many respects, it has not ended yet. For those of us who were raised in the concrete jungles of the temperate latitudes, these distant sylvan landscapes of Africa, Asia, Oceania, and the Americas continue to conjure images rooted in legends and preconceptions of the misty past, oscillating from noble and harmonious to chaotic and terrifying. For some, they are habitats of glorious biological beauty, nature's greatest evolutionary experiment expressing itself in magical and seemingly endless variety. Where else could surreally plumed resplendent quetzals ensconce in the trunks of dead trees, or rhinoceros hornbills haul their massive beaks and casques over the forest canopy, or birds of paradise perform their absurd dance of passerine passion? Where else could leaf cutter ants diligently tend their gardens of fungus, or caterpillars at one moment imitate squishy bird droppings and an instant later transform into a faux snake's head, complete with ballooning forked tongue? Where else do rhinoceros beetles grow to the size of bullfrogs, or do miniscule frogs fail to reach the size of a tiny button? And where else do so many species of birds and bats and trees and insects coexist and co-depend in such a restricted patch of real estate?

Closely shadowing these biological narratives are homages to the rainforest's fragile ecological harmony, plants and animals hopelessly intertwined through eon-aged, coevolved unions. Obscenely showy and aromatic flowers entice winged flower visitors, while succulent fruits lure sugar-craving birds and bats and monkeys. Green plants wage chemical warfare against the unending onslaught of herbivorous insects, while delicate symbiotic fungi form occult, subsurface networks of connections between rooted photosynthetic organisms. So specialized are some plants and animals in their behavior, and so mutually dependent are their food webs, that some researchers would describe these interactions as ecologically stable only so long as the individual elements of the system remain intact. Permit the extinction of one "keystone" species, many believe, such as a strangler fig or a specialized seed disperser, and the ecosystem experiences a domino effect of species loss, tumbling into simplified biological homogeneity. Add to these perceptions of forest instability the purported weakness and sterility of their soils, once cleared hardening into brick-red deserts of laterite, never again to sustain lofty evergreen forest, and the image of an equatorial earthly Eden, desperately in need of managerial intervention by well-intentioned outsiders, takes on a life of its own.

Images of native peoples and their forested homes are today as culturally constructed as they were in the time of Columbus. However primitive they may be technologically, indigenous people of the tropics are now most often depicted as noble stewards of their arborescent abodes, protecting their primordial homelands from the onslaught of loggers, planters, and cattle barons. Once derided as hairy-knuckled despoilers of useful tropical resources, eking out a meager subsistence through senseless slash-and-burn methods, indigenous forest folk are now seen as supremely adapted to the limitations imposed by the enervating climate and depauperate soils. Shifting cultivation (swidden) is now viewed as the most appropriate means of sustainably managing the forest and feeding families, and of contributing to the patchwork of successional habitats so crucial to species diversity. And the harvest of wild fibers and fruits and fuelwood for home and commerce is depicted increasingly by environmental scientists and aid agencies as a vital strategy for improving the nutrition and economic wellbeing of forest peoples, and encouraging meaningful forest conservation.

But there are other less sublime narratives forthcoming from tropical forests and fields, just as deeply ingrained, and just as culturally constructed. For while these landscapes manifest ecological wonder and protean biodiversity for some, for others they constitute landscapes of debilitating disease and nightmarish biotic afflictions. As the cradle of the planet's most lethal arsenal of microbes—malaria, smallpox, Ebola, break-bone fever, Zika, and so many others—the equatorial latitudes have long been considered the "white man's graveyard" as well as the merciless destroyer of untold millions of native peoples. Outside of the eradication of smallpox, most of these infectious maladies have proved immune to the efforts of Western biomedicine, afflicting twenty-first-century visitors and locals just as they did their millions of forebears.

The Jungle Medicine Narrative

Among the myriad stories and metaphors that have appeared over the centuries regarding the lands and peoples of the equatorial latitudes, none has proved more compelling in recent decades than the "jungle medicine narrative." It's a simple plot that evolved organically in the 1980s, with a compelling cast of heroes and villains, conflicts and noble causes. The story line goes something like this—tropical forests are pristine, largely unknown to science, and home to mysterious and wise native people who are privy to their great botanical secrets. Among these secrets are miracle-cure medicinal plants known and dispensed only by indigenous shamans and herbalists. Forest pharmacopoeias have the potential to cure society's most horrific diseases, but they are imperiled by the forces of globalization as well as unsustainable harvest to meet distant commercial markets. The plot thickens with the entrance of antagonists, especially foreign pharmaceutical corporations and their ethnobotanist minions, who are hell-bent on pilfering and patenting these tropical treasures. Ask ten reasonably well-informed adults why we should be concerned with saving tropical forests, and half are likely to mention the impending loss of medicinal drug plants. "What a crime it is," many will argue with passion, "that the world's natural medicine chest is being destroyed before humans can rescue its wondrous medical miracles." If this claim seems at all exaggerated, consider the pronouncement in a trade journal by a pharmacist and latter-day shaman that "besides being rich with an overpowering verdant fecundity and colorful wildlife, the rain forest holds secrets that could change the course of medicine as we know it" (Grauds 1997, 44). Or the words of a well-respected botanist, "Is it an impossible dream to hope that through medicinal plants the biodiversity of tropical forests might be able to save the world from cancer or AIDS and at the same time contribute to its own salvation?" (Gentry 1993, 21). Or a widely read ethnobotanist's pronouncement that "I believe they [shamans] are our greatest hope for finding cures to currently incurable diseases (cancer, AIDS, the common cold)" (Plotkin 1993, 14). Such enthusiastic prophecies are less often forthcoming from scientists nowadays, but they continue to be deployed by environmental groups, New Age books, and websites. One recently stated that "if there is a cure to cancer or AIDS, it will be found within the rainforest's biodiversity," and that the Amazon region harbors "new drugs still awaiting discovery—drugs for AIDS, cancer, diabetes, arthritis and Alzheimer's" (http://www.rain-tree.com/). Exaggerated, perhaps. But a host of useful and, yes, some miraculous medicinal drug plants have been uncovered in the tropical forests of Asia, Africa, and the Americas over the past few centuries, and these bioprospecting efforts have been going on more or less continuously since the early sixteenth century. But the pervasiveness of the jungle medicine narrative and its emotional resonance is a new phenomenon, having arrived and diffused rapidly throughout the developed and developing world beginning only in the 1980s. It has shaped a generation's perception both of the value of tropical forested landscapes, and of the urgency to preserve them.

The Biochemical Factory

There are five fundamental and complementary features of the jungle medicine narrative (Myers 1984). Some are derived from ongoing research and hypothesis building in tropical ecology, biochemistry, and ethnobotany. Others find their inspiration in ancient theories and preconceptions regarding the relationship between nature and society. The first involves the biochemistry of green plants and fungi. Like other possible prey, plants and fungi are in more or less constant jeopardy of being consumed by predators, such as mammals, insects, bacteria, and just about every other life form. But being rooted in the soil, plants and fungi are at a distinct disadvantage compared to animal prey, which can simply run or fly or slither away. Responding to their immobile (sessile) disadvantage, rooted organisms (in fungi, these are hyphae) have developed various predator-avoidance strategies, including mechanical (load leaves and stems with cellulose and lignin), phenological (time biological functions to outwit herbivores), and biotic (provide food and shelter for good insects). A well-known example of the latter is the bullhorn acacia (*Acacia cornigera*), whose hollow thorns provide homes and miniature fat-protein nodules (Beltian bodies) give sustenance for colonies of carnivorous ants (*Pseudomyrmex ferruginea*). The acacia provides food and lodging for the ants, which for their part tenaciously attack insects or other animals (including humans) that threaten the tree.

Other plants and fungi have developed over time a complex arsenal of chemical defense mechanisms (Coley and Barone 1996; Sumner 2000, 107–123; Waterman and McKey 1989). These compounds, termed secondary compounds (or allelochemicals) because they appear to serve no primary metabolic function in the organism, include saponins, cyanogenic glucosides, tannins, phenols, alkaloids, isoprenoids, and others.

Some of these, such as alkaloids, serve as toxicants against herbivory, whereas others like tannins act as digestive inhibitors. Others serve merely as feeding deterrents or pollinator attractants. Some use volatile compounds to help defend against a coterie of herbivorous assailants by "crying for help" with aromatic attractants. In this case, insect-eating birds, beetles, and other potential plant protectors are drawn by odor-plume distress calls put out by the plant to deal with one or another herbivorous attacker. Good for the plant; not good for the plant eater (Dicke 2010; Hare 2011).

People are not the intended target of these biochemical bulwarks, as most evolved long before humans or even our immediate ancestors appeared. But we are nonetheless intimately familiar with their properties. We encounter them with the chemical receptors on our tongues and noses every day, and in most instances, until we grow up and learn to respond otherwise, reject them readily (the gusto-facial reflex) as toxic and bad tasting (Hladik and Simmen 1996). But many are also, literally and figuratively, the spice of life—the source of the habanero chili's fiery features, spearmint's soothing taste and aroma, tea and coffee's bitter flavor, and cilantro's soapy consistency. They are why so few people can tolerate eating Brussels sprouts (due to phenylthiocarbamide), and why so many are passionate about chocolate (theobromine). And they are why we enjoy our potatoes boiled, mashed, and French-fried, but never raw (to eliminate glycoalkaloids and proteinase inhibitors) (Johns 1990, 69–70).

Among this armada of plant-derived defensive compounds, alkaloids are particularly relevant to the jungle medicine narrative. They are numerous, new ones are being discovered daily, and they are frequently bioactive—that is, they have positive or negative biochemical effects on humans. In 1950 there were about 1000 known alkaloids; by 2008, the number had burgeoned to over 21,000 (Raffauf 1970; Wink 2008). Their primary evolutionary role is to repel the attacks of plant or fungi eaters, but their bitter-tasting toxicity often produces marked effects on biochemical activities inside the human body, particularly disruption in the brain and nervous system (neurotoxicity) and disruption of cell membranes (cytotoxicity) (Wink and Schimmer 1999; Wink 2008). Many alkaloids have a lengthy legacy of recreational consumption for their stimulating properties, such as caffeine from coffee and tea, nicotine from tobacco, cocaine from coca leaves, and ephedrine from Mormon tea. Others are employed for their soporific effect, such as morphine from poppies, while others, such as capsaicin in chilis, are consumed for their wonderfully painful burning sensation, what psychologists refer to as a "thrill seeking" flavor (Rozin and Schiller 1980). Some intoxicating alkaloids help us communicate with our gods through hallucinogenic visions, such as psilocybin in *Psilocybe* spp. mushrooms, and harmine in Amazonian ayahuasca vines (*Banisteriopsis* spp.). Still others stir our passions, or at least we think they do, such as atropine from the massive root of mandrake (*Mandragora officinarum*), the infamous "love root" for early Persians, and the "testicles of the demon" for medieval Arabs (J. Mann 2000, 21–27; Simoons 1998, 101–134). Atropine is also contained in deadly nightshade (*Atropa belladonna*), which was administered by medieval women to their eyes to dilate their pupils and thus appear more sexually alluring (M. R. Lee 2007). We also employ alkaloids to hunt and to harm our enemies, in convoluted pathways that reveal the complexity of plant and animal relationships. Take the case of garishly colored Central and South American poison arrow frogs (dendrobatids), whose powerfully toxic skin alkaloids have long been employed to toxify arrows. These poisonous frogs don't manufacture their own poisonous alkaloids, however, but rather assimilate them from the arthropods they consume, certain ants and beetles (*Choresine* spp.) and the occasional millipede. These in turn acquire them by eating other plant-eating insects (V. Clark et al. 2005; Dumbacher et al. 2004). Thus, poisons that are deployed originally by green plants to do battle with one or another plant-eater manage to ascend through the food chain, arriving finally on the pointed projectiles of skilled Amazonian hunters.

Tropical forests became central to the evolving jungle medicine narrative with the discovery in the 1970s of a curious geographical dimension to the distribution of secondary compounds. Most occur in the tropics. And while it is true that these compounds occur in plants at all latitudes (about 20% of all flowering plants contain alkaloids) (Wink 2008), a significant inverse relationship exists between the proportion of a native flora that tests positive for alkaloids and the average latitude of the country. Thus, the native plants in mid-latitude countries like the United States and New Zealand register only 13.7% and 10.8% alkaloid presence, respectively, whereas countries closer to the Equator, such as Kenya and Ethiopia, register 40.0% and 37.2%, respectively (Levin 1976). Studies of individual plant groups yield similar results. For example, along a climatic gradient from temperate to tropical eastern Australia, five eucryphia tree species (*Eucryphia* spp.) exhibit increasing plant chemical defenses (phenols and tannins) against herbivory (Hallam and Reid 2006). In this case, the lower the latitude, the higher the proportion of species containing allelochemicals. This compound gradient, from temperate to tropical latitudes, is illustrated nicely by the altitudinal vegetation transition present on the island of New Guinea. Located in the far western Pacific Ocean, only 5–6 degrees south of the Equator, the island (covering the Indonesian states of West Papua and Papua, and the country of Papua New Guinea) presents a dramatic vegetation gradient from swampy mangroves at sea level through dense tropical forests in upslope areas to windswept alpine meadows on 4509 m (14,793 ft) Mt. Wilhelm. In this insular microcosm of a temperate to tropical climate transition, subalpine forest and alpine grassland each maintain 0% alkaloids in their respective floras, whereas lowland and montane rainforest maintain 21.5% and 14.6% alkaloid presence, respectively (Hartley et al. 1973). Alkaloids, it seems, like the rainforest. But why is this so?

The most reasonable explanation for this high-to-low latitude chemical defense transition is increasing levels of predation in species-rich tropical habitats, or the biotic interactions hypothesis (Coley and Aide 1991; Lim et al. 2015). Predator-prey and other species-species interactions, at various levels of the food chain, are simply more intense in the tropics than in other biomes. For example, ants appear to attack wasp larvae with increasing frequency along a latitudinal gradient towards the Equator. And bird's nests in the tropics experience significantly higher predation pressure than those in temperate regions (Schemske et al. 2009). Leaf consumption by insects is higher in tropical than temperate areas, as is the infection rate of plants by endophytic fungi (most of which benefit the plant) (A. Arnold and Lutzoni 2007). And the density and diversity of insects (butterfly larvae and others) feasting on two species of wild pepper (*Piper aduncum* and *P. aequale*) increases along a gradient from higher to lower latitudes. But because these two species do not exhibit greater leaf damage near the Equator, individual plants must have better antiherbivore defenses in lower latitudes (Salazar and Marquis 2012). In summary, tropical plants more so than their temperate zone counterparts marshal an assemblage of chemical defenses, including alkaloids, phenols, and many others, to repel insect, fungal, microbial, and other attackers.[1] Over thousands of years, people have learned to exploit many of these compound-rich plant parts, especially roots, bark, and leaves, where most of these toxins are stored, to produce a veritable cornucopia of healing and intoxicating formulas.

The significance of this high to low latitude biochemical gradient in the search for drug plants is enhanced exponentially by the legendary biological diversity of moist tropical forests. Mantling only 7 to 8% of the Earth's surface, these arboreal habitats sustain one-half to two-thirds of the estimated 300,000–450,000 plant species on Earth (Mora et al. 2011;

Pimm and Joppa 2015). A survey of the trees in a one-hectare plot (100 m × 100 m, about 2.5 acres) in old-growth tropical forest, for example, is likely to yield over 200 adult tree species. In Brazil's Atlantic Coastal Forest (roughly 13° S), a single one-hectare plot revealed a remarkable 450 tree species (Anonymous 1993). Compare this to the meager list of tree species on my own forested, five-hectare property in northern Oregon (45° N)—Douglas fir (*Pseudotsuga menziesii*), ponderosa pine (*Pinus ponderosa*), Oregon white oak (*Quercus garryana*), bigleaf maple (*Acer macrophyllum*), and a few scraggily red alders (*Alnus rubra*)—and the enormity of this biodiversity bonanza looms large indeed. A comparison of the difference in total vascular plant species richness (alpha diversity) between temperate and tropical countries proves equally insightful. For instance, the total number of identified species in Brazil as of 2012 was 32,364, making it the most botanically rich country on the planet. The United States, by comparison, which is about 8% larger than Brazil, is inhabited by only 18,737 vascular plant taxa. Similarly, Canada at nearly ten million square kilometers (just over six million square miles) boasts barely 5000 total plant species, whereas Indonesia, with less than two million square kilometers (about 1.24 million square miles), harbors almost 30,000 species (Forzza et al. 2012). And of course, in Brazil, Indonesia, Cameroon, and elsewhere in the equatorial realm, there are scores of species awaiting discovery and documentation, whereas in mid-latitude countries like the United Kingdom, Japan, and New Zealand, new species finds are considerably less frequent. Clearly, if mining the Earth for botanically derived medicinal compounds is the objective, then tropical landscapes represent the mother lode.

Pharmacy in the Forest

A second critical element in the jungle medicine narrative involves the wealth of modern drugs that owe their existence to plants. Pharmaceutical companies have tested and ultimately incorporated an amazing array of plant compounds, either as drug components or as templates for synthetic drug development. Some of the better known examples include cascara, a purgative derived from the bark of North American cascara buckthorn (*Frangula purshiana*), and diosgenin from winged yams (*Dioscorea alata*), which is used as a female contraceptive. Paclitaxel (brand name Taxol) from the Pacific yew (*Taxus brevifolia*) was developed into a treatment for lung, breast, and ovarian cancer. And in the most widely publicized example, the alkaloids vincristine and vinblastine extracted from the Madagascar periwinkle (*Catharanthus roseus*) were developed into chemotherapeutic treatments for Hodgkin's disease and childhood leukemia (acute lymphoblastic leukemia—ALL), the most common form of cancer in children. With the discovery of vincristine (used in combination with other compounds), the five-year survival rate from ALL in children climbed from 20% in the 1960s to nearly 90% at present (Hudson et al. 2012). Importantly, of the 121 clinically useful prescription drugs developed from plants, fully 47 were derived from tropical forest species (Soejarto and Farnsworth 1989). And the contribution of plants and other natural products (insects, arachnids, sponges, snakes, and others) to drug development has not abated. Over 68% of new antibacterial, antifungal, and antiviral drugs (anti-infective) are derived from or inspired by natural products. In the case of new cancer treatments, nearly 80% come from or were inspired by plants and other natural products (Newman and Cragg 2016). And some years back, it was estimated that 30,000 American lives were saved each year just by anti-cancer drugs derived from plants (Myers 1997, 224). Bioprospecting clearly has made and continues to make important contributions to public health.

The purported economic value of tropical medicinal resources, both to private pharmaceutical corporations and to society as a whole, further legitimizes bioprospecting efforts. In the 1980s, 25% of all prescription drugs sold in the United States contained compounds that were extracted from plants, totaling over US $8 billion in annual retail sales (Soejarto and Farnsworth 1989). By the mid-1990s, the value of plant-derived drugs was placed at US $15.5 billion/year (Principe 1996). Focusing just on the tropical realm, a 1990s study estimated that there were roughly 375 medicinal plant taxa with commercial value yet to be revealed. Assuming these undiscovered drug plants will have the average economic value of previous discoveries, each would be worth in the area of US $96 million to a pharmaceutical interest, with gross revenue for all drug plant discoveries yielding a total of US $3.2–4.7 billion dollars (Mendelsohn and Balick 1995). Finally, if social benefits such as the value of saving many thousands of human lives are folded into this calculation, the potential dollar value of plant-derived drugs climbs astronomically, to between US $200 billion and US $1.8 trillion per year (Principe 1991). James Miller of the Missouri Botanical Garden revisited this question by providing an estimate of the total number of drugs yet to be discovered from plants. Assuming that there are between 300,000 and 350,000 plant species in the world, and that only 60,000 have to date been screened for pharmacological activity, he reasoned that 540 to 653 new drugs, many of them medical blockbusters, remain to be developed from plants (J. S. Miller 2011). What are these potentially worth? Consider the estimated US $100 million per year annual profits to the Eli Lilly corporation from the development of the vinca alkaloid (vincristine) treatment for ALL from just a single plant, *Catharanthus roseus* (Duffin 2000, 176.).[2] And this was just one plant species.

Regardless of how speculative or inflated these figures may have been, their prominence in the scientific literature and their wide dissemination in the press led to at least two important outcomes. First, the notion entered mainstream environmental consciousness that bioprospecting in tropical forests represents a hugely lucrative enterprise, one that could be leveraged strategically in efforts to protect these endangered habitats. If tropical countries could not be persuaded to protect their precious forests because of their intrinsic environmental and aesthetic value, perhaps the knowledge that they are potentially worth billions of dollars would sway them. Certainly the pharmaceutical option was worth more than the value of saw logs and scrawny cattle. Second, and just as pivotal to the developing jungle medicine narrative, the governments of less developed countries, as well as the various well-meaning defenders of indigenous people's rights, came to the sensible but largely baseless conclusion that there were mountains of money being made by unscrupulous foreigners and pharmaceutical companies in a blatant rip-off of intellectual property. In a repeat performance of the ruthless biological exploitation that characterized several centuries of north-south colonial relations, Big Pharma and its scientist errand boys were once again pirating away the botanical booty of the equatorial zone. Thus was born the biopiracy narrative.

There is, however, a fundamental feature missing from this discourse, and that is the fate of forest people. Regardless of whether the aforementioned economic values of botanical medicines are deployed as justification for "saving the rainforests," or are used to conflate all ethnobotanical enterprise into some international biopiracy conspiracy, these discussions omit the day-to-day meaning of medicinal plants and fungi to traditional forest and savanna communities. For a huge proportion of the two billion people in the less developed world the value of roots, bark, and leaves for healing has no relation to the bottom line of multinational drug companies, or to the aspirations of well-meaning scientists and environmentalists. For the actual practitioners of medicinal plant resources, who cure bellyaches and bee stings, dizzy spells and diarrhea, and the myriad other health problems that bring disease and premature death, the real value of green medicine comes from its ability to heal what ails them. This fundamentally important dimension of people-plant relations in the tropical realm in many ways trumps in value the development of one or another remedy for the battery of (often) lifestyle ailments that afflict the modern world. Most are not the diseases of the poor. And of those diseases that are restricted to the tropics, it is inconceivable that a profit-driven pharmaceutical corporation would invest much time or many resources.[3]

Medicinal plants do, however, possess real financial value for the indigenous people who discovered them and use them. This is because herbal remedies have considerable "replacement value" as substitutes for what someone would need to pay if they purchased a pharmaceutical drug. What are the financial savings, for instance, to a mother who gathers some leaves in her garden and brews them into a tea to treat her daughter's cough, rather than taking the time to walk to town and buy some cough syrup? Quite a lot, it seems. In Madagascar, one community employs at least 241 different medicinal plants. And calculated over the course of a year, the economic value of local medicinal plants to this community accounts for 43–63% of the median annual income of local people (Golden et al. 2012). Unfortunately, the question of the value of medicinal species to the people who actually use them in their daily lives has never played a significant role in the jungle medicine narrative. The story has always been about us and our health problems, not them and theirs.

Bioprospecting is a modern term but an ancient concept. It concerns the pursuit of new compounds and materials from living organisms, especially plants, fungi, animals, and increasingly microbes. In the case of the search for new plant-derived pharmaceuticals, the process has included random and targeted approaches. The former, originally employed by the US National Cancer Institute (NCI), used a broad net to opportunistically collect and screen as many plants as possible for the presence of bioactivity—the shotgun approach. Although by 1974 some 78,882 plant extracts had been screened by the NCI, this strategy met with surprisingly limited success (J. S. Miller 2011).[4] Other investigators took up targeted approaches, based on the belief that some plant groups are better candidates for investigation than others. In the phylogenetic approach, for instance, the medicinal properties of closely related species are assumed to be similarly bioactive. In this case, the search for effective drug plants is focused on close relatives of species and genera with known therapeutic properties. For instance, because malarial-fighting species in the pantropical legume genus *Pterocarpus* have been found to be closely related, further bioprospecting efforts are assumed to be most successful if they are concentrated on closely related groups (Saslis-Lagoudakis et al. 2011). Alternatively, the search for new medicines can be framed with reference to greater understanding of the ecology of tropical plants—the ecological approach. For example, most collecting for pharmaceutical screening in the past was carried out with mature leaves of trees and shrubs. However, older leaves are more likely to employ thick cell walls and overall toughness to fend off herbivory, whereas young and tender leaves are in fact better candidates for investigation as they are more likely to employ biochemical defenses, such as alkaloids and other possibly bioactive compounds (Coley et al. 2003).

Among the targeted strategies, the one most germane to the jungle medicine narrative is the ethnobotanical method. In a revival of tactics mastered long ago during the European colonial period, modern bioprospectors mine the medicinal wisdom of forest shamans and herbalists. Although indigenous societies in the temperate zone have lost many of their traditional links with nature's medicinal properties, rural societies in tropical Asia, Africa, Oceania, and Latin America continue to cultivate these material and spiritual relations with the plant kingdom. The ethnobotanical approach follows from the important discovery that the bioactive components of many contemporary pharmaceutical drugs were in fact known and employed originally by traditional healers long before they were "discovered" and developed by Western scientists. Through trial and error methods likely carried out over many generations, tropical healers learned and applied many of the medicinal secrets of the plant kingdom. And of course, there is the prospect that many more secrets remain to be revealed (Cox and Balick 1994; Bedoya et al. 2001). This indigenous connection to drug plant identification, popularized by so many books and films, was reinforced considerably by the findings of the late pharmacologist Norman Farnsworth. He determined that fully 74% of the plant-derived compounds employed in Western drugs maintained a similar medicinal application by traditional healers in the past (Farnsworth 1988). Later, it was found that over 70% of the 122 plant-derived natural products with therapeutic uses had at least a partial ethnobotanical component (Fabricant and Farnsworth 2001).

The historical record abounds with examples of indigenous-inspired drug plant discoveries. Knowledge of the antimalarial properties of the alkaloid quinine, for instance, which is derived from the bark of Andean cinchona trees (*Cinchona* spp.), was acquired from indigenous Peruvian informants sometime in the sixteenth century. As we will see later (chapter 4), the species that produced the miraculous Jesuit's bark were so coveted by foreign interests that a covert campaign was initiated by Sir William Hooker, then director of Kew Botanical Gardens in London, to secret the seeds away to colonial India (Brockway 1979, 103–139). On the one hand, this represented one of history's most unscrupulous acts of genetic thievery, and set the stage for future charges of biopiracy. At the same time, the eventual outcome of this international conspiracy was global access to a drug therapy that has, outside of penicillin, probably saved more human lives than any other drug (Balick and Cox 1996, 27–30). In another example of the ethnobotanical method, the cardiac glycosides (especially digoxin) in European foxglove (*Digitalis purpurea*) were first adopted into English medicine by physician William Withering in the late 1700s. He learned that the herb was employed as a family recipe for the treatment of dropsy (pulmonary edema), but that its identity "had long been kept a secret by an old woman in Shropshire." By testing the effect of the concoction first on paupers, Withering brought modern medical legitimacy to a humble herbal remedy of the "common people," one he eventually found fit to use on his paying patients (Withering 1785, 3 and 4). Although digoxin is the oldest compound in the arsenal of cardiovascular medicine, it remains to this day an important drug for the treatment of heart failure and atrial fibrillation (Gheorghiade et al. 2006). Finally, pilocarpine is an alkaloid derived from several South American species of jaborandi (*Pilocarpus* spp.). Among indigenous Brazilians it had numerous uses,[5] including inducing sweat, perhaps for ceremonial purposes, as well apparently to encourage the flow of saliva (sialagogue). In the 1870s, Brazilian physician Sinfrônio Olímpio César Coutinho carried the leaves of the plant with him on a trip to France, as he routinely used them in his own practice. He introduced the species and its medicinal properties to local French scientists, who tested it against a host of ailments, including glaucoma (Holmsted et al. 1979). It was found to dramatically decrease intra-ocular pressure, the source of this eyesight-ending disease, and it continues to be used for this end today.[6] It is also administered as a treatment for chronic dry mouth (xerostomia), a frequent side effect of chemotherapy and radiation therapy. Unfortunately, intense harvest of Brazilian jaborandi populations in recent years (especially *Pilocarpus microphyllus*) for pilocarpine export has nearly driven the species to extinction in the wild (Pinheiro 1997).

Plants identified by ethnobotanical leads continue to provide novel compounds for prospective drug development. The fruit of *Brucea javanica*, for example, has several quassinoid compounds (named for a famous eighteenth-century Afro-Surinamese healer) with significant cytotoxicity against various cancers, especially leukemia (K. Lee 2010). Known as Ya-Tan-Tzu in China and buah Makassar in Indonesia, this tropical fruit is used by traditional healers to treat malaria and amoebic dysentery. Its close relative, *Brucea antidysenterica*, is employed by healers in Ethiopia to treat cancer, and has shown promise as a future treatment for various cancers (Fiaschetti et al. 2011). The star anise (*Illicium verum*) is an aromatic cultivated evergreen tree also native to Asia but now widely naturalized. As a component of traditional Chinese medicine since at least the Ming Dynasty, it is used to treat stomach problems, insomnia, and rheumatism. In the United States and Mexico, the species is employed mostly to treat stomachaches and colic in babies (Wang et al. 2011). One of its secondary metabolites, shikimic acid, served as a prototype for the synthesis of oseltamivir phosphate (marketed in the United States as Tamiflu), an antiviral drug used as a prophylaxis against seasonal influenza for adults (FDA-approved 1999) and for children (FDA-approved 2005). In South America, the Brazilian peppertree (*Schinus terebinthifolia*) has served as an antiseptic and anti-inflammatory in traditional medicine for centuries. In the battle against increasing disease resistance to drugs, recent research shows that a flavone-rich composition extracted from the peppertree's red berries "essentially disarms the MRSA [methicillin-resistant *Staphylococcus aureus*] bacteria" (C. Clark 2017). And the most publicized recent example comes from the identification of the antimalarial properties of artemisinin in Asian wormwood (*Artemisia annua*). Now one of the most effective treatments for malaria, this compound and the proper way to prepare it were uncovered after years of ethnobotanical and archival research by Tu Youyou, who won the 2015 Nobel Prize for Physiology or Medicine (Callaway and Cyranoski 2015).

How effective is the ethnobotanical approach compared to a more random strategy for potential drug-plant identification? This question was tested empirically in various locales. In Belize, plant extracts from folk medicinals were found to yield greater antiviral activity in anti-HIV screens than randomly selected species (Balick 1990). In Samoa, 86% of species used in traditional herbal medicine were found to exhibit pharmacological activity (Cox et al. 1989). In Peru, folk antimalarial plants used by the indigenous Aguaruna showed significantly greater inhibition of malarial plasmodia than randomly collected plants (W. H. Lewis 2003). And throughout Latin America, plants with a history of medicinal use were more likely to be bioactive against fungal pathogens than species collected randomly (Svetaz et al. 2010). Some of these studies used relatively small samples, however, and some recent results were less encouraging. Doel Soejarto and his colleagues, in an ambitious bioprospecting effort in Vietnam and Laos, reported mixed results for random vs. ethnomedical leads. They found higher hit rates for healer-identified botanical leads for anti-tuberculosis, but not for HIV, cancer, or chemoprevention (Gyllenhaal et al. 2012). And recently, Applequist et al. (2017) discovered that plants employed in Madagascar to treat malaria failed to outperform randomly chosen species in antimalarial assays.

The Environmental Claim

At this juncture, scientific evidence and theory transition into a persuasive environmental claim, one that has resonated deeply among a health- and environment-conscious public. For just as the contribution of folk healers and their mysterious plant pharmacopoeias captured the imagination of the Western world, the combined forces of destructive forest exploitation, unsustainable plant harvest, and eroding plant knowledge among rural folk has undermined their anticipated contribution to drug development. Deforestation has been linked to decreasing access to traditional plant medicines in Samoa, Kenya, eastern Brazil, and elsewhere (Cox 1999; Jungerius 1998). Among Afro-Brazilian healers in the city of Salvador, I often heard the complaint that urban sprawl was eliminating the last remaining vestiges of habitat necessary for medicinal and spiritual plant collection. The last bico-de-papagaio (*Centropogon cornutus*), a small tree and shrub used to heal stomach ulcers and in religious initiation ceremonies, according to one informant, had recently been cut down, and there were no known sources for future use (Voeks 1997, 98). In other locations, valuable medicinal taxa appear to be declining due to excessive extraction to supply urban and international markets (Van Andel et al. 2015; Stewart 2003). The Himalayan yew (*Taxus baccata*), for instance, faces extinction in its native habitat due to overharvest of its ovarian-cancer-fighting bark and leaves for export to Europe and North America (Lanker et al. 2010). And overharvest of the medicinal pepperbark tree (*Warburgia salutaris*) in southern Mozambique to supply regional markets has severely depleted the species (pers. obs., June 2015). The magnitude of this trade is exemplified by South Africa, where an estimated 70,000 tons of medicinal plant material is consumed per year from over 2000 medicinal species. Of these, two species are now extinct in the wild, and another 82 species are threatened with extinction (Williams et al. 2013). Whether a result of habitat destruction or unsustainable harvest to meet the demands of distant markets, commercial and biological extinction appears to loom for a growing list of healing and other commercially extracted plants.

The most pressing global threat to medicinal plants and their knowledge profiles, however, may be declining knowledge of the healing floras maintained by rural tropical communities (Cox 2000). Numerous studies point to a sharp decline in knowledge and use of traditional medicinal plants, especially among urban immigrants and the younger generation (cf. Edwards and Heinrich 2006; Voeks and Leony 2004). Understanding of the curative properties of nature, accumulated and transmitted over many generations, is seemingly being sacrificed in a single generation. Religious conversion, the entrance of modern medicine and Western-style pharmacies, economic improvement, and enhanced access to formal education, especially in rural settings, are all factors in this nearly universally reported pattern of cognitive "devolution" (Balick 2007; Steinberg 2002; Voeks and Sercombe 2000). Although the nuances of this decline are discussed in detail in chapter 8, surely the most compelling phrase summing up the dilemma was coined by Mark Plotkin in his popular book *Tales of a Shaman's Apprentice*. "Every time one of these medicine men (or women) dies," he lamented, "it is as if a library has gone up in flames" (Plotkin 1993, 236).

The jungle medicine narrative provides in so many ways a persuasive justification for preserving Earth's tropical forests and encouraging continuity of the traditional practices of its indigenous inhabitants. It achieves credibility, on the one hand, through the authority of Western science and the status of its educated practitioners. Gone are the imprecise ethical arguments for saving tropical nature and its wondrous bestiary, replaced by the empirical and quantitative precision of science and economics. The case is bolstered by contemporary predator-prey theory, as well as the promise of possibly thousands of novel bioactive compounds. Constructed on a record of pharmaceutical drug successes, the jungle medicine narrative salvages the long-maligned meta-sciences of herbalism and shamanism from the medical nether reaches of ignorance and occultism. Traditional knowledge of nature, which had long been seen as "a kind of ignorance . . . to be overcome" (Ellen and Harris 2000, 11) suddenly is seen as a possible savior for humanity. Indeed, the notion that miracle cures for society's most debilitating diseases are harbored in the fading memories of a few elderly forest shamans, the perceived protectors of tropical nature's medicine chest, has become firmly entrenched in Western everyday wisdom. Rainforest herbalists and healers, long derided by the scientific community as misguided purveyors of witchcraft and hocus-pocus, have been metaphorically resurrected as the earthly embodiments of ancient forest wisdom.

The various interconnected elements of the jungle medicine narrative—pristine forests, noble natives, mysterious shamans, altruistic ethnobotanists, blockbuster drugs, fantastic profits, greedy multinational
corporations, and creeping globalization—all make for a good story. And
people love good stories. Like most good stories, this one has a compelling cast of characters—jungle healers and rainforest innocents—as
well as rugged ethnoscientists, slogging through the rainforest primeval,
risking life and limb in their mission to bring to light nature's healing
secrets. There are conflicts between good and evil: virginal nature and its
environmentally conscious forest-dwellers versus loggers, ranchers, and
other environmental destroyers. And there is the stark contrast between
the wise indigenous medicine men working for the benefit of their community and Big Pharma scheming for the sake of corporate greed and
the bottom line of their stockholders. As a degradation narrative based
on empirically tested hypotheses, the jungle medicine narrative has its
legitimacy buoyed by the credibility of its many proponents. And because so many of the sub-elements in the story have been provided by
scientists, the whole story may well be true. But truth "often comes in
a distant second to well-constructed and well-executed stories" (Worth
2008, 49). And this story is no exception.

In the following chapters, I explore the primary elements in the jungle medicine narrative. Chapter 2 considers the question of the pristine
nature of tropical forested landscapes. I argue that our view of tropical forests, sublime and sensual, or pestilential and pernicious, where
"vegetation rioted on earth" (Conrad 2012 [1899], 39), has metamorphosed over time in response to varying spiritual quests, entrepreneurial
exploits, and the romanticized prose of naturalists and adventurers. In
chapter 3, I explore the origin of the noble savage archetype so often
deployed in the jungle medicinal narrative. I show that these perceptions are largely culturally constructed to meet the wants and needs of
outsiders, and that they have varied dramatically over the ages. Chapter 4
examines the colonial quest for tropical drug plants, in particular, the
means through which intellectual and genetic property were exploited
from indigenous peoples over the past several centuries. Chapter 5 explores the geographical and cognitive nature of tropical medicinal plant
species. In spite of the image created by issue entrepreneurs, I argue that
most tropical medicinal plants inhabit disturbed habitats rather than
primary forests. One of the key elements in the jungle medicine narrative is the iconic image of the mystical male shaman guiding his Western scientific apprentice through the forest primeval in search of life-
saving drug plants. In chapter 6, I consider the question of gender and

medicinal plant knowledge in the tropical realm. Chapter 7 looks at the medicinal plant knowledge and skills of immigrants who have moved to or from tropical forested countries. Focusing on the African diaspora in the Americas, especially Suriname and Brazil, I show that many immigrant groups have maintained profound connections with the healing traditions of their homelands. Chapter 8 examines the question of cultural devolution, in particular whether the knowledge and use of healing plants maintained by traditional societies is disappearing before it can benefit society at large. Finally, chapter 9 reflects on the origins and impacts of environmental storytelling. In the case of the jungle medicine narrative, I suggest that it was deployed by well-intentioned scientists and environmentalists to address a pressing environmental problem—the destruction of the world's tropical forests. It was successful in part because it appeared just as the West was beginning to take a more globalized view of environmental issues. In the end, however, although this compelling story helped spread the gospel of "save the rainforest," evidence suggests that no life-saving drugs were developed, no diseases were cured, no fortunes were made, and no tropical forests were protected.

TWO

Terra Mythica

Tropical nature symbolizes for many an imaginary heaven on Earth, a latter day Garden of Eden. It is mantled by pristine vegetation, inhabited by a harmonious bestiary, and peopled by the innocent stewards of the forest cornucopia. These three complementary images—virgin landscapes, exotic animals, and noble savages—represent the conceptual starting point for the jungle medicine narrative. But the lower latitudes have not always cast such a romanticized glow. The space between the tropics of Capricorn and Cancer was long known as the Torrid Zone, and for good reason. According to the ancient Greeks and Romans, the burning blaze of the tropical Sun in the equatorial zone bred boiling seas, scorched landscapes, and monstrous semi-humans. Embellished and codified by the ancient philosophers and geographers, these forbidding images held sway for many centuries. But over the last two thousand years, our perception of tropical lands and peoples has shifted wildly, from purgatory to paradise and back again. Then as now, tropical forests and their mysterious denizens constituted a mythical landscape, a terra mythica, supported by a powerful but often contradictory array of metaphors and culturally constructed images. And none of these myths lasted longer and had a more enduring impact on our view of the tropical latitudes than the biblical Garden of Eden (Driver 2004; Glacken 1967, 82–83, 103–105; Flint 1992).

Paradise

God created a perfect world, according to the Book of Genesis, and in it, for the benefit of man and woman, he planted a wondrous and well-watered garden, situated to the east of Eden. To Adam, God gave "every tree that is pleasant to the sight, and good for food," as well as "every beast of the field, and every fowl of the air." All of these Adam and Eve could eat of freely, excepting the Tree of Knowledge. Eve of course ultimately yielded to temptation and ate the fruit of the tree, which according to Jewish and Islamic tradition, would have been a sugary fig rather than the apple of legend (Goor 1965; Wilson and Wilson 2013). And as punishment for consuming the fruit of the tabooed tree, she and her husband were banished from Paradise to eat of "thorns" and "thistles," and to live in sorrow. The healing properties of the Garden were lost, as was the promise of immortality. The first couple was condemned to grow old and diseased, and eventually die. But the belief that there actually existed such a salubrious spot somewhere on Earth, where the air was always fresh and the fruit always abundant, and where God's medicinal leaves never withered and died, persisted in the imaginations of Christian and Jewish scholars. For some, the geographical location of the Garden of Eden became an obsession. But its most lasting effect was as a metaphor for the mysterious healing powers of tropical nature (Gen. 2 and 3, King James Version).

The Garden of Eden, according to tradition, was a place where humankind lived in harmony with nature, and where there was food and water aplenty. For the desert people of the parched Middle East, this must have been a truly glorious image (Delumeau 1995, 4, 15–17). Over the ages, scholarly interpretations of the Garden changed many times. What kinds of beings, for instance, had God chosen to place in the Garden? Some argued that because animals committed the sin of sexual intercourse, they were banned from the Garden. Plants, on the other hand, which were rooted in place and believed not to reproduce sexually, were seen as "mirror images of God" (Prest 1981, 10, 82). Others had no great opposition to stocking the Garden with animals. For although the essence of Eden was a walled enclosure overrun with useful plants, the ancient Greek concept of the Great Chain of Being, with all animate and inanimate objects connected, still found considerable favor among biblical scholars. Life was a gradient, it was argued, and so there were quite naturally intermediate life forms between disparate entities, such as the corals that linked the kingdoms of rocks and animals. Plants, it was supposed, must also be connected to animals, perhaps as Claude Duret depicted in 1605 as leaves falling from the trees, only to mysteriously transform into fish or birds (Fig. 2.1). The Garden was indeed a great and wonderful mystery. But then as now, humans prayed for long and healthy lives. And the Garden, it was said, was overflowing with God's healing leaves.

The Hebrew prophet Ezekiel was the first to articulate the healing properties of Eden. "Their fruit will serve for food," he prophesized, "and their leaves for healing" (Ezekiel 47:12). Over the coming centuries, scholars rendered their opinions on what form these plants took. In the twelfth century, one of the fabricated letters from the mythical Prestor John noted that the fragrance of apple flowers in the Garden had the power to heal. Later in the thirteenth century, French biographer Jean of Joinville said that Paradise was populated by all sorts of wondrous healing spices, including ginger, rhubarb, aloes, and cinnamon. The seventeenth-century poet Du Bartas reflected another prevailing opinion—that the healing flora of the Garden was much more powerful than that of surrounding spaces. For rather than heal what ailed people, the Garden's plant life maintained its privileged dwellers in a perpetual state of wellness. The healing herbs and trees, in his view, were more prophylaxis than medicine.

"The rarest simples that our fields present us
heal but one hurt, and healing too, torment us,
and with the torment lingering our relief;
. . . But thy rare fruit's hid power, admired most,
Salveth all sores, sans pain, delay or cost;
Or rather man from yawning death to stay;
Thou didst not cure, but keep all ills away.
—DU BARTAS 1613, 221

Adam understood the special power of the Garden's flora, according to the apocryphal Gospel of Nicodemus. And when he was old and gravely ill, he sent his son Seth on a journey to the Garden to obtain oil with special healing powers. Seth was unsuccessful in his quest, but he received instead from God the "oil of mercy." Adam was forgiven his sins, and so he died in peace (Delumeau 1995, 46–49).

Did the Garden of Eden constitute earthly soil and vegetation, or was the whole story simply an allegory for the battle between good and evil? Early Christian scholars leaned towards the latter interpretation. The

*Portraict de l'Arbre qui porte des fueilles, lesquelles tombées sur ter-
re se tournent en oyseaux volants, & celles qui tombent dans
les eaux se muent en poissons.*

2.1 Claude Duret. 1605. Woodcut "Trees that Generate Both Fishes and Birds." In: *Histoire
Admirable des Plantes et Herbes*, Wellcome Library, London, Creative Commons,
M0005642.

Jewish philosopher Philo argued in the first century CE (=AD) that to believe that God had busied himself planting fruit and vines in a garden was "mere incurable folly" (Delumeau 1995, 15–17). But in the coming centuries, more and more writers, such as Bishop Peter Lombard of Paris in the twelfth century, began to speak of Paradise in the present tense, as "is" rather than "was." But if the Garden was of this Earth, where was it located? Many opinions were forthcoming over the centuries. St. Thomas Aquinas believed the Garden was somewhere in the east, since this was "to the right" of where he resided, which was nobler than "to the left." Most agreed that it must be situated on an immense mountain or massif, for it would need to have stood above the waters of Noah's Great Flood. The eastern extremity of Asia was often mentioned, again at considerable elevation. Sometimes it was spoken of as surrounded by fiery walls, and therefore it was inaccessible to humans (Cosgrove 2001, 65–66). During the Middle Ages, most writers felt the Garden was to be found somewhere in the Mediterranean, where the climate was benign, as described in Genesis. Others thought it was closer to the Equator (Glacken 1967, 164–165, 273). India was also considered, as were other geographical candidates—the Holy Land, Mesopotamia, Armenia, and others.

If Paradise was indeed a geographical reality, then surely it could be discovered and mapped. At no time in human history were people better placed to begin such an odyssey than during Europe's period of colonial and mercantile expansion. But where to begin? All that was agreed upon was that the Garden was perpetually warm, and that the trees were evergreen and bore fruit year round. There was also some speculation about the presence of agriculture in the Garden, but among those who believed this, it was universally accepted that cultivation was accomplished with little toil (Delumeau 1995, 195). The first allusion that Paradise might be situated in the equatorial latitudes came from Columbus, who exclaimed in a 1493 letter to his friend Luis de Sant Angel that the forest "was most beautiful, of a thousand varied forms, accessible, and full of trees of endless varieties, so high that they seem to touch the sky, and I have been told that they never lose their foliage" (Mancall 2006, 210). Curiously similar observations were forthcoming from Amerigo Vespucci, who reported from America that the trees never lose their leaves, and are quite aromatic. "I must be near the Earthly Paradise," he proclaimed (Vespucci 1992 [1504], 31). These and other statements of awe by the first visitors to the New World not only were similar to each other, but also plagiarized the vocabulary that had been used for fifteen centuries in Europe to describe the Garden of Eden (Delumeau 1995, 109–112). The two Italian mariners, it seems, filled their logs with observations and not a few fabrications, but the Edenic narratives they provided were simply restatements of stories that had been told and retold for many centuries.

Circumstantial evidence for an earthly Paradise from this point forward positioned it firmly in the humid tropics, and most likely in the Americas. Only in the newly found world were the trees sufficiently aromatic and evergreen, was fruit available all year round, and was the climate forever warm and balmy. Signs were also forthcoming from certain flowers and fruit, and even animals. The obscenely ornamented passion flower (*Passiflora* spp.), with its three stigmas likened by Brazil's Jesuit brotherhood to reflect the three nails in Jesus' hands and feet, its threads believed to be a symbol of the Crown of Thorns, its tendrils likened to the whips, and its five anthers meant to represent the five wounds, was said to represent "a theater of the mysteries of Redemption for the world" (Vasconcellos 1668, 244). The Indians' name for the magnificent flower and fruit was *maracujá*, but the Portuguese by the late 1600s referred to the flower only as "the Passion." Likewise the pineapple, known early in Brazil by its indigenous name *ananás* (and still so in Portugal and Mozambique), was considered at the time the king of fruits. Its indigenous name was early associated with Anna Nascitur, the name of the Virgin Mary's mother, Jesus's grandmother. The presence of chatty parrots added further evidence that the Garden was nearby. With their ability to speak human words, they were considered by some as a connection between humans and the animal kingdom. Parrots' notorious longevity further cemented their association with the immortality of God's Garden (Delumeau 1995, 112).

Years after Columbus's nipple hypothesis, Spanish historian Antonio de León Pinelo carried out an exhaustive study on the geography of Paradise. He pinpointed its location in the Amazon Basin, a few degrees south of the Equator. More systematic in his methods than Columbus, he similarly ruled out tropical Africa as being too hot and parched for the Garden, as well as India, which, although a more suitable candidate, was plagued by expansive dunes and excessive temperatures. The biblical Paradise was in the American rainforests, Pinelo reasoned, as evidenced by the tremendous stature of her evergreen trees, the benign nature of her climate, and the presence of the four fabled rivers reported in Genesis— the Magdalena, the Platte, the Orinoco, and of course the mighty Amazon (León Pinelo 1943 [1650]; Scott 2010). Still other Spanish writers, believing that Eden harbored all the plants and animals on Earth, positioned it high in the Andes, since the mountain chain embraced such a varied array of life zones (Cañizares-Esguerra 2006, 116–121).

In spite of the enthusiasm forthcoming from a few Eden promoters, most colonial writers understood that the idyllic American, Asian, and later South Sea Island tropics constituted metaphors rather than material locations for Paradise. These "Edenic isles set in sparkling seas" indeed seemed heaven-sent compared to the overpopulated and often "hungry landscapes" of Europe (D. Arnold 1996, 142–146). But if there ever were a physical and biological Paradise on Earth, most agreed that it had long since vanished, or that it was in any case quite unapproachable. What emerged over time, at least in the American tropics, were carefully crafted "tropicality" narratives, each with its own particular historical preconceptions and entrepreneurial agendas. Thus, for example, Columbus sought to gain financial backing for further forays into the American unknown, no longer in the pursuit of precious spices, but rather in search of the mother lode of the Christian tradition. León Pinelo likewise was a booster for his native Peru—it was teeming with a bewildering array of marvelous fruits and animals, woods and precious metals—and he wanted the world to know about it. But they were the only two who seriously took up the call to search for the Garden. As the myth of Eden in the Americas faded, it was replaced by other fantastic and equally unfounded obsessions, such as the Fountain of Youth, where people could slake their thirst for immortality, and especially El Dorado, where a bold man could realize unimaginable riches.

The search for the lost city of El Dorado replaced and redefined the quest for a tropical Eden. Incalculable quantities of gold became the new paradise in the forest. The story of the Golden Man, or El Dorado, was first related by Fernandez de Oviedo, in Quito, Ecuador, in 1541. He was convinced, as were many of his contemporaries, that gold "grew" near the Equator, where the golden glow of the Sun was most intense. In a tale he said he gleaned from some itinerant Indians, de Oviedo described how the rich chief of a hidden jungle kingdom, located somewhere to the east of Quito, painted resin on his naked body each day, and then was dusted by his assistants with gold dust. It was a city of gold, and a fantastic fable. And at the time, the story was considered completely plausible. Amerindians were known to paint their bodies, and of course there had already been huge discoveries of gold by Hernán Cortés in Mexico and Francisco Pizarro in Peru. Surely there were other treasures waiting to be unearthed in the mysterious forest primeval. From this simple beginning, the legend grew and become embellished with each telling, each new and more fanciful version serving to reinforce and legitimize the former. By the eighteenth century, the dream of El Dorado was so hypnotic that it was said that "tufts of grass pulled from the ground hold gold dust in their roots" (Hemming 1978b, 101–102; Hecht and Cockburn 2010, 6–7).

The imagined location of El Dorado migrated geographically over time, eventually coming to rest somewhere between the Amazon and the Orinoco river basins. There its cause was taken up by the British explorer and buccaneer Sir Walter Raleigh. At the time (1596) languishing in the Tower of London for personal transgressions, he crafted a sufficiently persuasive argument for the existence and location of El Dorado to gain his release (Raffles 2002, 141). Requesting five ships and one hundred men to aid in his crusade, Raleigh argued that the riches of the golden man were simply an extension of the Inca treasure so recently taken by Pizarro, the governments of the two entities "not differing in any part." His narrative provided freshly invented ethnographic details on how the people covered their bodies with gold. "All those that pledge him are first stripped naked, & their bodies annoynted al over with a kind of white Balsamum . . . having prepared gold made into fine powder blow it thorow hollow canes upon their naked bodies, untill they be al shining from the foote to the head" (Ralegh 1596, 10, 16). But like the other tropicality narratives of the time, it was all just a fable. There was no golden man, no lost kingdom, and Raleigh's journey into the unknown was an abject failure. For offenses committed during the trip, Raleigh was executed upon his return to England, his head preserved in a jar by his mourning wife for many years. His personal quest for El Dorado led to his execution, but the story he so effectively embellished encouraged generations of believers in the great mysteries to be discovered in virgin tropical forests.

The Sexualized Forest

In addition to his quest for the legendary El Dorado, Sir Walter Raleigh also introduced a metaphoric sexualization of the tropics, one that has been retold and reformulated in countless writings up until the present. This tropical paradise, according to Raleigh, "had yet to have her hymen torn" (Ralegh 1596, 10, 16, 96). His identification of the chastity of the rainforest, articulated first to gain favor with the Queen of England, goes on to become one of the most compelling and overused tropes of tropical nature—a pristine and virginal landscape ripe for penetration by intrepid

2.2 Jan van der Straet. 1600? "Discovery of America." *Nova Reperta*. Antwerp. Public Domain.

outsiders. The unknown forests of the Americas, and later other tropical spaces, slid effortlessly into sexualization by European visitors as they scrambled to justify their objectives and defend their often brutal methods. The vastness of the great forests they invaded, for the first time entered by "civilized" men, their many erotic mysteries yearning for exploitation, all were images that lent themselves to feminization of newfound tropical lands. Popular early convention presented the American landmass as a female, symbolically ready and willing to be dominated by intrepid European men. Theodor Galle, in his 1580 engraving, depicts a stoic Amerigo Vespucci as he awakens, or is perhaps invited to lie with, a naked and willing female America (Fig. 2.2). Columbus wasted no time in feminizing the New World with his comparison of the Earth to a woman's breast. And Raleigh's untorn hymen statement graphically summarized the violent side of this emerging metaphor.

For others, America was not so much a maiden as she was a lady of the evening, offering Europeans her sexual services for little or nothing. One of Raleigh's men, Lawrence Kemys, described America as "whole shires of fruitful rich grounds lying now waste for want of people, do prostitute themselves unto us, like a fair and beautiful woman in the pride and flower of desired years" (Kemys 1596, 487). In the fevered imaginations of outsiders, equatorial Asia, Africa, and the Americas became "porno-tropics," where Europe "projected its forbidden sexual desires and fears" (McClintock 1995, 22). Virgin landscape suggested empty landscape, devoid of rightful owners, inexperienced, easily exploited, and ready for trespass. Such eroticized conceptions of pristine tropical forests and fields, about to be violated by consent or force, fulfilled an important and rather clichéd aspect of imperial discourse—the resources of these new lands were calling out for domination and exploitation, lacking only men of vision and guile to do their duty.

But the emerging stereotype of the tropics was not all sex and vio-lence. In time, even the most grizzled explorers and ardent naturalists went slack-jawed in astonishment upon first entry into the rainforest. Sensually warm and moist and aromatic, the tropical latitudes provided a panorama of exotic beasts and erotic vegetation with which European visitors were wholly unfamiliar. Feminine chastity was as close as many could come to expressing their innermost feelings. Others, lacking fa-miliar features with which to attach, filled their pages with the words and images they had read so many times in the past—the virgin forests, the giant leaves, the succulent fruit—and likened the whole to the great biblical Garden. The forest was simply too grand and bewildering for precise prose. So otherworldly was it, even for early subscribers to Dar-winian theory, that it must have been heaven-sent. For Darwin himself, newly arrived in Rio de Janeiro and trying to transmit his thoughts to paper, "sublime grandeur" was the best he could do, at least early on in his journey. How to overcome this verbal paralysis, he thought? "It would be as profitable to explain to a blind man colours" (C. Darwin to John Henslow 1832). For other travelers, it was not always possible to distinguish the lands and the people who inhabited them. From Tahiti, the island and its inhabitants were seen conjoined as one great tropical idyll. During Louis-Antoine de Bougainville's visit in 1768, he described a young girl "as Venus . . . having the celestial form of that goddess"; the land "was like Paradise before the Fall of Man." No need to toil in this island paradise, where bread grows on the trees, and milk pours forth from the palms (B. Smith 1985, 42–46). But if there was a water-mark after which cultural construction of the tropics took a permanent romantic turn, it was surely after the return of Alexander von Humboldt from his epic four-year journey through Latin America. A generation of scientists and artists, Alfred Russell Wallace, Henry Walter Bates, Richard Spruce, and the painter Johann Rugendas, were drawn to the tropics and saw nature through the literary legacy created by Humboldt. As Darwin famously exclaimed after his first forays into the tropical forest,

"I formerly admired Humboldt, I now almost adore him" (C. Darwin to Henslow 1832).

Humboldt's felicitous writing and rigorous field methods caused a metamorphosis of sentiment regarding the "riches" of the rainforest. Fantasies of immortality and material wealth persisted, and still persist today, but for many the true fortune of the forest was now measured in the birds and the trees and the butterflies. The tropical forest for Humboldt was a great void of biological mysteries, awaiting only people of science and conscience to reveal its inner workings. In the beginning, he too was challenged to put his emotions into words, as one "can scarcely define the various emotions, which crowd upon his mind" (Von Humboldt and Bonpland 1818, 36). But he prevailed like none other before him. Humboldt was the consummate empirical field scientist, lugging hundreds of pounds of scientific instruments through the torturous physical terrain to measure, collect, and identify. He was also influenced by the romantics of his age, and his science included considerable aesthetics and the search for natural relationships (Helferich 2004, 25–28; Stepan 2001, 36). He viewed humans as integral elements in the landscape, and commented at length on the inhumane treatment of Native Americans and enslaved Africans, as well as on the negative effects of destructive resource exploitation. His notes on the adverse effects of deforestation around Lake Valencia for many marked the beginning of the field of environmental studies. And like so many before him, he harbored Edenic and environmental determinist sentiments on the relation between humankind and tropical nature. Chilly temperate climates and vegetation excited humans to industriousness, he argued, but in the tropical forests "the richness of the soil, and the vigour of organic life . . . retard the progress of nations . . . in the midst of abundance, beneath the shade of the plantain and bread-fruit tree, the intellectual faculties unfold themselves less rapidly" (Von Humboldt and Bonpland 1818, 14; Helferich 2004, 175 and 226; Slater 2002, 40–42). Life in Eden was just too effortless.

Humboldt became wildly famous in his time. His persuasive writings in many ways shaped our current perception of tropical nature—a motley mix of literary romanticism, astonishing biodiversity, and environmental catastrophism. In the coming years, these images were further molded and refined by the writings and artwork of Humboldt's intellectual disciples (Egerton 2012; Slater 2002, 40–42; Stepan 2001, 18–48). Among the most influential was Englishman Henry Walter Bates. The result of eleven long years collecting and cataloguing insects in the Amazon forests, his *The Naturalist on the River Amazons* (1863) was widely

2.3   Henry Walter Bates. 1864. "Bird-killing spider." *The Naturalist on the River Amazons.*
RB 721518. Vol. 1. Pg. 161. The Huntington Library, San Marino, California. Digital Image
by Robert Voeks.

read by a jungle-hungry public. Bates was a commoner and scientific
entrepreneur, who, unlike Humboldt and Darwin, had to pay his own
way. He accomplished this by collecting and selling tropical animals,
totaling over 14,000 species during his Amazonian sojourn. He was also
a gifted storyteller, and his book was (and is) considered a classic trav-
elogue and natural history of the tropical rainforest. One of the most
memorable images in the book is the engraving of the bird-killing spider
(Fig. 2.3). The sight of this crab-sized arachnid capturing and killing a
pair of finches underscored for his readers the fantastic and sometimes
macabre dimension of tropical life, a place where nature rather than
humankind was in full command. There was also Victorian sensibility
in his narrative, as Bates shared with his readers his unsuccessful attempt to rescue the second bird by driving away the spider before the
bird could be smeared with "the filthy liquor or saliva of the monster"
(H. W. Bates 1864, 95–97). He provided timely evidence in support of
Darwinian selection as well. Bates was the first to describe one organism
mimicking another that was more dangerous, thus obtaining protection
from predation, which he termed "mimetic analogies," and which later
became known as Batesian mimicry. And there were environmental lessons to be learned from his narrative as well. After many years living and
working deep in the forest, he returned to his jump-off point in Belém,

at the mouth of the Amazon River, only to discover the forest he had known years before had been cut and burned. "The noble forest-trees had been cut down, and their naked half-burnt stems remained in the midst of ashes, muddy puddles, and heaps of broken branches . . . Only a few acres of the glorious forest . . . remained in their natural state" (H. W. Bates 1864, 458–459). But Bates knew his audience. And at the conclusion of his book, he confirms the cultural superiority of England by informing the reader how elated he is to be home after such a long and difficult absence. Having survived "in gipsy fashion" for so many years, he preferred the civilized life and "intellectual nourishment" of Britain to "the spiritual sterility of half savage existence, even if it be passed in the Garden of Eden." Only in the "bleak north" can superior civilizations develop, he reasoned, but regarding nature, "it is under the equator alone that the perfect race of the future will attain to complete fruition of man's beautiful heritage, the earth" (H. W. Bates 1864, 460).

Dark Eden

The pristine paradise depiction of the American tropics may have been the most persuasive of the tropicality narratives constructed by travel writers and naturalists. But it was not the only one. Scientists and travelers and increasingly nature tourists continue to be astonished at what they see and feel upon entrance into the forest primeval, and continue to deploy Edenic reference to express their awe. But a host of other myths and metaphors developed over time as European men and women travelers struggled to describe and define tropical lands and people for their readership. One of the most persuasive early themes was the tropics as landscapes of deadly disease and nightmarish creatures. In sub-Saharan Africa, where Joseph Conrad's "merry dance of death and trade" was played out, the "dark continent" and "white man's grave" metaphors were built in part around the awful array of infectious and parasitic afflictions that plagued European visitors (Conrad 2012 [1899], 18). Although these specific phrases did not appear until Henry Stanley's description of his journey down the Zambezi River, the notion that Africa was awash with disease has been widely reported since the earliest European traders and missionaries began mingling with the local populations (Jarosz 1992). Until the widespread availability of quinine to battle malaria, disease was believed to be Africa's best defense against domination by Europeans. As Charles Darwin noted, "When civilised nations come into contact with barbarians the struggle is short, except where a deadly climate gives its aid to the native race" (Crosby 1993, 138; C. Darwin 1871, 229). The construction of an uncivilized and diseased Africa was seen as justification enough for European imperial ambitions in the continent. The late nineteenth-century scramble for Africa by colonial powers, especially Belgium, Great Britain, Portugal, France, and Germany, turned on Africa's "need" for a civilizing force. In the twenty-first century, the disease metaphor has been powerfully reinforced in the minds of Westerners with the knowledge that HIV/AIDS originated in the rainforests of Africa, as did Marburg, Zika, and Ebola hemorrhagic fever. Indeed, as grisly accounts of Ebola's symptoms metastasized in the Western press and films, sub-Saharan Africa for many became synonymous with Ebola (Haynes 2002).

Given the justified preoccupation of European visitors with premature death in the forests and savannas of sub-Saharan Africa, the Eden metaphor might seem to be a tough sell. But disease has also been a counterpoint to tropical American and Asian paradisiacal narratives as well. And the list of afflictions is nearly as horrific. For wherever the climate was perpetually hot and sticky, the vegetation damp and decomposing, disease and death seemed to follow, at least for outsiders. In Latin America, Chagas disease leads to heart failure and intestinal blockage. The disease, which is transmitted by kissing bugs (*Triatoma* spp.), likely disabled Charles Darwin for much of his post-voyage life. Leishmaniasis, which is contracted by the bite of female sand flies (subfamily Phlebotominae), reveals its presence by great seeping skin ulcers and an enlarged spleen. Malaria, yellow fever, dengue, schistosomiasis, and many other animal-transmitted diseases (zoonotic), all introduced early in the colonial era to the American tropics, quickly became the scourge of indigenes and expatriates alike (Colfer et al. 2006). The situation was equally grim in South and Southeast Asia. Malaya (present-day peninsular Malaysia and Singapore), for instance, was early described in the usual Garden of Eden epithets—fertile soils, luscious fruits, healing herbs, and a perpetually balmy climate. By the nineteenth century, however, disease and deprivation narratives replaced the idyllic. By then the region was depicted as filthy and overrun with disease, and hence ripe for rescue by colonial saviors (Ahmad et al. 2011). In early colonial Batavia (Jakarta, Indonesia), the intolerable climate was believed to cause bloody flux, cholera, dropsy, jaundice, and killing fevers. Mortality among incoming Dutch was said to reach 50% some years (Eekelen 2000). In India, the death toll was even worse. According to a sixteenth-century Portuguese proverb, "Of the hundred who go to India [from Portugal], not even one returns" (Pearson 1996, 241). Given India's huge latitudinal expanse, English physicians conceptually divided the colony between the healthy temperate regions and the unhealthy tropical lands. The latter, for Europeans, could be improved only by change of food and clothing to emulate native Indians, or by escape to cooler hill stations, where salubrious mini-Britains could be established (Harrison 1996). The British also believed that India's tropical forests were the source of the infectious diseases, and advocated in 1830 that 52,000 square kilometers (about 20,000 square miles) of trees be removed (D. Arnold 1996, 153). The forest region could be made safe if the forests were simply removed. Indeed, since the earliest days of exploration and colonization, the tropics have been thought of in terms of debilitating and civilization-stifling disease. Every textbook on tropical medicine is replete with frightful photos of the symptoms of tropical illness, and none are complete without the obligate photograph of some victim of filariasis (elephantiasis) "carrying his testicles in a wheelbarrow, or at least sufficiently enlarged so that a wheelbarrow would seem a convenience if available" (M. Bates 1952, 134–136).

Less lethal, but the source of even more hair-raising images, are those creatures great and small so often dramatized in Hollywood films and travelogues. A perennial favorite is the *candirú* (*Vandellia cirrhosa*), a tiny Amazonian catfish reported by travelers since the eighteenth century to swim up and become permanently lodged in the penis (or rarely rectum or women's urethra). During my first visit to Amazonia, I was encouraged to wear a condom if I went for a swim. At the time, I thought it was a joke. I was also told never to urinate into the water, as the *candirú* could easily navigate the warm stream. In fact, the scientific name of the little fish was at one time *Urinophilus diabolicus*—the devilish urine-lover. This supposed scourge of the Brazilian watershed is said to be extricated only by slicing open the affected appendage to remove the offender. Although this all makes for a great story, there is not to a single verified case of the *candirú* actually doing the evil deed. Nevertheless, no travel account or adventure story to this day fails to feature the little beast (Bauer 2013).

Other equally creepy creatures are said to abound in the tropics. Tarantulas, bullet ants, and foot-long centipedes are constantly underfoot, and venomous snakes dangle like vines or lay coiled and impossibly camouflaged: the Burmese krait, the bushmaster, the black mamba, and many others too numerous and too awful to mention. And the parasites, my god, the parasites! Who hasn't cringed at the thought of the Guinea worm, the two- to three-foot, noodle-thick nematode that grows under the epidermis of people's legs and feet? Or the human botfly, whose eggs swell into plump larvae under the skin of its human host? I had the pleasure of having a bicho-de-pé (*Tunga penetrans*), a parasitic flea, extracted from my foot after my last visit to Brazil. The female burrows under the skin and lays her eggs, which develop into an incredibly itchy sack of larvae. And who can forget the famous scene in *The African Queen* when Humphrey Bogart lifts his shirt to reveal a blanket of leeches? These days there is a new nightmare from the jungle making the rounds on nature programs and travelogues—the Brazilian wandering spider (*Phoneutria nigriventer*). No myth, this aggressive arachnid inflicts an excruciatingly painful bite, leading to tremors and possible paralysis. What has received the most press, however, is that the neurotoxin causes a painful and long-lasting erection (priapism) that can lead to permanent potence and even death.

Of all the dreadful creatures inhabiting the tropical biome, none surpasses in reputation the legendary piranha (Family Serrasalmidae). Indeed, no tropical narrative is complete without an obligatory encounter with a school of these supposedly bloodthirsty fish, rumored to be able to de-flesh a human body in minutes. Native to the Amazon, Orinoco, and several other South American river systems, the various species of piranha can be aggressive, and attacks on people occur about as regularly as shark attacks do in the open ocean, in other words, rarely. But the notion that they attack in feeding frenzies appears to be mostly a fabrication, possibly initiated by President Theodore Roosevelt's book *Through the Brazilian Wilderness* (1914, 42). In it he describes his personal observation of the little beast's "savage fury" and "embodiment of evil ferocity" as the fish tore to pieces a cow that had been pushed into a small pond. Unbeknownst to Roosevelt, this scene appears to have been cooked up by his Brazilian hosts who were eager to impress the American luminary. According to one version, the locals had captured schools of piranha, and kept them captive and starved in a small backwater until Roosevelt's appearance. Their hunger-fueled ferocity as they tore into a bloodied bovine was duly recorded by the President, and repeated by nearly every subsequent depiction of Amazonia. In the end, and for as long as outsiders have been transmitting their culturally constructed conceptions of the equatorial zone, paradise and purgatory have been obliged to coexist (D. Arnold 1996, 142–145).

The Illusion of Virginity

Among the numerous tropicality narratives that have persisted to the present is the notion that tropical forests are pristine or virgin (Hecht and Cockburn 2010, 14). The terms have been used so generously as epithets for "tropical forests" that anything less than "pristine tropical forests" or "virgin tropical forests" seems somehow incomplete. The concept has circulated in the writings of travelers and naturalists since the time of Humboldt, and for many was code for unpillaged, unowned, and available for exploitation. But the concept of a biblical Eden in the jungle, a wondrous geographical comfort zone of health and happiness where God's creatures exist in harmony with humankind, has endured since the colonial era. And if anything, the exciting biological discoveries made over the past century or so, the protean biodiversity and its complex ecological relations, have only reinforced the West's preconceived perception of the tropical forest primeval. No matter that the fantastic flora and bestiary is the product of natural selection and evolution rather than special creation, what matters is that Earth's greatest natural garden exists, relatively untouched and unaffected by the actions of humankind, deep in uncharted tropical jungle, far from the soul-crushing uniformity of post-industrial society.

But field research has not been kind to the concept of pristine tropical forests. For most scientists, the idea of barely touched tropical landscapes has yielded to the view that most are in large measure cultural reflections of past and present human actions. Indeed, considering the pedigree and pervasiveness of human impacts in the equatorial zone, the distinction between natural and cultural landscapes appears for many to be rather arbitrary. As long as humans have occupied and made their living in tropical forests, they have been cutting, burning, trimming, transplanting, sparing, mulching, and otherwise contouring their surrounding forests and fields to meet their needs. The effects are well documented in the Americas, much less so in Africa and Asia. But wherever scientists have concentrated their efforts, tropical forests have been revealed to be complex mosaics of coupled human-natural systems.

The pristine myth in the Amazonian tropics followed from the widespread belief that these landscapes were sparsely populated when first encountered by Europeans. Indigenous forms of agriculture, it was assumed, were insufficient to produce the sorts of surpluses necessary to sustain large numbers of people. And the environment was simply too challenging. The red soils (principally oxisols) were believed to be fragile, acidic, and infertile, and so their ability to provide for food surpluses, and thus maintain dense populations of people, was severely constrained. Nor was the climate cooperative, being considered too warm and wet for the sorts of intensive agriculture that developed in the temperate zone. In addition, Amerindian technology was technologically primitive. Unlike the cultures of tropical Africa and Asia, which had forged or at least had access to iron axes, knives, and other metal tools, and who husbanded cattle, goats, and other livestock, the people of the Americas were still mired in the Stone Age. For animal protein, they subsisted on sparsely distributed wildlife—pacas, agoutis, monkeys, and tapirs. The ability of indigenous Americans to modify the great tropical forests of Central and South America, it was long supposed, was nearly as limited as that of other forest primates.

But research over the past four decades has turned these ideas upside down, starting with the "low population hypothesis." Rather than a few tens of thousands of technology-limited indigenous folk struggling to survive in a hostile environment, the American tropics were in fact teeming with people, and likely had been so for several thousand years. From the seasonal forests of southern Mexico to the semi-arid grasslands of southern Brazil, the pre-Columbian population was likely forty million or more. Even the so-called green deserts of the lowland tropics probably sustained ten million or more people, concentrated in particular along the large, resource-rich rivers (Clement et al. 2015). As Native American populations increased their numbers and influence over the centuries, they reshaped the forest landscapes in myriad ways. The forests were not, by the time of the arrival of the first Iberians, pristine in the least, but rather humanized and profoundly domesticated, according to William Denevan, "nearly everywhere." It was only after the arrival of horrific epidemics, especially Old World viruses during the early decades after the European landfall, that the indigenous population began its epic collapse. Within a century, smallpox and influenza and other infectious diseases had reduced the numbers of Native Americans by 90–95%. It was the most extensive and least known genocide the Earth had ever witnessed.

Ironically, the ecological consequence of the demographic collapse was that by 1750, two-and-a-half centuries after the arrival of Columbus, the continent-wide anthropogenic changes in forest distribution and composition several millennia in the making had long ceased. The forests had recovered many of their pre-cultural features, and many of the workings of the great indigenous Amazonian populations were smothered by trees and vines and undergrowth. For a short time at least, before European immigrants and enslaved Africans began to leave their own footprints in the forest, the landscapes of the American tropics were as Edenic and pristine as they had been in at least 10,000 years (Denevan 1992; 2011, 578). But all of the evidence for high population figures raises the question: If agriculture was indeed severely constrained by climate, soils, and technology, how did indigenous people manage tropical forested landscapes to produce so much surplus? And are their impacts still evident?

When we think of early humans in the rainforest, we usually imagine them carrying out some form of shifting agriculture—cut, dry, burn, plant, replant, abandon. However, this mode of subsistence required domesticated crop plants, such as maize and manioc in the Americas, yams and millet in Africa, taro in the South Pacific, and rice in Asia. And these were not developed until some 5,000–10,000 years ago, depending on the region. Thus, whatever impacts resulted from the actions of shifting cultivators were predated many millennia by small-scale societies that pursued hunting and collecting for their subsistence. When these people first arrived in tropical forests is still debated. In sub-Saharan Africa, a foraging mode of subsistence similar to that of present-day hunter-gatherers is traced back at least 44,000 years. In northern Borneo (Malaysia), occupation and sophisticated plant preparation date back 50,000 years. And in Amazonia, which lacked humans until much later, palm and other fruit remains of species still consumed today point to the presence of peopie by 11,000 years ago, that is, the late Pleistocene and early Holocene (Barker et al. 2002; d'Errico et al. 2012; Mercader 2003; A. C. Roosevelt 2013).

However long their forest occupancy, ancient human foragers had minimal but by no means insignificant impacts on tropical forest biota compared to their cultivating counterparts. Most groups were small in number and nomadic, moving from one location to another after game and other forest resources had been depleted. The most dramatic means of forest modification would have been tree clearance, but this was rare because, lacking crop plants, there was little reason to expend the energy to remove trees. Indeed, the Penan (or Punan) hunter-gatherers of northern Borneo until recently maintained prohibitions against cutting trees that were larger in diameter than a man's forearm. All trees were considered to have spiritual entities, and only those few destined to become blowpipes, their principal hunting weapon, were felled. But the Penan do "manage" the forest in more nuanced ways, and these likely affect species composition patterns. The defining resource philosophy among the Penan is *molong*, which translates loosely to "don't be greedy." This ethic addresses both the necessity of sharing in a small and interdependent foraging group, and the need to conserve resources. The Penan exercise this principle in particular with respect to wild sago palm (*Eugeissona utilis*), their main source of carbohydrates, and perhaps the main starch source for Bornean people going back 40,000 years. The sago is removed by harvesting the stem and removing the pith. They *molong* sago by carefully husbanding and not overexploiting it; the older stems are harvested and processed, but the younger stems, *uvud*, are marked and protected for future exploitation. These actions likely increase seed production and thus affect the palm's abundance and distribution in the forest. Like so many other useful species, sago is not completely wild, but neither is it domesticated (Brosius 1991; Voeks 2007b; Voeks and Sercombe 2000).

The manipulation of individual wild plant species to produce more food is a common feature of hunting and gathering peoples. The effects on forest structure and species distribution are subtle, but they are certainly recognized by local inhabitants. One of the surest signs of past hunter-gatherer presence in the forest is concentrations of fruit trees. These are often floristic signposts marking old hunting camps, where people have discarded or defecated the seeds of wild fruit trees. The habitat disturbance created by the temporary presence of people allows light-demanding fruit tree seedlings to grow quickly after the site is abandoned. And the seedlings are fertilized by the nutrient influx from organic waste. In a few years, the old camp is overgrown with useful fruit trees, a distinctly unnatural occurrence. Foraging groups also affect useful species patterns by weeding, seed scattering, and replacement planting. Wild yams are particularly suggestive of forager forest management. In places as distant as Sumatra, the Philippines, Malaysia, and India, hunter-gatherer societies manage wild yams (*Dioscorea* spp.) by replanting the head after harvest. For the Baka Pygmies of Cameroon, this process of "protocultivation" of a wild food source encourages reproduction and contributes to the unnatural geographical distribution of species. Overall, for the dwindling numbers of tropical forest hunter-gatherers, the distinction between wild harvest and species management "becomes almost impossible" (Dounias 2001; Ichikawa 2001; Puri 2005; Rambo 1985, 70; Rival 2002, 81).

Cultural Rainforests

The gradual development of domesticated food plants in the tropical realm, or their introduction from afar, signaled the beginning of a profound structural and floristic transformation of Eden. With few exceptions, the presence and patterns of tree species, especially those useful to people, became less and less a function of ecological processes—soil properties, precipitation regimes, topographic aspect, and seed dispersal syndromes—and more the product of human intention. Some forests were transformed to savannas, as humans employed fire to improve forage for wild and domestic animals, and to kill off snakes and scorpions. Others could best be described as "cultural forests," as people employed myriad strategies to modify species and manage habits (Balée 2013). A few of these strategies were nearly universally applied; one was the management of palms with fire. There are over 2000 species of palms in the world, mostly tropical, and nearly all are useful to people, for food, oil, timber, medicine, thatch, and many other ends. Temperate zone people, according to Alexander von Humboldt, exist on cereal and meat, but tropical people are basically "palmivorous" (Von Humboldt and Bonpland 1852, 336).[1] And many palms, perhaps most, serve as landscape signatures of the prior actions of collectors, cultivators, and deforesters. In eastern Brazil, for example, a palm known as piassava (*Attalea funifera*) is restricted to sandy soils near the coast. It is neither tall nor majestic, but it does produce a valuable fiber, used during the colonial era as anchor rope, and later as brooms, brushes, and roofing thatch. Like many other palm species, piassava has the habit of sprouting immediately and abundantly after a fire. This is due to an unusual germination pattern, known as saxophone growth (cryptogeal germination), in which the terminal bud of the seedling initially burrows down into the soil, rather than reaching towards the sky. At about 20 cm in depth, it reverses course and pushes its leaves above ground. As a consequence, if the forest is cut and burned, the heart of the palm (apical meristem) is protected under the soil from the flames, allowing it to resprout a short while later. This feature is key to understanding how the species has been managed to the present day. When the density of the palm in an area is low, small patches of rainforest are cleared, sparing only adult piassava. The slash is allowed to dry, and then it's burned. Within a few weeks, a near carpet of piassava seedlings emerge, having survived the heat of the flames safely ensconced below the surface. Within a few decades, the seedlings grow and form an almost monotypic stand of adult, fiber-producing palms that people are able to harvest for many decades to come (Voeks 1988) (Fig. 2.4). Other useful palm species were in the past managed in a similar manner: the babassu (*Orbignya martiana*) of Brazil, the palma real (*Scheelea liebmannii*) of Mexico, the conga (*Welfia georgii*) of Panama, the African oil palm (*Elaeis guineensis*), and many others. Today, however, many palms in the American tropics are overharvested and/or managed unsustainably. Saxophone germination in palms now often points to deforestation and irrational exploitation, with isolated and unreproductive palms in pastures and farm clearings marking "virtually dead" populations (Bernal et al. 2011; Biwolé et al. 2015; Montúfar et al. 2011; Sanín et al. 2013; Sauer 1963; Voeks 1988).

2.4 A. Piassava palm (*Attalea funifera*) growing behind a home, Maragojipe, Bahia. B. Piassava fiber brooms for sale in Rio de Janeiro. Photo credit: Robert Voeks.

In addition to fire management of individual species, the presence of concentrations of individual trees in otherwise biodiverse forest assemblages can also be attributed to the movement of plants by people. In the course of their trading networks along rivers, indigenous people (and current residents) move planting materials, such as seeds, suckers, and cuttings, along informal associations of friends and family members. These networks enhance the diversity of garden and swidden plants, and contribute to the overall distribution patterns of useful forest species (Ban and Coomes 2004; Sander and Vandebroek 2016). Transplanting of useful wild species in swiddens and along trails, sparing valuable trees during clearing, and other forms of encouragement further modify species composition and distribution. In Papua New Guinea, for example, there are a wealth of so-called wild tree species, including marita (*Pandanus conoideus*) or tulip (*Gnetum gnemon*), which owe their presence to the husbanding activities of local people (Guix 2009; Kennedy 2012). In Serem, Indonesia, concentrations of the valuable nut-yielding iani trees (*Canarium* spp.) point to selective discarding of seeds near settlements (Ellen 2007). And across Borneo, the presence of patches of betel nut palms (*Areca catechu*) in the forest attest to the past kitchen gardens of indigenous forest folk (Fig. 2.5).

Entire landscapes of supposed virgin tropical forest may well be cultural artifacts of past (and some ongoing) human modification (Clement et al. 2015).[2] The primary tool is the seemingly primitive process of forest farming with fire. Sedentary cultivators in nearly all tropical forests employ a form of shifting horticulture to cultivate their crop plants. The vegetation in a relatively small area, certainly less than two hectares, is cut down, allowed to dry, and then burned. The fire incinerates most of the leaves, twigs and branches, where most of the ecosystem's nutrients are stored, leaving the trunks and large branches to slowly decompose. The charred soil, now enriched with minerals from the burned plants, especially phosphorus, magnesium, and potassium, is planted with the annual crops particularly prized in that particular region, often manioc (cassava), maize, chili peppers, hill rice, yams, sweet potatoes, taro, or millet. The crops are not planted in tidy rows, like those on temperate farms, but are rather haphazardly distributed, conforming to the vagaries of the site's soils and micro-topography. This is an overly simplified version of shifting cultivation, as there are in fact many variations on the theme. These include short and long fallow cultivation regimes, homegardens, timber and fruit stands, intensive cultivation, and grazing lands (Padoch et al. 2007). In all cases, at least to the uninitiated, swiddens appear rather chaotic and disorganized. But, assuming low

2.5   Row of betel nut palms (*Areca catechu*) in second-growth forest, Sarawak, Malaysia. Photo credit: Robert Voeks.

population numbers, shifting cultivation is the best adapted system for supporting people in an otherwise challenging environment. And the long-term impacts of these actions are crucial to understanding the biogeography of tropical forested landscapes.

Until roughly the 1970s, it was generally assumed by researchers that swidden plots were abandoned at the end of a few years of cultivation. By this time, they were believed to be overrun with weeds and pests, and most of their soluble nutrients had been leached from the soils by torrential rainfall. Consequently, forest farmers moved on to a new plot to start the process anew. This is, however, only part of the process. Upon further examination, it was discovered that these abandoned sites (fallows) are seldom completely abandoned, but rather represent a fundamental tool for long-term human management of the forest. For example, cultivators nearly always spare and protect valuable species during the swidden process, such as useful palms, fruit trees, rattans, or medicinal plants. They are useful, and so they are protected. People further enrich the site by collecting seeds and seedling of wild species (wildlings) or semi-domesticated species and planting them, such as Brazil nut (*Bertholletia excelsa*), durian (*Durio* spp.) rubber (*Hevea brasiliensis*), Ramon tree (*Brosimum alicastrum*), rambutan (*Nephelium* spp.), illipe nut (*Shorea* spp.), and many others. In this case, with little immediate competition from other trees and bathed in sunshine due to the removal of the overhanging forest, these otherwise wild but now "cultivated" trees grow and eventually came to dominate the abandoned plot. In some cases people will clear away competing species in the swidden, prune the useful plants, mulch near the especially valuable species, and otherwise improve the site. The result is that as the fallow matures over the coming years, it comes to appear more and more to the untrained eye like untouched forest. In reality, however, it is an artifact of the long-term process of cultural management of nature (Biwolé et al. 2015; Brookfield 2007; Denevan et al. 1985; Peluso 1996).

Cultural forests are heterogeneous in form. They resemble jigsaw puzzles of more or less humanized habitats, grading from barely affected to nearly completely products of human agency. For the Ka'apor of Brazil, for instance, some patches of the forest are frequently disturbed, especially the current swidden (*kupiša*) and the house garden. Others were swiddened and managed many years in the past (*taperer*), while still others are barely managed at all (*ka'a-te*). All are crisscrossed with a network of trails. Recent fallows are where domesticated crops like manioc and sweet potatoes are intensively cultivated. Nearby house gardens (*kar*) are planted with valuable old-growth forest trees, like wild cacao (*Theo-*

broma speciosum), and vegetables. Around the edges of the swidden, non-domesticated species are encouraged or planted to attract game, such as brocket deer and collared peccaries. For the Amazonian Kayapó, their old fallows provide food and medicine, bait for fish and bird traps, thatch, paint, oil, insect repellents, and many other uses (Balée and Gély 1989; Posey 1985). Similarly in northern Thailand, most of the 295 recorded wild food plants are derived from old fallows (Ellen 1998). In Indonesia, a single Iban community registered a total of 214 species of wild vegetables, most growing in long fallows (Christensen 2003). Old fallows may appear natural to the untrained eye, but for indigenous people they are a quintessential part of living biological history.

Tropical forest management is not restricted to old fallows. Useful species, especially wild vegetables and medicinals, are encouraged to grow along trails and around the edges of villages. Useful trees and herbs are transplanted into backyard gardens, where they are tended and genetically modified over time. Mesoamerica (southern Mexico through Central America) in particular is a hotbed of past and present species management. Some 600 to 700 useful species are managed in the wild by myriad traditional methods. Many are trees, shrubs, and cacti, but many others are *quelites* (weedy and herbaceous greens) that are crucial additions to the diet of Mesoamerican people. *Quelites* are managed in place, transferred to backyard gardens, or left wild. Mexico alone has upwards of 3000 weedy species, and almost all are useful for food or medicine or both (Vibrans 2016). People generally distinguish tasty *hembra* (female) varieties from less flavorful *macho* (male), although neither of these designations relate to the actual sex of the plant. People encourage the growth and reproduction of the *hembra* types, while weeding out the less desirable *macho* types (Casas et al. 2007). Management of wild trees and herbs has likewise been documented in the Republic of Benin. A total of 22 wild herbs and 18 trees are consumed and managed in three villages. Some are maintained in old fallows, while others are otherwise protected, nurtured, or tended in nature, or they are transplanted and tended near the home (Avohou et al. 2012). Even in more intensively cropped areas, wild species appear in the repertoire of local people. In Laos, *phak* (edible weedy herbs) are encouraged around the edges of paddy rice fields. Some of the 25 edible species are pruned and protected, and most enter into a popular local dish called *sup phak* (boiled vegetables) (Kosaka 2013).

Although they are made up of useful trees, shrubs, and herbs, the vegetation mosaics associated with tropical cultural landscapes are more than simply material resources. They also represent the living legacies of indigenous people, past and present. They are spaces where the hunting was remembered as good or bad, where personal relationships were renewed and memories were made, where battles were fought, and where ancestors passed on. The Waorani of the Ecuadorian Amazon, for instance, have 33 "family landmark" species that record the presence of habitat time categories. Among these, unmanaged and least anthropomorphic sites are known as *titeiri kewënko*, and can be identified by the presence of showy heliconia (*Heliconia episcopalis*). Some ancient fallow sites, known as *pikenani bai*, can be identified by the presence of the peach palm (*Bactris gasipaes*), a remarkably useful palm with delicious and nutritious fruit. These and other time-vegetative categories for the Waorani show both that their forest home is a function of past and present management activities, and that each category represents a signpost

2.6   Kenyah burial pole, Sarawak, Malaysia. Photo credit: Robert Voeks.

of the group's living and nonliving social history (Zurita-Benavides et al. 2016).

The social significance of vegetative landmarks was brought home to me many years ago, as I traveled up the Baram River by longboat in Sarawak, Malaysia. As we rounded a bend, a tall carved wooden pole stood barely visible in the forest. My colleague, a Kenyah villager from the area, explained that in the old days, when a village headman died, his remains were placed in a jar at the top of a burial pole. There his corpse remained for several years, decaying and being devoured by forest creatures. After some years, the bones were taken down and cleaned, there followed an elaborate celebration, and the bones were properly buried. In this particular case, my friend could not recollect the name of the deceased, as it had been more than a century since he passed, but he made it clear that this particular vegetative landmark held special significance for local people[3] (Fig. 2.6).

Footprints in the Forest

The current use and management of tropical forested landscapes reveals much about the processes that make them cultural. But these actions, including seed dispersal, species sparing, and shifting cultivation, have been going on for centuries if not millennia. In some cases, researchers have been able to discern current forest formations and species distributions that are the outcome of ancient landscape manipulation. In the rainforest zones of the Congo River Basin, for example, people have been carrying out agriculture for at least two millennia. Locally domesticated oil palm (*Elaeis guineensis*) was an early subsistence crop, as was Southeast Asian water yams (*Dioscorea alata*), taro (*Colocasia* spp.), and plantains (*Musa* spp.—starchy bananas), which may well have arrived in Central and West Africa over 3000 years ago. The arrival of these latter three exotic crop plants—the "tropical tool kit"—have been implicated in the long-term penetration and management of ica's tropical forests by early Bantu peoples. Prehistoric burning of forests and the widespread presence of charcoal in African soils during particularly wet periods attests to the ubiquity of human forest use and management. The pollen record from lake sediments clearly shows major changes in species presence several thousand years ago, as useful species and other light-demanding plants increasingly dominated the landscape. Species of Irvingiaceae (such as *Irvingia gabonensis*—bush mango), an important forest food tree, and *Pausinystalia* (such as *Pausinystalia yohimbe*—an

47

2.7 Brazil nuts. Shepard, G. and Ramirez, H. "'Made in Brazil': Human Dispersal of the Brazil Nut (*Bertholletia excelsa*, Lecythidaceae) in Ancient Amazonia." *Economic Botany* 65 (1): 44–65. Photo credit: Glenn Shepard.

herbal aphrodisiac), may well have been spared and husbanded for the past 1000 years or more. In Cameroon, apparently pristine forests are frequently underlain with charcoal, indicating previous shifting agriculture. The species composition of central African rainforests "still echoes historical large scale disturbances," mostly of human origin (Blench 2009; Brncic et al. 2007; Van Gemerden et al. 2003, 1388).

Brazil nut represents one of the best examples of human influence on so-called wild species distribution (Fig. 2.7). This nutritious and delicious seed has been appreciated by native South Americans since the earliest human occupation, and it is today the only globally traded seed crop that is harvested entirely from nature. For many, Brazil nut is iconic for the pristine nature of Amazonia. However, although the species is widely distributed over the lowland forests of South America, from Bolivia and Peru to the Guianas, it has a suspiciously patchy distribution that has long intrigued scientists. Where Brazil nut occurs, it is usually in groves of 50–100 individuals, which are separated by many miles of suitable habitat that entirely lack the species. Using a combination of genetic and linguistic evidence, researchers have determined that Brazil nut was dispersed by indigenous people thousands of years ago, as Arawak speakers expanded their range and cultivated large sections of South America's tropical forests. Today, even Brazil nut trees in seemingly pristine rainforests are there largely due to the actions of ancient Amerindians (Shepard and Ramirez 2011).

At the landscape level, the best documented example of humanized tropical landscapes may well be those inhabited by the Mesoamerican Maya. The Classic Lowland Maya civilization prospered from about 250–900 CE and then declined precipitously over one thousand years ago. Although the cause of the classic Maya collapse continues to be debated, the botanical impacts of their actions are still in evidence. In their time, the Maya built imposing temples and cities, often surrounding *cenotes* (limestone sinkholes), as well as massive irrigation systems and earthworks. Their population was large, and the demand for food by urban elites was considerable. They built raised fields to avoid flooding, and fertilized them with nutrient-rich soil from swamplands. They practiced long fallow forest management, through culling and planting and tending. In nearly every way, the forests of the Yucutan, Belize, lowland Guatemala, and Honduras were modified to meet the needs of a growing human population. The great civilization that built the cities and managed the forests is gone, but their descendants remain, as does the aftermath of their ancient actions. One of these was the creation of orchard gardens, or *pet kot* (*pet kotoob* plural). The ancient Maya planted wild, old-growth fruit trees in old fallows, or perhaps discovered useful fruit trees in the forest and began planting around them. They then stacked limestone around the periphery. *Pet kotoob* ranged considerably in size, from 20 to 24,000 square meters. They were so widespread and in existence for so many centuries that today the otherwise old-growth forests of Guatemala and Belize are dominated by the descendants of useful wild species that were planted and tended by people over one thousand years ago. So humanized are the forests of Mesoamerica that some refer to them as "artificial rainforests" (Clement 1999; DeClerck et al. 2010; Gómez-Pompa et al. 1987; Peters 2000; Ross 2011; Wiseman 1978).

The footprints of age-old forest management in lowland South America are equally in evidence. There, entire watersheds have been domesticated over thousands of years, to the point that many plant and animal populations today owe their density and distribution to the efforts of ancient peoples. Hidden from view under the canopy of old-growth forests is evidence that people built major earthworks to improve the productivity of the forests, fields, and waterways. Canals were dug, roads and aqueducts were constructed, and soil was heaped into mounds and long fishbone-shaped ridges. Fortified and densely populated human settlements were built. Ancient peoples modified the soils as well by heaping their organic trash and charcoal into peripheral dump heaps. Over time, they turned otherwise nutrient-poor soils into nutrient-sufficient and even nutrient-rich dark earths, known as *terra preta do indio*, or anthrosols. Five hundred to over two thousand years later, *terra preta* covers some 10% of Amazonian territory. Old secondary forest inhabited by unusually large numbers of useful species suggests a form of soil and vegetation coevolution (Clement et al. 2015; N. J. Smith 1980).

The significance of these early land use actions is highlighted by the fact that although Amazonia is hyper-diverse—with a total number of tree species estimated at 16,000 or more—just 227 tree species account for over half of the individual species encountered in any single forest area. The fact that many of these dominant species are particularly useful to people makes for a "compelling hypothesis" that the floristic dominance of the entire Amazonian forest ecosystem is at least partly due to the actions of pre-Columbian people. This hypothesis is further supported by the recent discovery that the 85 domesticated woody species native to Amazonia are five times more likely than non-domesticated species to be hyperdominant in "natural" forest formations. The only sensible explanation is that the composition of these forests is largely cultural (Levis et al. 2017). Even further evidence comes from the observation that useful trees increase in density as a major or secondary river is approached. There is no reason for this to be true if species distributions are controlled entirely by environmental factors, such as soils and aspect. But because human populations in the past were highest near rivers, this useful species gradient is exactly what you would expect if the forests were anthropogenic in origin. Clearly, at various scales and in different tropical forested landscapes, the distinction between fallow and forest, virgin and disturbed, anthropogenic and natural, in the end all seems rather arbitrary. For the people who have lived, toiled, and died in the tropics for millennia, the virgin forests of the Western imagination are profoundly humanized, with roots penetrating deeply into the cultural histories of its native peoples (Clement et al. 2015; Junqueira et al. 2011; Levis et al. 2017; Ter Steege et al. 2013).

The ethnobotany of Eden has drawn outsiders to the tropical latitudes since before the Columbian landfall. Hidden deep in the virgin jungle, they imagined, were the answers to some of humankind's greatest desires—unimaginable wealth, healing plants, and perhaps immortality. Over time, as the early biblical quest turned secular, new myths and metaphors arose to capture their imagination, each as culturally constructed and impervious to facts as the previous. The notion of an Amazonian El Dorado persisted well into the twentieth century, fueled by the disappearance of British adventurer Percy Fawcett in his search for the great lost city. And the recent discovery of fantastic 2000-year-old geometric figures (geoglyphs) under the forest along the border between Brazil and Bolivia, rivaling in complexity and mystery Peru's Nazca lines, threatens to rekindle interest in a new generation of rainforest dreamers and schemers (Watling et al. 2017). Pop culture has had a hand in preserving and twisting these myths as well. Between 1900 and 1980, over two hundred American films were released with the word "jungle" in their title, and many more appeared with the subject of lost tropical worlds as their setting—*The African Queen, King Kong, Tarzan,* and *Congorilla* to name a few. The most resilient myth, however, seems to be the existence of virgin nature. In spite of decades of evidence to the contrary, the notion retains "a pernicious grip on the Western popular imagination" (Enright 2009; Schaan et al. 2012; Sluyter 1999, 380). Its persistence is rivaled only by the myth of the noble savage.

THREE

People in the Forest

How we envision tropical forests—wild and ungodly, pestilential and pernicious, or virgin and pristine—has metamorphosed over time in response to varying armchair theorizing, travelers' embellishments, and the environmental claims of outsiders. But just as our understanding of these sylvan landscapes is the product of centuries of cultural construction, so too are our views on the indigenous folk who call the forest their home. At present, many envision the warm and wet equatorial latitudes as inhabited by mysterious and wise native peoples. They are the caretakers of tropical nature's hidden healing secrets. And they are instinctive conservationists. Far removed from the frenetic pace of urban existence, indigenous forest-dwellers move in harmony with the rhythms of the natural world, taking only what they need, exploiting plants and animals and soils, but never to excess. They walk gently in the forest, and they are mindful of its physical and biological limitations. In their apparent wildness and liberty, they are foil to so much that is wrong with industrial and postindustrial society. Through trial and error, they have learned the limits of their resources, carving out a sustainable and life-affirming subsistence through their conscious and ritually reinforced relations with nature. They are unencumbered by vicious politics, mind-numbing bureaucracy, and corrupt religions. The forest canopy is their cathedral; the beasts and birds are their gods, the streams and soils their goddesses. They harvest with reverence the succulent fruits of natural wisdom. In their innocence and simplicity, the archetypal indigenous men and women of the mythical pristine tropical rainforests represent everything good and meaningful that we do not.

This idealized and over-romanticized representation of tropical forest people is a centerpiece of the jungle medicine narrative. But unlike the other strands of the story that are grounded mostly in history and science, this depiction of the denizens of the forest is a miscellany of short-term observation, sentimental conjecture, and wholesale fabrication two thousand years in the making. And although much of the forest-dwelling archetype was constructed and deconstructed over the course of European colonial expansion, it is impossible to fully appreciate how indigenous people have come to be seen as the mysterious gatekeepers of nature's botanical treasures without first plumbing some deeper history.

Tropical Monsters

For Aristotle, Hippocrates, Strabo the Geographer, and other Greek and Roman scholars, the unknown equatorial Torrid Zone was at best uninhabitable by humans and at worst a scorched earth or boiling sea, bristling with monstrous human races and a polymorphic bestiary. Among these were the Amazons, warrior women who chopped or burned off their right breasts to improve their bowmanship, the Antipodes, whose feet faced backwards, the Androgeni, who had both male and female genitals, and the Cynocephali, dog-headed monsters who lived in caves, communicated by barking, and were skilled hunters (Friedman 1981, 9–21). Although the mystical lands of India were well within the known world of the ancients, their perception of its inhabitants was surprisingly no less bizarre. Pliny the Elder, the great Roman natural historian, reported that the Astomi, who resided near the mouth of the great Ganges River, were a people without mouths, "who subsist only by breathing and by the odours." There were also the Sciapods, the one-legged tribe who lie on their backs and use their monstrous foot to shade themselves in hot weather (Fig. 3.1) (Schedel 1493). If these depictions seem exaggerated or beyond reasonable belief, recall that common people of the period believed that the world was flat, that disease resulted from bile and bad air, and that the Sun charted its course around the Earth, not vice versa. The belief that the Torrid Zone was inhabited by fiendish beasts and sub-human beings would not have seemed at all farfetched.

In the view of the ancient Greeks, the inhabited physical world (ecumene) was centered on the region around the Aegean Sea. This is quite

A

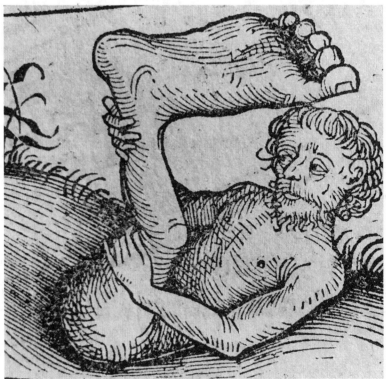

B

3.1   Hartmann Schedel. 1493. A. The headless Blemmye. B. The giant-footed Sciapod. *Liber Chronicarum* (or *Nuremberg Chronicle*), Folium 12. Nuremburg: Anton Koberger. RB 92586. The Huntington Library, San Marino, California. Digital Images by Robert Voeks.

sensible, as people generally employ space and proximity as a distinguishing feature of Us and the Other. As distance from this peopled core area grew, they reasoned, the chance of encountering proper human inhabitants declined progressively. Aristotle further conceived of the polis (the central community) as the domain of rational men (women and slaves were considered irrational beings), and that those who lived outside of the polis were either non-humans, or were barbarian monster-people suited only for enslavement. Indeed, the term barbarian was derived from the Greek term "barbaros," named for the onomatopoeic sound of the babbling foreigner who could not speak Greek (Nail 2015, 188). The barbarian stereotype, unsurprisingly, legitimized later colonial and imperial enterprises among the distant and different "primitive races."

Food choice for the ancient Greeks represented a fundamental index of humanity. The gods, it was said, dined on the sublime aromas of food, civilized men ate domesticated animals, uncivilized men subsisted on raw meat and insects, while the lowliest of beasts were cannibals, who fed on their own kind. "Man is not an animal," noted Greek philosopher Porphyry, "that eats raw flesh" (cited in Jahoda 1999, 98). The primary distinction between man and beast was the rule of law; man had justice, animals did not. And because animals obeyed no laws other than brutish instinct, they were condemned to consume raw food and each other. There was, in effect, a perceived core to periphery spatial gradient—from civilized man, to raw flesh eaters, and finally, in distant and unknown spaces, to cannibals. "The kingdom of cannibalism begins at the frontier where justice ends" (Detienne 1981, 218). The seeds of our later colonial-era obsession with cannibalism, it seems, were sown early.

The core-periphery conception of inhabitable spaces also translated neatly into a greater astronomical cosmography, with the globe cleaved precisely along mathematical lines: poles, longitudes and latitudes, Equator, tropics, and arctic zones. Most importantly, the short distance from Sun to Earth in the horizontal swath between 23.5° N and 23.5° S, or between the Tropics of Cancer and Capricorn, seemed to render it uninhabitable by humans due to its extreme heat and blistering solar radiation. Common sense dictated that any adventurer foolish enough to journey to such perilous latitudes could expect "a great and burning stream of ocean" (Flint 1992, 33). The word Ethiopia, which for the ancients was shorthand for black Africa, was derived from the Greek term for "burned face," referring to the belief that proximity to the Sun was the cause of darkly pigmented skin. At the other geographical extremes, the high latitude Frigid Zones, lack of sufficient sunshine reaching the

Arctic and Antarctic to the north and south of 66.5°, respectively, made them similarly lands (or seas) uninhabitable by people.

Mediterranean cosmographers developed elaborate schemes around these underlying geographical principles. Although they often carved the Earth into dozens of "klima" zones, it was the elegant simplicity of Aristotle's five climate zones—two Frigid Zones, two Temperate Zones, and a single, equatorially situated Torrid Zone—that assured its continuity through the ages. The term "torrid" seemed intuitively rational given the limited experience of the Greeks. Knowing that the Earth was a sphere (common knowledge among the educated), and that the Sun migrated increasingly overhead as the Earth's midline was approached, it stood to reason that the great latitudinal swath equidistant from the poles must be torturously hot, or torrid. Indeed, the hottest place with which Mediterranean people were familiar was nearby Libya, at 32° north latitude, which still holds the world's heat record (136°F, 58°C). Imagine how hot it must be at lower latitudes, they reasoned (Glacken 1967; James 1972, 37)! So logical and compelling was this geographical division of the Earth that writers well into the nineteenth century still referred to the tropics as the Torrid Zone, and of course, the Temperate Zone term continues to be used to this day.

This cleavage of the Earth into equally spaced zones, often in groups of 3 or 5, reflected as much the early Greek's love of symmetry as it did known terrestrial space. Herodotus divided the barely known regions of the upper Nile Valley into three zones: the "inhabited region," the "wild beast tract," and finally an unforgiving region to the south too fiery to be inhabited by living things. What little the Greeks and Romans knew of the equatorial latitudes convinced them that skin color was the result of solar intensity. Pliny the Elder remarked that "in the regions which lie to the south of the Ganges, the people are tinted by the heat of the sun, so much so as to be quite coloured, but not burnt black like the Ethiopians" (Pliny the Elder 1890 [77–79 AD], 46).

These and other "environmental determinist" theories, wherein human morphology and behavior, disease and healing, were assumed to be influenced by the Sun, Moon, constellations, winds, and air, continued to dominate Western thought for nearly two millennia. They were keenly on the minds of Iberian argonauts as they planned and executed their fifteenth- and sixteenth-century expeditions into terra incognita (Bunbury 1959, 274–275; Cosgrove 2001, 36–39; Flint 1992, 33; Friedman 1981, 9–21; Vespucci 1992 [1504], 45). Columbus was well schooled in the theories of the ancients, as well as the opinions of more recent thinkers. He relied heavily on the ideas of French Cardinal Pierre d'Ailly's cosmography *Imago Mundi*, written in the fourteenth century, which theorized on what to expect of unknown people to the west. And he carried his own dog-eared copy of *The Travels of Marco Polo* (Flint 1992, 53 and 72–73). From these and other sources, Admiral Columbus was fully prepared to encounter, and did in fact report, the existence of dog-headed humans on the island of Hispaniola, as well as people waving their tails about on the newly discovered isle of Cuba. Ferdinand Magellan was also well versed in Pliny's fanciful ruminations. During his inaugural circumnavigation of the globe, he expected to meet African men who morphed into women on the night of their wedding, as well as tropical women who "rejoice at the rebirth of their deceased infants" (Bergreen 2003, 74 and 81–83).

The European Age of Exploration disproved the direst and most macabre speculations advanced by the early philosophers and cosmographers. Amerigo Vespucci, for instance, who navigated the Caribbean and most of the eastern seaboard of South America during his (likely) four voyages (1497–1500), disputed out of hand the old theories about the boiling lands and oceans of the equatorial latitudes, and about the presence or absence of humans. With little fanfare, given the monumental significance of his proclamations, he recorded in a 1502 letter to Lorenzo di Pierfrancesco de' Medici that "we entered the Torrid Zone and passed south of the equator and the Tropic of Capricorn." From first-hand experience, he stated matter-of-factly that "this opinion of theirs [the ancients] is false and contradicts all truth, since I have discovered a continent in those southern regions that is inhabited by more numerous peoples and animals than in our Europe, or Asia or Africa" (Vespucci 1992 [1504], 29 and 45). Columbus would have agreed, believing that he too had safely entered the Torrid Zone during his pre-1492 voyages to northwest Africa. However, he strategically failed to publicize this particular opinion, as silence regarding the habitability of equatorial regions reinforced his argument that the only plausible route to the riches of the Indies was to the west across the vast Atlantic rather than the proverbial sizzling seas to the south (Flint 1992, 1–41).

These and subsequent discoveries of equatorial human populations forced a reconsideration and reconstruction of the long-held theories about the lower latitudes. The uninhabitable intertropical world was, in fact, quite habitable. But were these beings really human? Dark skinned and barbarous by the standards of the time, and exposed to the influence of solar excess and astrological imperfections, newly encountered tropical people challenged Mediterranean conceptions of what constituted the human condition. Whatever their God-given lineage, these human-appearing creatures were clearly brutes as seen through the medieval European lens. And so chroniclers of the age, marshalling once again the malevolent influence of the tropical Sun and other geographically associated environmental influences, reformulated their earlier ideas regarding equinoctial habitability. What had been conceived of as spaces too climatically harsh to sustain people, morphed over time into landscapes inhabited by beings that were physically human, but were culturally closer to jungle beasts.

According to observers of the time, newly encountered people in the American tropics were distinguished by a suite of distinctly un-European features—licentiousness, indolence, godlessness, and worst of all, cannibalism. From the point of view of the medieval Spanish and Portuguese, these were beings whose God-given powers of innovation and invention were cobbled by climate and astrology, and who were in every way inferior to themselves. Much like the exotic and unique plants and animals that were being discovered and documented in these new lands, the idealization of tropical people was molded and stretched over the centuries by the preconceptions of explorers and settlers, the ecclesiastic expectations of the Catholic clergy, the Machiavellian behavior of distant monarchs, and the philosophical musings of European scholars. More so than any of the cultures encountered in equatorial Africa, Asia, and Oceania, the newly encountered "Indians" in the Americas, a landscape unimagined by the ancient philosophers and geographers, would be the cornerstone of these changing views and, ultimately, the "noble savage" paradigm.

New World Natives

There was little to suggest in the early explorer and traveler's accounts of Amerindian behavior that they would achieve the ennobled cultural status they are often afforded today by Westerners. Indeed, almost from the initial New World landfall, reports of animalistic behavior and sub-human traits were the rule. Foremost among these, cannibalism appeared conspicuously and with grisly detail in the journals and letters of explorers and settlers in the Caribbean and South America (Petersen and Crock 2007). New World indigenes were man-eaters, it appeared, surely the vilest of all bestial behaviors. Guillermo Coma, a sailor on

Columbus's second voyage, reported, "These islands are inhabited by Canabilli, a wild unconquered race which feeds on human flesh" (letter from Guillermo Coma 1494, cited in Sale 1990). Amerigo Vespucci, in his third voyage along the northeast coast of Brazil, chronicled the scene of natives clubbing and dismembering one of his young seamen, and finally "roasting him before our eyes, showing us many pieces and then eating them." In other correspondence, he described seeing "salted human flesh hanging from house beams, much as we hang up bacon and pork" (Vespucci 1992 [1504], 50 and 89). Hans Staden, a German gunner captured and held captive by Brazil's indigenous Tupinambá for nine months in the 1550s, transcribed in lurid detail the process leading up to the act of people-eating (anthropophagy) by his captors. During the months that the natives fattened him for his own day of reckoning, Staden witnessed firsthand the ritual slaughter and consumption of other prisoners. The Tupinambá were, on the one hand, "a well shaped people," he recorded, but they were also "cunning in all wickedness, and it is their custom to capture and eat their enemies" (Staden 1928 [1557], 129). Jean de Léry, who established the first Protestant mission in the New World, corrected the widely publicized opinion that the Brazilian Indians cooked their enemies on a spit. They were "barbequed," he observed, on a wooden grill called a "boucan" (De Léry 1990 [1557], 79). Later relying on these vivid, first-hand narratives, European geographers compiled the earliest maps and wood-cut illustrations of the New World, often with human legs and torsos skewered on a barbeque or grill (Bry 1592) (Fig. 3.2).

Whether these early reports represented factual depictions of events actually witnessed, or rather were exaggerations intended to justify colonization and enslavement of newly encountered peoples, has long been debated (Whitehead 2000). Kirkpatrick Sale calls Admiral Columbus's report total myth and "fabrications born in the fable-heated mind of Cristobal Colon" (Sale 1990, 130). By demoting indigenous people to the bottom of Aristotle and Porphyry's cultural food chain, early chroniclers negated any sense of customary ownership and intellectual property regarding local resources, and legitimized whatever level of violence was necessary to wrest them from the occupants. Local men were "written as animal and hence deprived of the rights of his lands and entitlements" (Rodríguez 2004, 14). In the case of Columbus, he most certainly expected to discover cannibals among the Indians he encountered from his reading and personal annotations on human flesh eaters in his copies of the *The Travels of Marco Polo* and the writings of

3.2   Theodor de Bry. 1592. *Americae Tertia Pars*. 1st ed. Part 3. Pg. 179. RB 70614. The Huntington Library, San Marino, California. Digital Image by Robert Voeks.

Pope Pius II. Nevertheless, the veracity of these observations, and whatever cultural significance the capture and consumption of their human neighbors may have had for aboriginal Americans, were less important than the twisted idea it planted in the minds of distant Europeans of the American native's bestial manner of subsistence (Dean 1995, 31–32; Flint 1992, 140–142; A. C. Metcalf 2005, 47–49; Staden 1928 [1557]; Vespucci letter to Pedro Soderini 1992 [1504], 89; Vespucci letter to Lorenzo di Pierfrancesco de' Medici 1992 [1504], 50).

The lowland tropics are by definition hot, wet, and humid. In such conditions, covering the body with clothing is not only uncomfortable, but also unhealthy. Consequently, the forest-dwelling people encountered by Europeans were almost entirely naked. What clothing they did wear was more ceremonial than protective; it certainly was not worn for the sake of modesty. Amerindian nudity, it should come as no surprise, was a topic of considerable interest to the all-male cast of European visitors. And considering the moral codes of medieval mariners, native nudity was easily equated with carnal liberty. From the perspective of early explorers, these latter-day sirens wantonly exposed even the most private parts of their bodies in shameless efforts to entice them and their men into illicit sexual unions. Their husbands, according to many, were willing accomplices. Amerigo Vespucci noted with his usual directness, "They showed themselves to be very desirous to copulate with us Christians" (Vespucci 1992 [1504], 64). Christopher Columbus, in a letter to Queen Isabella after his fourth voyage, described two young Indian girls who were sent to him, aged roughly seven and eleven, who were "both so abandoned that they were not better than prostitutes" (Columbus 1932 [1498], 100). Pêro Vaz da Caminha, the scribe on Pedro Cabral's April 1500 maiden voyage to Brazil, offered the first Portuguese accounts of eastern South America. In Brazil's "birth certificate," as this famous letter is often termed, Vaz da Caminha offers an encyclopedic and rather uninspired account of the lands and resources encountered in the Terra de Vera Cruz. The notable exception was his steamy sketch of the charms of the native women. In numerous passages, Caminha could barely contain his excitement as he described the women who "go about naked, with no clothing. They do not bother about to cover or to uncover their bodies, and show their private parts as readily as they show their face." Later "One of those young women had the whole body painted, from bottom to top with that tincture [achiote—*Bixa orellana*], and sure she was so good shaped and so rounded, and her private part so graceful that most women in our land, if had seen those features would feel ashamed for not having their own like she has hers" (Caminha 1963 [1500], 40; also see De Léry 1990 [1557], 67). At least two sailors who participated in this inaugural voyage to Brazil were so enamored that they jumped ship, forsaking forever a return to mother Portugal.

Amerigo Vespucci similarly recorded the provocative nature of the women in nearly all his correspondence. "Their women, being very lustful, make their husbands' member swell to such thickness that they look ugly and misshapen; this they accomplish with a certain devise they have and by bites from certain poisonous animals" (Markham 1894 [1503], 46). Missionary Jean de Léry had extensive contact with Brazil's indigenous Tupinambá. In regards to their morals, he reported "they have no other law than that of nature," they are free to have sex before marriage, and the girl's fathers have "no scruples about prostituting them to the first comer" (De Léry 1990 [1557], 153). For Europeans intent on emphasizing the animalistic character of newly encountered tropical people, such obvious sexual deviance was nearly as damning as cannibalism. And for young and restless would-be adventurers back in Spain and Portugal, the arousing depictions of physical beauty and moral lassitude added mightily to the draw of the newly found world. Most importantly, the sexualization of Amerindian women created a distorted stereotype that would endure in various forms well into the twentieth century.

Another point upon which nearly all early visitors agreed was that Amerindians were addicted to idleness. Compared to their industrious Mediterranean conquerors, New World Indians were almost criminally lazy. And much like their skin color and purported lack of ingenuity, environmental influences drawn from the texts of the ancients, especially the effects of celestial bodies, were used to explain these character shortcomings. The Spanish physician Francisco Hernández, known in his time as the "Third Pliny," spent seven years in early sixteenth-century Mexico compiling an astounding census of the healing flora, some 3000 new species of medicinal plants. Working closely with Nahua indigenous herbalists and healers, he tested the efficacy of local medicinal plants at the colonial hospitals throughout Mexico. His great herbal (not published until several decades after his death) offered numerous observations on the behavioral attributes of his patients and native assistants. The local people were docile and idle, he argued, resulting from the weak star arrangement and the humid air. Similar views were put forward by Samuel Purchas, the early seventeenth-century English cleric who published travel narratives in his widely read *Purchas his Pilgrimes*. He suggested that the stars in the Americas were smaller and dimmer than those apparent in Europe, and that these astronomical features as well as other environmental influences left the natives dull and dim-witted (Cañizares-Esguerra 2006, 64–95; Chabrán 2000, 21).

The construction of the "phlegmatic Indian" stereotype was a necessary element in colonial efforts to rationalize the forced labor of tens of thousands of native people in the mines and sugar plantations. The solution to their idle and "humid nature" was the harsh regimen of forced labor and iron-willed discipline (Cañizares-Esguerra 2006). Accordingly, colonial authorities in New Spain and Peru redeployed the old "mita" system in 1605, a pre-Columbian process of counting and conscripting laborers for the military and public works projects previously employed by the Incas. In what amounted to a thinly veiled form of slavery, a given percentage of men and boys over the age of fifteen years from every village were obligated to toil for two to four months for the Spanish. Similarly in Brazil, enslavement of the native population was officially banned from the beginning of Portugal's colonization effort. But sugar plantation (*engenho*) owners desperately needed laborers, and so early on the answer was to barter Indian labor for iron pots and axes. For the local Indians, this represented a revolutionary leap in technology. But after they had acquired sufficient iron tools for cooking, killing, and cutting trees, they simply lost interest and went back to hunting and luxuriating in their hammocks. How to deal with this native idleness? One remedy was the acquisition of Indian slave laborers during "just wars" with intransigent indigenous groups who refused to acquiesce to Portuguese control (Schwartz 1985, 53). Another was the infamous "ransom law," or *resgate* (rescue). This provision, earlier employed in the Mediterranean and Africa, allowed for the lawful purchase of natives who had been taken captive in inter-tribal wars. In the case of Brazil, acquisition and enslavement of hostages was taken as an act of mercy, since the poor souls, if not "rescued," were assumed destined for death and the cooking pot of neighboring tribes (Hernandez cited in Cañizares-Esguerra 2006, 64–95; A. C. Metcalf 2005, 177–79). Better slave than supper, the thinking went.

The clash of European and Amerindian cultures was characterized by profound differences, such as cannibalism and standards of modesty. But contrasts in character and behavior were revealed as well by unusual day-to-day customs. Many of these habits seemed peculiar to visitors, but not necessarily menacing or heretical. They did, however, give early explorers and missionaries pause to reflect on their own lifestyles. For instance, the native people bathed incessantly compared to the colonists, especially the women, who were said to jump into a lake or river to bathe upwards of twelve times a day (De Léry 1990 [1557], 66). Moreover, these bare-skinned native people were robust and healthy, a conundrum for Europeans since domesticated wheat and livestock, such as cattle, pigs, goats, and sheep, were absent from the American cuisine. The primary source of animal protein for indigenous tropical Amerindians was wild game, such as tapirs, capybaras, pacas, and other forest biota, which was not especially abundant in the tropical forested zones. Perhaps it was the cannibalistic feasts that sustained them? Still other groups consumed the rudest of husbanded creatures, such as guinea pigs (*cuy*) in the Andes, hairless dogs (*xolo*) in the Yucatan, and river turtles in Amazonia (*arrau*). These New World "barbarians" were strong and sturdy beings, yet their venary and mostly vegetarian subsistence confounded and challenged the nutritional preconceptions of newly arrived Europeans.

These peculiar behaviors—frequent bathing and a naturalistic diet— contributed to fantastically long life spans among the native peoples of tropical America. Or so it was reported. How the earliest visitors would have come to these conclusions, with little or no command of the various languages, is an open question. In any case, Amerigo Vespucci recounted that the natives "do not sicken either from pestilence or corruption of the air," and live long lives, "one of the oldest men explained to me that he had lived seventeen hundred lunar months . . . 132 years" (Vespucci 1992 [1504], 31). Jean de Léry supported this view, reporting that Brazilian Indians regularly reached 100 to 120 years, and few of the elderly have gray or white hair (De Léry 1990 [1557], 56–57). In a 1549 letter, Brazil's first Jesuit missionary, Padre Manuel Nóbrega, encouraged elderly and infirm Portuguese to immigrate to Brazil, which is "very healthy and has good air" (Cartas do Brasil 1886 [1549–1560], 63). Years later, on the island of Guadeloupe, French missionary Du Puis reported that the native people "are subject to few illnesses and live long lives; it is not uncommon to see five generations alive at one time" (Du Puis 1972 [1652], 243).

These and other observations suggested to some that in spite of their considerable cultural shortcomings, the lifestyle of indigenous "savages" was in some ways preferable to that of their European invaders. For in addition to the salubrious atmosphere that blessed these sublime latitudes, a prerequisite to good health within the ancient humoral medical system, Léry contrasted the politically complicated lives of Europeans with those of Amerindians. He observed that they suffer none of the "avarice, litigation, and squabbles, of envy and ambition, which eat away our bones, suck out our marrow, waste our bodies" (De Léry 1990 [1557], 57). Such comments were not the pervading view of the time, but they would serve as a preamble to the noble savage paradigm that would emerge in later generations. In addition, the apparent existence of primitive super-centenarians ran contrary to long-held environmental determinist notions about longevity and climate, such as those put forth by British natural historian John Ray. He argued in 1692 that "the Inhabitants of Cold countries are longer live'd than those of the Hot because the cold keeps in the natural heat." However, when he was confronted with first-hand reports out of Brazil and elsewhere in the Americas, Ray was forced to admit the utility of empirical observation, noting that we are "deceived if we trust to our own Ratiocinations [logical reasoning] . . . and consult not Experience" (Ray 1692, 187).

Religion was a primary catalyst for European colonization of the Americas, at least in principle, and the clash of cultures in the spiritual realm was intense. The very earliest reports suggested that converting the godless natives would be an easy exercise. In Columbus's first description of Caribbean Amerindians, he maintained that they were a people without gods or idols, and that they were easily persuaded to kneel before the Christian cross. They were spiritually flawed, in his initial opinion, but salvageable. Léry likewise reported from coastal Brazil that aside from Satan, the natives "neither acknowledge nor worship any false gods" (De Léry 1625, 1336). Later, as longer-term observations came to light, it became clear that the forest-dwelling people of the American tropics were in fact not spiritual blank slates. The coastal Tupinambá speakers retained a complex cosmology, perhaps as rich as that of their bearded invaders. These included a high god, Monan, who was said to have introduced mankind to horticulture (Hemming 1978a, 55–60; Leite 1938, 14–19). Coastal people also recognized and feared "a prophet called Toupan," the deity of lightning, thunder, and rain, who "revealeth secrets" (Thevet 1625, 916).[1] During religious ceremonies, according to captive Hans Staden, "They go first to a hut and take all the women . . . and fumigate them. After this, the women have to jump and yell and run about until they become so exhausted that they fall down as if they were dead . . . [after which the women] are able to foretell future things" (Staden 1928 [1557], 150). The "heathens" maybe have been creatures of the Christian God, but their behavior, it was agreed by European scholars, was controlled by the Devil (Glacken 1967, 360–361). There were of course more technologically advanced civilizations in the Americas, with complex systems of religious belief and practice. The Incas, the Aztecs, and earlier, the Mayas constructed great edifices, elaborate cosmologies, complicated divination systems, and intense managerial relations with nature. But these exceptional civilizations contributed neither to the emerging outsider conception of indigenous forest dwellers nor to the later jungle medicine narrative.

The way of life of newly encountered peoples of the American tropics challenged in so many ways the European conception of what it meant to be human. In the early years, outsiders constructed a humanized equatorial zone that was "foil to the temperate, to all that is modest, civilized, cultivated" (Driver 2004, 3). Yet in spite of the profound Otherness stereotypes that were emerging, both observed and constructed, there were breaches in the narrative that left room for fresh interpretations. In the eyes of outsiders, Amerindians were cannibals, heathens, slackers, and sexual deviants. But they also exhibited a pervasive innocence and symmetry with the natural world that intrigued many visitors, as well as armchair adventurers who gained their experience vicariously through the travel accounts of others. If the natural habits of tropical peoples could highlight the flaws in European society, was it possible that these lessons could be marshaled for the improvement of all (Gareis 2002)?

Noble Savages

The barbaric images that emerged from early encounters with New World Native peoples, whether from observation or fabrication, or somewhere in between, are conceptually distant from the elevated cultural and environmental standing they later inherited. From lazy and long-lived, pagan and pleasure-loving, to wise stewards of the pristine tropical paradise, the Amerindian metamorphosis from jungle barbarian to noble savage seems nothing short of a miracle. But like earlier narratives that were so dependent on human preconceptions and conditions of the times, these were similarly constructed by Europeans around their own recent history and general social malaise. The lines from Alexander Pope's 1734 poem "Essay on Man" are often used to transmit a sense of this sentiment:

Lo, the poor Indian! whose untutor'd mind
Sees God in clouds, or hears him in the wind;
His soul proud Science never taught to stray
Far as the solar walk or milky way;
Yet simple Nature to his hope has giv'n,
Behind the cloud-topp'd hill, a humbler heav'n;
Some safer world in depth of woods embrac'd,
Some happier island in the wat'ry waste,
Where slaves once more their native land behold,
No fiends torment, no Christians thirst for gold!
—POPE 1828, 6–7

Moving from the sixteenth to seventeenth centuries and beyond, the noble savage concept was interjected increasingly as antithesis to the decadent and degraded state of European society. At its foundation, it was both a critique on European life and on the belief that Native Americans were immune to such problems. Wars, religious strife, and political oppression in their homeland challenged Europeans to reconsider who indeed were the civilized men and women, and who were the savages. Jean de Léry had in his earlier years in France witnessed some of its darkest hours. He was present at the St. Bartholomew's Day Massacre, where many thousands of Protestants were massacred and "Frenchmen were to be seen roasting and eating other Frenchmen's hearts," (De Léry 1990 [1557], xvii). And so he wondered aloud, in the presence of these wild native folk, whether their relaxed and libertine ways weren't somehow preferable to the horrors that Europeans were capable of meting out to each other.

The image of the noble savage became over time inseparably entwined with the romanticized depiction of tropical nature that prevailed into the mid-nineteenth century. Exalted nature and natural man, the latter living in sublime harmony due to the luxuriant fecundity provided by the former. Alexander von Humboldt did as much as anyone to crystallize this view and bring it to the attention of his vast rea ˙ ˙rship. Caught between the cool rationalism of the Enlightenment and the intense aestheticism of the Romantic Period, Humboldt was at once sympathetic to the circumstance of indigenous (and enslaved) people in America's tropical lands, and the careful cataloguer of its verdant plant life. His overriding concern was to understand the unity of nature, climate, landforms, vegetation, and humans (Helferich 2004). His sumptuous and widely read narratives that followed his great Orinoco-Amazonian expedition (1799–1804) spread his empiricist-romantic vision of tropical lands and primitive peoples, and influenced a generation of naturalists and artists.

Humboldt's impact cannot be exaggerated. A legion of naturalists, including the likes of Charles Darwin, Alfred Russell Wallace, and Henry Walter Bates, and artists, including Johann Rugendas, Philip Henry Gosse, and Frederic Edwin Church, feasted on his stories and followed in his footsteps, seeing and recording with pen and brush tropical nature and people through the Humboldtian lens (Stepan 2001, 35–48). (Fig. 3.3). Upon his first entrance into Brazil's tropical rainforest, Charles Darwin confided to his friend the Reverend John Henslow, that only Humboldt "gives my notion of the feelings which are raised in the mind on entering the Tropics" (C. Darwin letter to Henslow 1832, 141).

The concept of the noble savage passed through various permutations over the centuries, waxing and waning in relative acceptance depending on the interests of its advocates or detractors. The stereotype withered for a spell in the mid-nineteenth century, as artists and naturalists opted to depict the gloomier side of nature and culture in the tropics. Artists increasingly concentrated their gaze on the dangerous and bizarre face of the tropics—the venomous snakes, biting spiders, and life-threatening disease. Darwin's publication of *On the Origin of Species* in 1859, coincidentally the year of Humboldt's passing, signaled for many the death of romanticized tropical nature (Stepan 2001, 35–59), at least as it was conceived in the nineteenth century.

But the mid-twentieth century witnessed a resurrection of the noble savage concept by social and natural scientists. As they searched for

3.3   Henry Walter Bates.1863. "Interior of Primeval Forest on the Amazons." *The Naturalist on the River Amazons.* RB 721518. Vol. 1. Pg. 72. The Huntington Library, San Marino, California. Digital Image by Robert Voeks.

explanations for the success or failure of small subsistence-based societies, they increasingly employed the model of ecological sustainability
pioneered by biological scientists. And much like the vitality of predator-
prey interactions in natural ecosystems, small-scale societies were examined in terms of their ability to adjust their cultural practices to the
rhythms and limitations of nature. A particularly fine example from the
1970s was Gerardo Reichel-Dolmatoff's *Amazonian Cosmos: Sexual and
Religious Symbolism of the Tukano Indians.* Working with informants from
an isolated tropical forest society in Amazonian Colombia, the author
related the various means through which the group regulated their own
reproduction. Among these factors were pervasive sexual taboos related
to the hunting of game, in which the success of male hunters was believed to be a function of their ability to sexually attract (symbolically)
female game. These included long periods of abstinence on the part of
hunters, before and after the hunt, so that the female quarry would not
become jealous and consequently difficult to locate. The cosmos of the
hunters and the hunted was conceived as a single cauldron of sexual
energy; if hunters consumed too much, that is, were too sexually active with their wives, then there would be insufficient energy for the
procreation of their game. These and other ritually enforced strategies,
including infanticide, allowed this technologically primitive group to

maintain a stable population in a challenging physical and biological environment (Reichel-Dolmatoff 1972).

By the 1960s, the notion of indigenous ecological harmony resonated perfectly with scientists and environmentalists as they struggled to make sense of the many looming environmental challenges. Because so many of these problems were the result of population pressure, agro-industrial modernization, and unconcealed contempt for natural processes, the idea emerged again that perhaps the simpler lives and livelihoods of tropical forest dwellers, who seemed to live in a state of environmental stasis, held lessons for the more developed world.[2] Research with small-scale tropical forest communities, if not supportive of the more romanticized rendition of the noble savage concept, nevertheless revealed the highly nuanced nature of human-environment interactions, as well as the role of traditional ecological knowledge in forest management and conservation, however that is conceived (Berkes 2012). Rather than the destructive activity described by earlier observers, shifting cultivation of tropical forest was shown to be supremely adapted to the many challenges imposed by tropical ecosystems, including a perpetually hot and wet climate, leached soils, and lots and lots of pests. Sparing of useful plants, as well as transplanting wildlings to the swidden plot, was shown to enhance the long-term utility of fallows, and to contribute to overall species diversity. The cultivation of dozens of varieties of staple crops—cassava in Amazonia, millet in Africa, maize in Mesoamerica, and rice in Southeast Asia—translated to agrobiodiversity insurance against droughts, storms, pests, and other natural calamities. Homegardens added to the sustainability of tropical life by providing ready supplies of food and medicine, as well as experimental stations for trade and incipient domestication of useful forest plants (Coomes and Ban 2004). Rituals and community-enforced rules for management of plant and animal resources protected them from overexploitation. And religious structures such as sacred groves served as "priceless treasures" of threatened ecosystems in West Africa, India, and elsewhere (M. O. N. Campbell 2005; Gadgil and Vartak 1976).

For others in academe, however, the noble savage concept found less support, at least when defined by Western scientific conceptions of conservation. By the 1990s, scientific publications began to challenge the notion that hunter-gatherers and small-scale agriculturalists deliberately conserved their biological resources. Careful study of indigenous forest communities revealed that many of their actions were both environmentally unsound and ecologically unsustainable. This was particularly the case when one or a very few ecosystem elements were examined critically. In the case of Peruvian forest dwellers, for instance, the most sustainable strategy would seem to be for hunters to avoid killing mammals and birds that were vulnerable to overexploitation. However, rarity and likelihood of local extinction did not cause hunters to exhibit conscious restraint as they harvested vulnerable species. They did not hunt sustainably (Alvard 1993). There are in fact few documented examples of indigenous groups acting as noble savages in regards to wild game. When Western scientific methods of inquiry are used to measure their impacts, "conservation by Native peoples is uncommon" (Hames 2007, 186).

The long-term, landscape-level effects of indigenous relations with nature are even more damning. In both North and South America, for example, the profusion of wild game that was noted soon after the arrival of Europeans was long interpreted as a reflection of the harmonious relationship that existed between Native Americans and their environment. But rather than a byproduct of pre-Columbian indigenous conservation strategies, the cornucopia of American big game appears now to be a consequence of the demise of Native peoples following the arrival of Europeans (chapter 2). Wildlife that had been kept in a steady state of scarcity due to heavy hunting pressure by indigenous people was suddenly able to rebound when Indian populations were depleted. "Where there were large Indian populations," according to William Denevan (2016, 389), "most large wild animals became rare or locally extinct."

The demise of Easter Island's lands and peoples is a particularly compelling example of the lack of ecological harmony between a local population and its natural resource base. The original Polynesian colonizers of Easter Island encountered an isolated environment that was rich in terrestrial and marine resources. The founder population grew quickly and in the process the island's vegetation and soils were heavily and unsustainably exploited. The most visible sign of the prosperity of early residents is the presence of more than 200 massive stone statues, up to ten meters tall (33 feet) and 80 tons, known as *moai*. It is thought that palm trunks were employed as rollers to move the huge stone blocks, in some cases up to ten kilometers (six miles), and that there was dramatic wastage of this arboreal resource. This along with construction of canoes, clearing vegetation for agriculture, and the insidious impact of introduced fruit-eating rats, caused untold environmental damage. Counted among these losses were two dominant tree species, an endemic palm (*Paschalococos dispersa*), which was driven to extinction, and the toromiro tree (*Sophora toromiro*), which now exists only in botanical gardens. Over time, with no trees to build water craft, marine resources could no longer be harvested. All of the island's birds were driven to extinction, food became extremely scarce, and the human population descended into cannibalism (Diamond 2005, 79–119).[3] The residents of this isolated island may well have been noble and marvelously creative in their iconic stone works, but they apparently were not natural conservationists.

There are in addition numerous examples of regional and even continent-wide faunal extinctions engineered by early humans. The extermination of most animal species larger than 44 kilograms in North, Middle, and South America at the end of the last glacial period has been widely attributed to unrestrained hunting by Ice Age humans. According to the Pleistocene Overkill Hypothesis (Martin 1973), recently arrived big game hunters from Siberia mowed down dozens of "inexperienced species" as people migrated south and east across the Americas. Similarly in the Pacific Islands, the arrival of Polynesian peoples several thousand years ago signaled the eventual demise of perhaps 800–2000 species of birds, particularly large and flightless species (Duncan et al. 2013).

What these examples of "un-noble savages" have in common is that they all represent invading rather than native peoples. With no social history or ecological feedbacks to constrain them, newly arrived humans exploited ecosystems to the full extent possible. Only over time, as resources become less abundant and people learn from their early mistakes, does coevolution and longer-term, culturally reinforced ecological adaptation occur (Berkes 2012, 245–246).

Having been discredited by many in the scientific community, the noble savage concept discovered a permanent refuge in popular culture. The theme of innocent and environmentally conscious indigenous peoples has been deployed by a long list of popular magazine stories, Hollywood productions, and adventure novels. Among these, the 1992 film *Medicine Man* was an early example of the cinematic edification of indigenous rainforest residents and their knowledge of nature's healing mysteries. In this case, a tribe of peaceful and knowledgeable Amazonian natives holds the botanical treatment, later discovered to be a species of ant, for "the plague of the century"—the cure for cancer. The film included the usual cast of characters: the curmudgeonly and jungle-savvy male ethnobotanist employed by a pharmaceutical company, who has hidden himself for several years in the rainforest, the female American scientist whose main role was to be constantly shocked by creepy rainforest life, the technologically simple, child-like, and ever good-natured indigenous folk, the wise but shrewd shaman, and the sinister loggers and road builders. The standard parade of physical and biological villains makes their appearance: torrential rains, venomous snakes, bloodsucking insects, and those nasty parasitic fish (*candirú*) that swim up the human urethra (a myth that won't seem to die). The nearly naked indigenous people are mostly background props for the story, but the overriding messages of the film—the mysterious and pristine jungle, the innocent but environmentally knowledgeable indigenous folk, the altruistic ethnobotanist, the greedy pharmaceutical corporation, and the environmental destroyers from the outside—make it a perfect model for the jungle medicine narrative (Fig. 3.4).

A more recent literary installment of the jungle medicine narrative with a nearly identical storyline is the adventure novel *A State of Wonder* (Patchett 2011). In this tale, a curmudgeonly and jungle-wise female ethnobotanist, long missing in the Amazon, has discovered the cure for reproductive senescence in women. The forest-dwelling women in the fable are able to bear children well into their 70s and 80s, and the scientist discovers that their secret is to suck and chew on the bark of a rare rainforest tree. But unlike the noble savage image portrayed in *Medicine Man*, this offering legitimizes the forest wisdom of the native women while at the same time portraying them as apish jungle beasts. While standing about gnawing on tree trunks, "the others [women tree chewers] hooted lightly and slapped their trees to make a sort of tree-plus-human applause." There is no evidence that the intellectual property of the people is respected or even acknowledged, or that the basic human dignity of this simple-minded forest folk needs to be considered. "As for the Lakashi, they were patient subjects, submitting themselves to constant weighing and measurement, allowing their menstrual cycles to be charted and their children to be pricked for blood samples." They submitted to such indignities, the ethnobotanist explains, because "I tamed them." Later she reveals that her discovery is also a cure for malaria, thus solving another plague of the century, and that she tests its curative properties by asking the indigenous men to be bitten by infected mosquitoes. For their willingness to contract malaria, the good doctor rewards each with a Coca-Cola (Patchett 2011, 214, 258, 294). There is the usual lineup of threatening jungle creatures, venomous snakes, vampire bats, bullet ants, and those pesky parasites that swim up the urethra. And of course, the pharmaceutical firm that funds the project stands to make a fortune by developing and patenting the drug. What is most instructive about this book, however, is that although it is clichéd and totally lacks moral compass, the public and reviewers seemed to love it. And why? Because serving up a simplistic jungle bromide, with all the stereotyped elements that have been trotted out again and again in films and fiction, satisfies the preconceived expectations of a misinformed public. However poorly presented, the myth of slow-witted

3.4 *The Medicine Man,* 1992, starring Sean Connery and Lorraine Bracco.

forest stewards is reinforced, and because the book sold well, it becomes the standard for the next installment of the story.

The *National Geographic* magazine is the international flagship for reporting on the lands and peoples of exotic landscapes. With its earliest issues appearing in the late nineteenth century, the journal has weathered changing public perceptions of the Otherness of these distant domains. And these transitions are reflected with crystal clarity in the ways the authors of the articles present their indigenous subjects. From the nineteenth into the mid-twentieth century, forest folk were most often depicted as savage and intellectually backward, and as impediments to rational development of the equatorial zone. Early accounts portray the travel writers as heroically entering one or another unknown terrain, a sweltering "green hell" overrun with physical hazards and biological enigmas, all for the sake of satisfying their armchair readership. The native peoples in these early pieces are "portrayed either as powerful hunter-warriors who glare wild-eyed at the camera or as backward savages who are dazzled by the modern devices of the white man" (Nygren 2006, 510). In either case, the native people are primitive, and they are desperately in need of modernization.

Beginning about the 1970s, as international environmental concern for the plight of tropical lands and peoples emerged, the *National Geographic* increasingly depicted forest dwellers as noble denizens of the forest primeval, wise subsistence farmers living sustainably in their Edenic abodes. Green hell was transformed into green paradise. The native peoples metamorphosed into the natural guardians of the forest and its myriad secrets, who "live in the rhythm of the forest and feel a sense of oneness with nature." Non-indigenous waves of colonists—the "ignoble villains" and "mindless destroyers of tropical biodiversity," are now depicted as part of the problem, not the solution (Nygren 2006, 516 and 518). For the *National Geographic* magazine's seven million plus readers, the notion of the noble savage is alive and well.

Are Africans Noble?

If early European representations of Amerindians were mostly crude and judgmental, those reserved for their sub-Saharan African neighbors to the south, as well as enslaved Africans in the Americas, were even more offensive. Unlike Anglo-European perceptions of Native Americans, which mostly softened over the ages, those directed at Africans did not. In the eyes of outsiders, they began as unsalvageable degenerates, barely more civilized than jungle beasts, and there they remained. And as in the case with New World indigenous people, their cultural construction began many centuries before the Portuguese reached Senegal and the Cape Verde archipelago in the 1450s.

Africa and Africans were geographically situated within the known universe of Europe since the classical period, and had been the subject of speculation for at least a thousand years prior to Columbus's voyages. Aristotle, Pliny, and others wrote that numerous monstrous races[4] existed in Africa, including the Androgeni, who possessed the genitals of both men and women, the Pygmies, ebony-skinned dwarfs believed to inhabit interior Africa, and the Ethiopians, mountain people whose skin had been scorched by the intense rays of the Sun. These races were believed to be degenerated humans, the product of an early fall from grace with God (Friedman 1981, 9–21). Consequently, the earliest written observations of tropical Africans restated and reinforced long-held notions of primal, biblically ordained savagery and inferiority. Amerindians, it was believed, had the possibility of redemption through European ecclesiastical influence. But Africans had purposely strayed from God, and in the eyes of most Europeans, were beyond salvation.

This early derision of African people stemmed from religiously informed geographical concepts as well. The thoughts of Saint Isidore of Seville have been deployed for this purpose often since his seventh-century encyclopedia *Origines* was written. In his interpretation of Genesis, the world after Noah's great flood was divided into three unequal parts, Europe, Asia, and Africa. The remainder constituted one great ocean. Only eight humans survived the deluge, and they served as the seeds of all subsequent civilizations. The continents were subsequently repopulated by the offspring of Noah's sons. Sem traveled to Asia, Japhet journeyed to Europe, and Noah's sinful son, Cham (or Canaan), went to Africa. According to Isidore, because the African continent was populated by the offspring of Noah's wicked offspring, the entire population was to that day awash with sin. Added to this biblical preconception, Africans were disadvantaged further by the aesthetic prejudices associated with their skin color by Europeans. Black was associated with the occult, with deception, and with death. Their ebony skin color embodied perfectly the medieval notions of evil and monstrosity sketched long ago by the ancient philosophers. Taken together, the tale of Noah's wayward son and the pigment of their skin were more than sufficient justification for exploitation and enslavement of Africans (Cosgrove 2001, 65; Duquet 2003).

First-hand physical descriptions by explorers, traders, and missionaries of sub-Saharan Africa were often less flattering than the charm and sensual beauty attributed by the first visitors to tropical America and the South Pacific. John Atkins, a surgeon in the Royal Navy, reported of the women from Guinea in 1720 that "their Breasts always pendulous, stretches them to so unseemly a Length and Bigness . . I believe, could suckle over their Shoulders" (Atkins 1737, 50). The merchant Richard Rainolds reported from the coast of Africa in 1591 that "their women esteeme it the greatest part of goodly feature, to have large breasts, which by Art and industrious stretching of them, they enlarge, and some of the have them hanging to their Navill" (cited in Purchas 1613, 538).

But there were exceptions, and not every visitor and merchant painted such brutish images of Africa's women. Some found considerable physical beauty, rivaling that of Europeans. From the Gold Coast, Dutch trader Pieter de Marees remarked in 1602 that the women "are very conscious of the way they walk and of their clothes. . . . They have firm bodies and beautiful loins, outdoing the womenfolk of our lands in strength and devotion" (Marees 1987 [1602], 37–39). In the late seventeenth century, Portuguese navigator Francisco de Lemos Coelho likewise commented on the considerable physical attractiveness in the local women, noting, "The black females [of Cacao on the Gambia] are the most beautiful women on the whole river; and are unsurpassed by any on the Coast of Guinea . . . the Fulo [Fulani] women [farther up the river] . . . are extremely beautiful and very neat, but not at all chaste" (Lemos Coelho 1684, 299 and 304). French botanist Michel Adanson was also very admiring of the comely appearance of Gambia's women, stating, "There are some of them perfect beauties. They have a great share of vivacity, and a vast deal of freedom and ease, which renders them extremely agreeable" (Adanson 1759, 39). And Englishman William Smith offered his own intimate appraisal of his West African lover: "Her lovely Breasts, whose Softness to the Touch nothing can exceed, were quite bare, and so was her Body to the Waste . . . though she was black, that was amply recompenc'd by the Softness of her Skin, the beautiful Proportion and exact Symmetry of each Part of her Body, and the natural, pleasant and inartificial Method of her Behaviour" (W. Smith 1774, 253).

Conflicting body images of African women were frequently accompanied by reference to their alleged promiscuity. Like the native women of the Americas, Africans were often portrayed in travel biographies as lacking sexual scruples, and as ever willing, even eager participants, in trysts with European travelers. And each new observation, or claimed observation, served to reinforce the stereotypes that had been produced before it. Jean Barbot, a commercial agent in West Africa in the late 1600s, described the women as of "a warm temperament and [they] enjoy sexual pleasures, European men being more pleasing to them than their own men. Few of them will refuse the final favour if offered the most trifling gift" (Hair et al. 1992, 85). Immediately after European visitors arrive in Fetu [Ghana], according to Lutheran pastor Wilhelm Johann Müller in the 1660s, "unchaste women soon offer themselves and, for a small present, even as little as a bottle of brandy, prostitute their body. They have intercourse with him as long as he lives in the country . . . and then they look around for another" (A. Jones 1983, 153). Whether attractive or not, African women were almost always depicted by early observers with an "L"—lewd, lustful, lascivious, lecherous, licentious, and libidinous. Sexual deviancy, as reported by early observers, signaled the lack of natural laws, of the sort that separates the civilized from the savage. Of course, the near or total nudity of women never ceased to influence the opinions of visitors. Prostitution, which is what most observers were in fact reporting, and not the morals of married women, most likely existed before the arrival of Europeans. But the ubiquity of the "oldest profession" without question dramatically increased following the arrival of Europeans, especially in the towns and around the fortifications (pers. comm. Alpern 2012).

Sexually charged narratives served to fortify ancient bestial representations of African woman. On the one hand, like the sub-humans in Pliny's early ruminations and the Ethiopians in *The Travels of Sir John Mandeville*, African women were said to bestow their sexual favors on as many African men as they chose, and to freely give up their offspring to one or another of the possible biological fathers without the least care. It was widely reported that African women experienced childbirth without any pain, much like livestock. In this case, by bypassing God's curse on Eve of painful childbirth—"I will make your pains in childbearing very severe; with painful labor you will give birth to children" (Genesis 3:16)—the association of black people with lower forms of life was further cemented. By the time Europeans had colonized the Caribbean and implemented the slave trade, most of their grotesque perceptions of Africans were already firmly in place (Morgan 1997).

These early and contradictory depictions of African women, at once hideous and licentious, breasts dragging in the dirt, but also comely and willing to employ their perverted eroticism to seduce innocent Christian men, reinforced perfectly ancient European imagery of witches and medieval wild women. And these stereotypes fed effortlessly into the racist and dehumanized stereotypes that would emerge in the eighteenth and nineteenth centuries. By then, Africans were being depicted, even by men of science, as having sexual liaisons with chimps or other primates, and some were believed to be the biological outcome of such unions. By 1826, French professor Jules Virey wondered aloud how many chimpanzees "have prostituted themselves to the ardour of Africans?" and that the result of "the love [that] goes on in these ancient forests," has produced "these monsters half way between humans and apes" (cited in Jahoda 1999, 45–46).

As for African men, they were judged by Europeans in terms of their capacity for brute labor as well as their notorious dishonesty. Almost every description paints them as physically capable, lazy to excess, and pathological thieves and liars. One of the very first descriptions of African men was provided by Venetian trader Alvise Cadamosto in 1455. He described the men at Budomel, south of the Senegal River, as "great Idolaters, and have no Law; but are very cruell." He also reported cannibalism, "They eate their enemies which are slain [in] the warres" (cited in Purchas 1613, 540–542). According to Barbot, the men are made lazy due to "the bounty and fertility of the land," but also "sensual, knavish, fond of lying, gluttons, abusive . . . foul eaters, drunkards" (Hair et al. 1992, 84). From his 1720 visit to Senegal, John Atkins reported that the men were idle, planted scarcely enough to feed themselves, and spent most of their waking hours "smoaking all day in long Reed-Pipes together, unplagued with To-morrow, or the Politicks of Europe" (Atkins 1737, 50). The men of the Gold Coast had especially bad reputations. Pieter de Marees reported that the men were "great liars and not to be believed . . . I know of no Nation in the world which would be their Master in stealing" (Marees 1987 [1602], 32 and 105). The West African Director of the Dutch East India Company, Willem Bosman, agreed, noting in 1700 that Gold Coasters "are all without exception, Crafty, Villanous and Fraudulent, and very seldom to be trusted; being sure to slip no opportunity of cheating a European, nor indeed one another" (Bosman 1721, 116). Michel Adanson in Senegal was more interested in their physical attributes, noting "the negroes of Senegal are . . . generally above middle sized, well shaped, and well limbed. . . . They are strong, robust and of a proper temperament for bearing fatigue." In terms of temperament, however, he adhered to earlier stereotypes, describing them as "negligent and idle to excess" (Adanson 1759, 38 and 117). French naturalist and biopirate Pierre Poivre (see chapter 4) noted similarly from West Africa that he witnessed nothing but agricultural privation, testimony to the lack of industry by the natives. The "unhappy negroes," he observed, "cultivate just as much as prevents their dying of hunger." So limited was their agriculture expertise, he argued, that

"I have never been able to discover in their industry any process which could in the least improve our own" (Poivre 1770, 4–6).

The question of super-longevity of tropical people, so often forthcoming from the New World, was quite absent in reports from tropical Africa. Bosman stated, "Most of the Negroes live healthful Lives, but seldom arrive to a great Age . . a Man of fifty (a good old age here) seized by any Sickness, generally leaves this World" (Bosman 1721, 110). Michel Adanson concurred a few years later, noting that a man of fifty is an old man here, "because the negroes of Senegal are really old at the age of forty-five, and oftentimes sooner" (Adanson 1759, 48–49). Whether this perceived difference was a result of the notoriously debilitating disease conditions in tropical Africa compared to tropical America, or was simply the product of verbal flights of fancy, is not known. But the impression disseminated to the outside world was that Africans were robust but not at all long-lived.

Most importantly, the reversal in stereotype regarding indigenous Americans that occurred over the centuries, from lazy, libidinous, subhuman cannibals to innocent and noble stewards of nature, never made the journey to sub-Saharan Africa. The reasons are many, but certainly the exorbitant death rates that prohibited whites from penetrating Africa's interior must be considered. There was little missionizing in West Africa in the early centuries of exploration compared to the Americas because of the heavy death toll among whites. Few missionaries were willing to post in these regions, and those who did "usually died of some tropical fever before they could do much good" (Boxer 1963, 7). Some 25% to 50% of Portuguese traders working in Africa in the fifteenth century died before they could return to their homeland. And the situation did not improve much over the coming centuries. The hardships and early mortality experienced by pre-colonial European visitors to lowland Africa, "led to the construct of the African native as the quintessential and permanent embodiment of savagery" (Duquet 2003, 23). Surely only beasts could survive these diseased and enervating conditions. In the New World, long-term and intimate association with naïve peoples fostered cultural paradigm shifts, especially the emergence of the ecologically noble savage model. But the continued fleeting familiarity of Europeans with sub-Saharan Africans served to calcify the worst of their early prejudices and stereotypes.

These centuries-long negative portrayals of Africa and its peoples were reinforced in the nineteenth century, especially by the Victorian "dark continent" metaphor introduced by Henry Stanley in his travel accounts

In Darkest Africa and *Through the Dark Continent*. This and other powerful metaphors—"The Land of Death" and "The White Man's Graveyard"— were used to encompass Africa's hostile, disease ridden environment, and bestial and idolatrous people (Jarosz 1992). The simianization by whites of racial and ethnic minorities, including Irish and Japanese and Jews, has certainly occurred at various times in history. But the ubiquity and persistence of the characterization of Africans as "ape-like beings, a hair's breath away from the nonhuman primates on the Great Chain of Being" is unparalleled. The early twentieth-century exhibition in the Bronx Zoo's monkey house of Ota Benga, a Mbuti pygmy, testifies to the persistence of these beliefs (D. L. Smith and Panaitiu 2015, 78). The myth of the noble savage, such a crucial element in the jungle medicine narrative, is a stereotype that never included tropical Africans. And it continues to exclude them to this day.

Environmental Determinism

In the centuries following the Columbian landfall, the native peoples of the New World, as well as those in Africa, Asia, and Oceania, were subject to paradoxical interpretations of their human nature: good or evil, noble or savage, enlightened conservationist or country bumpkin. Indeed, like the exuberant biodiversity of their forested homes, tropical people were the object of diverse and often conflicting stereotypes. The climate, for example, was said to produce great natural abundance and variety, yet it was also associated with human poverty and hunger. The visual elegance of the forests and fields contrasted with the ferocious wildlife, the destructive hurricanes, and especially the obscene assortment of infectious and parasitic diseases. These conflicting and often synchronous tropicality tropes, at once both "pestilential and paradisiacal" (D. Arnold 2000, 8), saw their direst expression beginning in the mid-nineteenth century, and reaching into the 1920s. During this period, colonial officials, scheming entrepreneurs, and misguided missionaries extended their influence ever deeper into the world's tropical spaces. By then they were aided in their efforts by the widespread availability of quinine, the miracle plant-based remedy for malaria. Earlier literary depictions of noble savages living in a pristine paradise evaporated temporarily, appearing again not until the environmental crises of the mid-twentieth century. Centuries of European exploitation and evangelism had finally convinced most writers that the positive attributes of the tropics were mostly illusions. The purported fertility of the soils was replaced by the belief that tropical substrate consisted of sterile clays subject to compaction and hardening into brick-hard laterites. There were precious few breeds of indigenous livestock, and those that were introduced from Europe and the Americas generally did not thrive. As a physical geographical reality, the tropics were in every way inferior to the temperate zone, pronouncements that served to flatter Europe's sense of cultural superiority.

The most pernicious of these twentieth-century revisionist theories, later known as environmental determinism, found its inspiration in Charles Darwin's *On the Origin of Species*. Anglo-European writers, arguing that culturally backward peoples, like inferior biological species, are eventually outcompeted by superior strains, adopted a social Darwinian approach to bolster their theory. The lands and peoples of the equatorial reaches, they argued, were degenerate and backward due to the devitalizing effects of the benign climate. Some, like Alfred Russell Wallace, cofounder of the theory of natural selection, argued that tropical climates were too enervating, and tropical resources too naturally abundant, to ever challenge humans to improve their lot. In his Indonesian study area, for example, he reported that sago palms (*Metroxylon sagu*) produced all the carbohydrate that people needed, retarding any further cultural development. "In ten days," he reported, "a man many produce food for the whole year." Because of this natural abundance, "no civilized nation has arisen within the tropics" (Wallace 1862, 136).

This Edenic theory, that the environment was just too fruitful and naturally abundant to encourage further cultural development, had appeared at various times and places in the equatorial orb. Humboldt mused over the question in Amazonia. The harsher climate of the northern latitudes, he reasoned, encouraged human labor and industriousness. But not so the humid tropical climates, where "in the midst of abundance, beneath the shade of the plantain and breadfruit tree, the intellectual faculties unfold themselves less rapidly than under a rigorous sky . . . where our race is engaged in a perpetual struggle with the elements" (Von Humboldt and Bonpland 1818, 15). Likewise in the South Pacific, the supposed tropical abundance was employed by various authors to further the degenerate savage ideal. According to John Hawkesworth, a biographer of Captain Cook's voyages into the South Pacific, there was considerable agreement that Tahiti's bountiful nature provided all of mankind's needs. Why bother to till the soil in a land where bread grows on trees [breadfruit] and palms supply milk? The land, he proclaimed "was like Paradise before the Fall of Man" (cited in B. Smith 1985, 42 and 48). An anonymous poem penned a few years after the voyage (1774) expressed this sentiment perfectly:

Refreshing Zephyrs cool the noon-tide Ray
And Plantane Groves impervious Shades display.
The gen'rous Soil extracts no Tillers Aid
To turn the Glebe and watch the infant Blade;
Nature their Vegetable Bread supplies,
And high in Air luxuriant Harvests rise.
No Annual toil the foodful Plants demand,
But unrenew'd to rising Ages stand;
From Sire to Son the long Succession trace,
And lavish forth their Gifts from Race to Race.
Beneath their Shade the gentle tribes repose;
Each bending Branch their frugal Feast bestows:
From them the Cocoa yields its milky Flood;
To slake their Thirst, and feed their temp'rate Blood.

Others took up the environmental influence cause, in most cases to further elevate the vaunted status of the favored European races. Writing in the mid-1700s, English historian Thomas Salmon negated the much-cited racial and cultural effects of the intense tropical Sun, arguing that if this were the only relevant factor, then the tropical peoples of the Americas should look and act much more like Africans. Rather, he suggested that the well-documented inferiority of the tropical races was due to a combination of Sun and soil and the overall tropical climate. Indeed, he pondered, if African and European men were to exchange geographical locations, "Blacks in a few generations would become white . . . and possibly white men would become black" (Salmon 1746, 59). French physician Julien-Joseph Virey, a supporter of the theory that blacks and whites had different origins (polygenism), similarly thought that solar influence could not be the primary cause of dark skin. Nevertheless, he argued that the benign climate and plentiful resources retarded the development of Africans: "It is this [intense sun] that causes negroes to be so lazy, so indolent and weak; it is this which incessantly assisting a luxuriant and rich vegetation, permits those nations to live without any trouble" (Virey 1837, 45).

This connection between tropical abundance and cultural retardation was expressed at various times in history. But in the hands of social scientists it took on a new and particularly racist turn. Thus, in his 1863 *Introduction to Anthropology* text, Theodor Waitz put forth his views on the impact of abundant resources and enervating climate: "in the Torrid Zone, nature yields her gifts too freely in supporting man, labour, and mental activity naturally languish." Bananas are so abundant, he suggested, producing twenty times the yield of temperate zone cereals, that people who evolved in the tropics were subject to "a relaxing influence" that required neither intelligence nor ingenuity (Waitz 1863, 331).

The most influential proponents of the theory that culture was a product of the environment were American geographers Ellen Churchill Semple and Ellsworth Huntington. It was in their efforts to make the field of geography somehow more scientific and less descriptive—"the new science of geography" (Huntington 1915, v)—that they produced a series of widely read books and articles on the subject of culture, race, and environmental control. The gist of their argument combined social Darwinism, long-held prejudices concerning race, and the influence of climate on human behavior and character. In one of Semple's seminal essays, "The Influence of Climate" (1911), she argued that mankind is the most adaptable of animals, who "makes himself at home in any zone." She then went on to articulate the behaviors that are caused by the different climatic influences. The people of sunny Andalusia (southern Spain), for example, are "light-hearted, gracious peasants," contrasting markedly with "the reserved, almost morose inhabitants of cool and cloudy Asturias (northwestern Spain)." She goes on to provide numerous examples of atmospherically determined behaviors and customs. But her sharpest criticisms were reserved for the "backward" races of the equatorial latitudes. Here lived the clearest proof of the differential influence of climate: "As the Tropics have been the cradle of humanity, the Temperate Zone has been the cradle and school of civilization. Here Nature has given much by withholding much. Here man found his birthright, the privilege of struggle." And finally, regarding the cultural and intellectual development of humans in the tropics, "Where man has remained in the tropics . . . he has suffered arrested developments. His nursery has kept him a child" (Semple 1911, 611, 622, and 635).

Ellsworth Huntington advanced the climate-culture connection even further. He argued not only that the rise and fall of the great civilizations could be traced to variations in climate, but that racial differences could be explained by climate zonation, especially between blacks and whites. Drawing on the flimsiest of personal experience, he suggested that the cultural differences between temperate and tropical people were due to the level of "climatic energy" which could be mapped on a global scale (Fig. 3.5). The enervating climate caused blacks in the equatorial latitudes to be lazy and shiftless. As for the evolutionary outcome of this debilitating influence: "Every exact test which has been made on a large scale indicates mental superiority on the part of the white race, even when the two races [black and white] have equal opportunities."

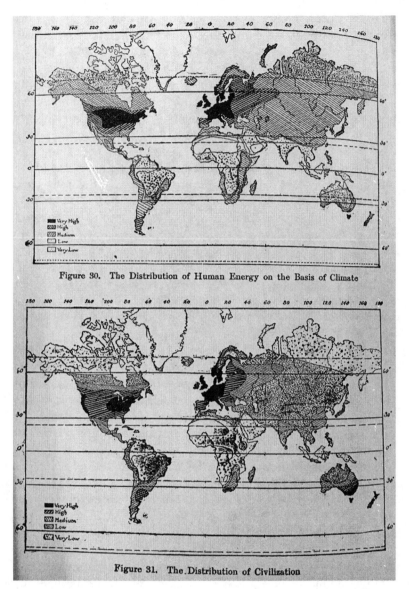

Figure 30. The Distribution of Human Energy on the Basis of Climate

Figure 31. The Distribution of Civilization

3.5 Ellsworth Huntington.1915. Fig. 31. "The Distribution of Civilization." Pg. 208. *Civilization and Climate*. GF31.H8. The Huntington Library, San Marino, California. Digital Image by Robert Voeks.

His evidence was based on personal visits to the Bahamas. There he observed that the only industrious workers were those who made frequent visits to the United States, where the effect of the "bracing climate was astonishing." Later, he commented on the creeping laziness that takes over whites living in the tropics, quoting a local white informant: "Until I came to the Bahamas, I never appreciated posts. Now I want to lean against every one that I see." White men who had migrated to the Bahamas, he noted, took on the same stupefied demeanor as the blacks, becoming over time the "cracker" type of "poor white trash." My personal favorite was Huntington's proclamation that the hot and wet tropical climate affected even the mental capacity of whites. When visiting the American South, for example, he found it impossible to find "higher grade magazines," only "trashy story magazines" (Huntington 1915, 48, 49, 51, and 52).

By the 1940s, environmental determinist explanations for human cultural development had largely disappeared from the social sciences.[5] In their continuing inquiry into human-environment relations, geographers turned away from racially motivated determinist theory to focus on the core of the discipline, the reciprocal relationship between humans and the non-human world. However, the intuitively appealing notion that human behavior and even cultural development is affected by ambient air temperature—who hasn't felt lethargic on a hot and muggy day?—is firmly cemented in the mindsets of many temperate and tropical peoples.

Instinctive Ethnobotanists

The stereotypes employed to venerate or denigrate forest-dwelling tropical people have varied dramatically over the ages. During some periods, multiple stereotypes coexisted simultaneously, and they do so today. Scientists are less inclined now to deify indigenous people as the original environmental crusaders. But their traditional ecological knowledge and associated management of nature is orders of magnitude more respectful of ecosystem services than that of modern ranchers, miners, loggers, and planters. Environmental determinism has died out (mostly) in academia. Yet these preconceptions of climatically determined cultural behavior are alive and well among the general public. The media can likewise be contradictory in their depiction of tropical folk, delivering on the one hand the noble savages of *Medicine Man* and James Cameron's *Avatar*, or on the other hand the deranged brutes of Peter Jackson's

King Kong. But whether forest-dwellers are portrayed as noble or bestial, the characteristic that has endured all paradigm shifts is that they are knowledgeable about the inner workings of nature. In this, everyone seems to agree. And for the jungle medicine narrative, this is a crucially important feature—the more technologically atavistic the people, the more likely they are to have mastered the healing properties of botanical nature.

There are more than sufficient studies from the lowland tropics to convince us that subsistence hunter-gatherers and shifting cultivators maintain a profound ethnobotanical inventory of useful plant species. And as we see later (chapter 8), this knowledge can vary significantly in terms of age, sex, and degree of modernization. Yet the notion that indigenous forest people are able to tap into nature's medicinal mysteries transcends rational scientific inquiry. Rather, it seems to be deeply ensconced in our subconscious. There is in fact a considerable pedigree to the belief that marginalized societies are somehow closer to nature, and are therefore better equipped to understand its healing properties. However technologically backward they may appear, indigenous rainforest healers and herbalists are seen as touchstones for the vitalist energy of nature that we in the industrialized world have mostly lost. They have harnessed the "magic of primitivism" (Taussig 1987, 216).

This perceived primitivist connection with nature's healing properties has been an important guide for bioprospecting efforts, now as well as during the colonial period. Edward Long, in his 1774 *History of Jamaica*, provided an important spin on this idea when he declared that "brutes are botanists by instinct" (Long 1774, 180 and 181). Long was simply articulating what was widely accepted in his time, that is, that beasts, savages, and other lower forms of life had an instinctive understanding of nature's medicinal virtues. Civilized races, however, having climbed out of this natural depravity, had forsaken instinct for intellect (Schiebinger 2004, 82). At least in this one aspect, Africans have joined their indigenous American counterparts through their inborn knowledge of the workings of nature. For colonial bioprospectors, hell-bent on uncovering tropical "green gold," the path to fortune and fame ran directly through the instinctive insights of indigenous tropical folk.

FOUR

Green Gold

Staff Sergeant Francisco de Melo Palheta was a dashing eighteenth-century Portuguese officer stationed in far northeastern Amazonia, in the Brazilian state of Pará. Having organized and led a harrowing journey up the Amazon River in 1722 and 1723, he had proved himself a courageous navigator and a keen negotiator. But the sergeant's primary claim to fame was not that he was a skilled military man, but rather that he was "an incorrigible seducer of women." This character flaw, or virtue depending on your perspective, was well known to his superiors, including the governor of the state of Maranhão, who in the early 1700s was embroiled in a struggle over the colony's geographical delineation with French Guiana, their neighbors to the north. Fearing encroachment by French interlopers into Brazil's rich mining districts, the governor ordered Sergeant Melo Palheta on a reconnaissance mission to resolve the border dispute by mapping out the poorly known territory. The true nature of his journey, however, had little to do with political boundaries; its primary objective was to secret away some seeds of Europe's latest botanical craze—coffee. The popularity of this stimulating bitter brew was growing exponentially, and colonial powers zealously guarded their coffee plant monopoly against thievery from rival states. But despite numerous attempts, Portuguese Brazil had yet to cultivate a single coffee bush. Their competitors to the north, however, had been more successful. Agents in French Guiana had managed to steal some seeds from their Dutch neighbors in Suriname who, in turn, had imported them from their possessions in distant Indonesia. Melo Palheta's mission was simple: to acquire the highly coveted propagules from the French and convey them back to Brazil. And so, according to legend, having finished his cartographic calculations, Sergeant Melo Palheta traveled to Cayenne, the capital of French Guiana, and was there able to strike up a friendship with the wife of the governor, a Madam d'Orvilliers. True to the Portuguese lothario's reputation, the governor's wife was quite smitten with Melo Palheta, and the relationship quickly turned amorous. And as a sign of her great affection for her Portuguese paramour, Madam d'Orvilliers presented him upon his departure with a bouquet of flowers, within which were hidden a cache of prized coffee seeds. The beans were spirited back to Brazil, where they likely became the earliest source of Brazil's eventual global dominance in coffee production (Hecht 2013, 111–112; Magalhães 1980; Pinto 2012).[1] Melo Palheta for his part became one of the colonial world's first biopirates. But he would not be the last.

The tale of Brazil's gambit to acquire the precious coffee seeds, by whatever means necessary, was a minor footnote in the frenzied quest for green gold in the Torrid Zone. Men of science and the cloth, not to mention adventurers, travelers, and drifters, often with scant understanding of the perils that lay ahead, scoured the moist tropical latitudes of Asia, Africa, Oceania, and the Americas in search of mysterious and potentially blockbuster crops and spices and herbal remedies. They were a multinational cast of characters, including Spanish, Portuguese, English, French, Dutch, and many others. In their day, many were considered heroes, either patriotically advancing the cause of national economic expansion or seeking to solve one or another of the great medical maladies that vexed the increasingly globalized world. Today, however, these historical figures are seen through a quite different lens, either as fame and profit-seeking fortune hunters or as minions for the colonial mercantile powers as they jockeyed for position against their rivals. Some of these characters, the Frenchman Pierre Poivre and the Britons Charles Ledger and Henry Wickham, fit the biopirate archetype perfectly. Each stole hugely profitable genetic and intellectual property from tropical regions with the full complicity of their colonial sponsors. Unfortunately, at least with regards to ongoing ethnobotanical efforts, these historical figures and their patrons represent for many the yardstick against which current medicinal plant research is measured. "If foreigners were stealing our genetic and intellectual property in the distant past," one can argue, "is there any reason to believe that they are not doing exactly the same today?" But then as now, people were not so predictably linear, and avarice and power were not the only motives that inspired these early ethnobotanists on their odysseys into the unknown. Chivalry, simple scientific curiosity, and, yes, even altruism, the same features that spur current ethnobotanical inquiry, were also at play. Among the latter, the search for new medicines stands out. As conquest and colonization of new lands and peoples progressed, diseases of the Old World found their way to the newly contacted lands of the east, west, and south with devastating consequences. An armada of hitherto unknown viruses, bacteria, and parasites emerged from the microbial-rich tropical forests and savannas to plague colonial settlers, traders, and their enslaved labor force. Colonial medicine was mired in archaic and mostly useless theories and practices that had persisted since the time of Galen and Hippocrates. Never in Earth's history was the need for new botanical cures more acute, and never was the quest to discover them more intense.

First, Do No Harm

When Christopher Columbus and Pedro Álvares Cabral first encountered the warm and tropical idyll of the Caribbean and Brazil, respectively, both waxed poetic on the soft air and salubrious winds. Surely, these lands without seasons, where succulent fruit hung languidly on the trees year-round, must be close to the Garden of Eden and its spiritually charged healing flora. Five decades later, when the northeast of Brazil was being settled, Jesuit priest Manuel Nóbrega could proclaim, "It is very healthy," and those few who fall ill, "we cure quickly" (Cartas do Brasil 1886 [1549–1560], 63). The good brother's enthusiastic claim would prove sadly premature, however, especially in light of the massive microbial onslaught that was beginning to grip the newly encountered tropical world. Europeans were no strangers to ill health and epidemics, having suffered through the fourteenth-century Black Death, in which over one hundred million souls had been lost. Leprosy was distressingly common, malaria had been killing Europeans since Roman times, and a host of other lethally infectious maladies plagued cities and rural landscapes (Hoffmann 2014). But the breaching of the intercontinental germ barriers that started with the Columbian landfall unleashed an unprecedented invasion of pain and pestilence on the peoples of the world. And none of the afflictions would have a more immediate impact and encourage the coming herbal quest more than an innocuous genital chancre carried away from the West Indies by one of Admiral Columbus's crew.

The origin of syphilis is still contested, but it was likely introduced to Europe in 1493 when Columbus returned from his maiden voyage. German author Ulrich von Hutten, himself suffering from the affliction, wrote extensively on syphilis in 1536, noting that it was unknown to "the forefathers," and that "the French Poxe arrived in 1493, or thereabouts." He said that the doctors of the day were not sure what caused it, but they agreed that it was related to "Unholsome blasts of the aire," that "the sees were corrupted" and "venemus vapours to come downe from the ayre" (Von Hutten 1536, fol. 1 and 3). The epicenter of the first outbreak was Italy, but when the French and their multinational mercenaries marched on Naples in 1495, their reward for rape and pillage was the unseen spirochete bacterium. Carried to their respective homes by the victors, the French pox spread like wildfire throughout central and northern Europe. Asia would shortly come to know the illness as well, as Vasco da Gama, in his 1498 circumnavigation of the globe, likely transported syphilis to India. Soon enough, sailors and soldiers had spread the dreaded disease around the world. An equal opportunity ailment, syphilis infected the low- and high-born, including kings and popes, covering their bodies in great seeping tumors before it drove them to dementia and eventual death (Desowitz 1997). Syphilis was the first of the great new plagues, but it would be dwarfed in virulence and sheer killing capacity by those soon to follow.

The infectious invasion of the Americas and Oceania by a flotilla of Old World diseases resulted in nothing short of a demographic catastrophe for indigenous peoples. Figures continue to be debated, but what is clear is that the biological quarantine of the Americas, enforced since the last (Wisconsin) Ice Age advance, translated to almost zero immunity on the part of New World inhabitants to Eurasian and African microbes. These "virgin soil epidemics" reduced Native American populations by upwards of 90–95%. Some particularly vulnerable locales, such as the island of Hispaniola, may have declined from a pre-contact population high of one million people to only a few hundred or so souls a mere fifty years later. Guatemala witnessed a collapse from some two million people in 1520 to less than 100,000 a half century later. The Incas and the Mexicas (Aztecs) and myriad lesser-known societies similarly witnessed decimation of their original numbers from a barrage of exotic ailments, including influenza, measles, yellow fever, and malaria. But smallpox was the worst. In the Antilles and northern South America, smallpox was joined by leprosy and, especially among the enslaved African population, elephantiasis, yaws, guinea worm, and river blindness in Guatemala. Among the Caribbean enslaved African population,

French observer Nicolas Bourgeois reported that they "bring illnesses we don't know and therefore don't have medicine for" (Bourgeois 1788, 425). For the century and a half after the Columbian encounter, the arrival and impact of foreign germs and worms represented "the greatest human catastrophe in history" (Cook 1998, 13; Denevan 1992; Leblond 1813; Lovell 1992; Orellana 1977).

The Jesuit Order during these early years of encounter was often depicted as the great protectors of the native population. But in many ways, they were also their most efficient exterminators. For while they initially marched their message of salvation through Christ directly into the Brazilian hinterlands, over time they realized that such a conversion strategy was both ineffective and ephemeral. So they abandoned their peripatetic approach, and began rounding up thousands of Tupinambá men and women, crowding them into mission villages. The consequences were devastating. Smallpox and other contagious diseases burned through the concentrated population within a few short years. Some 30,000 mission Indians perished during a single smallpox epidemic of 1562–1563 (Leite 1938, 574). "Obsessed with their personal soul-count . . . the padres resolved that it was better for the native peoples to be baptized and buried than pagan and alive" (Hemming 1978a, 145).

The spread of disease among vulnerable indigenous populations represented a demographic catastrophe. But it was not in any way a catalyst for the search for tropical nature's healing secrets. For outside of their role as laborers, the lives of native peoples were not much valued. Rather, it was the expansion of Europeans into the inhospitably hot and humid tropics that stimulated this medicinal quest. Whether in Africa, Asia, or the Americas, the health of arriving Europeans simply did not seem up to the challenge. Portuguese Goa on India's west coast had a very high mortality rate for Europeans, and in West Africa (as noted in chapter 2), there was little missionizing due to the heavy death toll among whites. According to Willem Bosman, who spent many years on Africa's Gold Coast, the worst diseases were smallpox, which killed thousands, and the "grievous pain" of leg worms, "some as long as Pikes." The black Africans were less prone to illness than the Europeans, however, because "they were born in this Air, and bred up in this Stench" (Bosman 1721, 108–109). In the West Indies, where most of the native population had already died off, the situation for Europeans was similarly dire. "Newly arrived people," according to Nicolas Bourgeois writing in the 1780s, "are less adapted to live here; their pores never open to let out the heterogeneous humors with sweat." It was, he argued, the malignancy of the air in these "hot countries," which is "acidic like the fruit" that grows there, and "which ordinarily pu[t]s the victim in his tomb as soon as the first bout of fever attacks" (Bourgeois 1788, 136 and 415; Boxer 1963, 7; Pearson 1996, 241; Rutten 2000). Fortunately, as Jewish physician Garcia de Orta informed his sixteenth-century readership from his post in Goa, India, "each day brings new diseases . . . and God is so merciful that in each land He gives us medicines to cure us. He who causes the illness provides the medicine for it" (Orta 1913 [1563], 105). God may have been up for the task ahead, but were the medical professionals of the era?

The corpus of healers, or those who pretended to be, was considerable during the early modern period (16th–early 19th centuries). These included physicians, surgeon-barbers, bloodletters (phlebotomists), bone-setters, herbalists, priests, sorcerers, witches, diviners, and astrologers. Some were university-educated; most were not. Cause and treatment was based largely on the ancient humoral system of Galen and Hippocrates, which had been adhered to on and off for fifteen hundred years. Illness was interpreted as a function of an imbalance in bodily fluids, including blood, phlegm, black bile, and yellow bile, and these were in turn interpreted as hot or cold, moist or dry. This state was usually identified through an examination of the urine. To treat these illnesses, whether caused by sin, bad air, or the alignment of the stars, physicians leaned heavily on the five hundred or so plant species (as well as some animals and minerals) described in Dioscorides' ancient *De materia medica*, formulated around 50–70 CE. An imbalance in the humors could be redressed through various means, but most physicians relied on diet, medicines, purging, vomiting, and of course bleeding (Newson 2006, 382; Pearson 2011).

Phlebotomy was an especially popular practice in humoral medicine. The idea was that the stomach digested food, and sent it on to the liver, which then produced blood. But the liver sometimes produced more blood than was necessary, known as a plethora, a condition that "corrupted the blood" and led to a number of ailments. The draining of excess blood from the body was nearly always recommended. Bloodletting was also considered a form of prophylaxis against illness. For instance, during the change of seasons, people were encouraged to get a good bleeding. Bloodletting was accomplished by opening up a vein with a lancet, or with small cuts and heated glass cups, and also by the application of leeches (Kusukawa 2004). Unfortunately for their hapless patients, doctors of the time (at least through the seventeenth century) were under the false impression that the human body contained 24 liters of blood, and that 20 of these could be safely bled away. But because the average person actually contains only about 5 liters, the immediate impact of excessive bleeding was often much worse than the ailment (Pearson 2011, 3). Patients frequently bled to death, including in all likelihood President George Washington. Suffering from a sore throat and hoarseness (later identified as inflammation of the epiglottis), his attending physicians drained over 2.4 liters (about 2.5 quarts) of blood over the course of 12 hours. Washington died 7 hours later (Morens 1999).

The concept of removing human blood to achieve health must have resonated widely, as the practice caught on among many non-Europeans. Perhaps because their Ayurvedic medicinal traditions recognized a similar balance between vital elements as that in the humoral system, bleeding was eventually adopted by Indian physicians, although to a lesser extent, where the two cultures came into contact (Pearson 2011, 8). In colonial Jamaica, Sir Hans Sloane noted that "the Negroes use very much bleeding in the Nose with a Lancet for the Head-ach" and other ailments. "In colonial Brazil, Indians stood in long lines waiting for the phlebotomists to open their veins, reddening the streets with the result of their treatment" (Hemming 1978a, 468). Guatemalan Indians may well have practiced phlebotomy prior to contact with Europeans, and were so good at it that many Spanish relied on them for their bloodletting (Orellana 1977, 141).

Bleeding may have been the most perilous medical practice, but it was not nearly the most macabre. Europeans and others clung to centuries-old "traditional" ingredients for their medicaments, some disturbing and grisly. For instance, in spite of their revulsion at cannibalism among New World native people, moneyed Europeans resorted to ground up human skulls and fresh human blood, as well as a slimy liquid known as "mummy" in their medical formulas. Portuguese apothecary Tomé Pires, who composed the famous *Suma Oriental* from Malaysia (1513 to 1515), set the record straight on the medicine known as mummy. "Mummy is not human flesh," he commented, "for the true [mummy] is an exudation from corpses . . . the moisture of the body drips out and . . . this liquor is called mummy" (T. Pires 1967 [1512–1515], 515). The deceased should have met with a "sudden, preferably violent death," in order that the substance maintained its occult healing qualities. "In a dead thing," as Leonardo da Vinci reasoned, "insensate life remains which, when it is reunited with the stomachs of the living, regains sensitive and intellectual life" (cited in Noble 2011, 17 and 20). And so mummy medicine was all the rage among those who could afford it.

Human bone and body fluids were not the only imaginative healing therapies of the time. Irishman Robert Boyle, one of the early founders of modern chemistry, wrote a short treatise in 1696 on local cures with the unlikely title *A Collection of Choice and Safe Remedies*. Although he admitted that "physick be not my profession" he nevertheless felt compelled to record a considerable list of unusual remedies. "For Convulsions, especially in Children. Take Earth-Worms, wash them well in White-wine to cleanse them, but so as that they may not die in the Wind . . . [then] dry the Worms with a moderate heat . . . reduce to a Powder." "To Clear the Eyes . . . Take . . . human dung of a good colour and Consistence, dry it slowly till it be pulverable: then reduce it into an impalpable Powder: which is to be blown once, twice, or thrice a day . . . into the Patient's Eyes." "For Obstructions . . . Let the Patient drink, every Morning fasting, a moderate Draught of his own Urine newly made, and . . . whilst 'tis yet warm." And "for colick, use balls of fresh horse dung . . . mixed in white wine" (Boyle 1696, A3, 11, 20, and 27). These from the man for whom Boyle's Law would be named. From Cypress, which was widely known in the seventeenth century for its love potions (philters) and male tonics, one virility recipe recommended to "Take some drops of your own blood one Friday in the spring . . . dilute it with fresh sperm, and dry it next to a fireplace. Put it then in a pot containing two testicles of a hare and the liver of a female pigeon. After the compote turns into fine powder, add some powder of hibiscus flower and dried leaves of endive." Also, in order to stir your manhood, Italian physician Giambattista Morgagni recommended adding "Earthworm oil . . . together with viper's fat, powdered human cranium, woman's milk, viper flesh, and mouse blood" into a warm chicken soup (cited in Finucci 2008, 531 and 539–540).

Early modern period Europeans did not hold a patent on bad medicine. From China, there was a considerable pedigree to the employment of what are referred to as "human drugs," that is, urine, hair, bones, teeth, and other body parts (Barnes 2005, 48). Brazilian Indians, according to botanist Karl Von Martius, "employed smegma as an antidote to snake venom and to the bite of huge ants" (Von Martius 1939, 224). In Santo Domingo, Nicolas Bourgeois records the eighteenth century use of fat from cockroaches to induce sweat (sudorific) and to treat chest illness (Bourgeois 1788, 468). And from the early seventeenth-century Greater Antilles, Jean-Pierre Moreau tells us that for stomach ache, a man "cuts the pubic hair of his wife, then chops it very finely and drinks it in potato wine (the exact inverse is done for a woman)" (Moreau 1994 [1618–1620], 224). If there is a message here, it must be that colonial physicians as well as their patients had nothing to lose by experimenting with the cornucopia of new medicinal plants that were being discovered and experimented with in the Old and New World tropics. It also instructs us not to over-romanticize the traditional healing practices of the distant past.

Colonial physicians and the clergy were desperate to find cures for the catalog of new fevers, pustules, and parasites that tormented their patients in the Torrid Zone. And they were in luck. For according to sixteenth-century Spanish physician Nicholás Monardes, God had provided "new medicines, and new remedies, which, if we did lack them, [the new diseases] would be incurable" (Monardes 1580, 1–2). But there was a minor dilemma. These new remedies could not be incorporated into the various standard pharmacopoeias if they in any way legitimized the heathen beliefs of the "savages" from whom they were acquired. This issue was especially important for Spanish and Portuguese men of the cloth, who were often on the front line of disease identification and treatment. The Holy Inquisition was in full battle mode, even in distant Asia and the Americas, and those unfortunate enough to be accused of heretical beliefs or practices, including innocent herbal treatments, were imprisoned, tortured, and occasionally burned alive. And none would have attracted greater attention than activities linked to the godless tropical natives. Included among their ranks were shamans, priests, mediums, and other spiritually guided healers, who administered their healing leaves, roots, and bark with a healthy measure of supernatural rigmarole. Even respected East Indian priests and healers (Brahmins) were occasionally persecuted (Gunn 2003, 104–109). As a result, like it or not, physicians and priests felt compelled to incorporate newly discovered botanical remedies into the ancient humoral system (Harris 2005, 75; Huguet-Termes 2001, 361).[2] Shedding their questionable association with native medical systems, newly encountered American drug plants were administered according to their supposed hot-cold, dry-wet properties, at least in the early years. American chili peppers (*Capsicum* sp.) were likened to hot and dry Asian black pepper, and thus were considered suitable for treating weak arms and legs, and to settle stomach problems (Estes 2000, 113). Caution was required, however, because the excess heat of the chili pod, according to Padre José de Acosta, "provokes to lust" (Acosta 1970 [1604], 240). Sixteenth-century Brazilian planter Gabriel Soares da Sousa noted that the herbaceous medicinal capeba (*Pothomorphe umbellata*) was very hot in nature and good for relief of kidney stones (Soares da Sousa 1971 [1587]). From India, Garcia de Orta suggested that the humoral temperament of cannafistula (*Cassia fistula*), a notoriously effective purgative, depended on where it grew; sometimes it was cool, other times it was hot (Orta 1913 [1563], 193). Guaiacum

(*Guaiacum officinale*) resin, the alleged miracle cure for syphilis, was a hot plant and therefore effective at treating syphilis' cold qualities. It also reached spots in the body that mercury, the standard treatment, could not (Rutten 2000, 47).

European physicians of the period employed a fairly standardized pharmacopoeia of spices, herbs, and minerals in their practice. Many had been in usage since the time of Galen and Hippocrates, and many were perishable. But rather than shrinking in popularity as new botanical discoveries were being made in the sixteenth century, many of these ancient medicaments found renewed popularity, especially those contained in *De materia medica* (Stannard 1999). But transporting these botanicals, many of which were collected from lands around the Mediterranean, to East Asia, West Africa, and the Americas, was problematic. Spoilage was common, and many medicinals simply lost their effectiveness after long voyages. So physicians residing in lands distant from Europe were encouraged to experiment with local medicinal plants. Pedro Arias de Benavides, a Spanish surgeon who practiced in America in the mid-1500s, explained, "The simples [single drug plants] were damaged by the long period of storage in Spain, followed by a long sea voyage and then the time it took for them to be sold to apothecaries in the Indies. The merchants would hold on to them until there was a demand, and then the shops too kept the simples a long time; inevitably they were ruined" (Huguet-Termes 2001, 372). In 1640, Jacques Bouton similarly noted from the island of Martinique, "The majority of remedies that we bring here from France lose all or part of their virtue" (Bouton 1640, 94). The situation in the East Indies was even worse. Hendrik Van Reede reported from Malabar (Kerala, India) in 1678 on the ludicrous requirement that all official medicines pass first through the Netherlands. This meant that local Asian drug plants were collected, shipped to Europe, packaged into pills, and then re-exported back to Asia for sale, often in an advanced stage of decomposition (Váczy 1980). The answer to the dilemma was obvious, according to Nicolas Bourgeois, "Why resort to foreign drugs while here [the Caribbean] they grow so healthy and in such abundance?" (Bourgeois 1788, 460).

Horrific new diseases, bad doctors, and rotten medicine; all of these factors triggered the demand for new medicines, or at least for cheaper sources of long-imported species. Demand created opportunity, and opportunity was met by a multitude of strategies and motives. In the Spanish New World Empire, medicinal plant discoveries followed four primary trajectories. First, they were identified and recorded as part of large metropolitan expeditions headed by leading scientists and physicians. The results were to be collated and published as great tomes. The best example was the herculean effort of Spanish physician Francisco Hernández. Working in Mexican hospitals in the late 1500s with indigenous healers, he compiled detailed ethnomedical data on some 3000 new species of useful plants. The enormity of this effort is best appreciated when it is compared with 500 total species reviewed in the classic *De materia medica*, a pharmacopoeia considered so complete that it remained the definitive source for herbal treatments for over fifteen hundred years. The second motive was a patriotic impulse. In this case, clerics investigated new botanical medicines for the purpose of elevating the status of the empire. For ecclesiastics, this obligation was accepted as a form of holy crusade, whose weapons were their compasses, maps, and astrolabes. Third were enterprising settlers, who used their own entrepreneurial initiative to identify new species and export marketable quantities to Spain, often accompanied by extravagant claims regarding their "extraordinary curative properties." Fourth, large government-coordinated campaigns to document colonial resources were set in motion. Chief among these were the comprehensive *relaciones geográficas*, continent-scale surveys of everything that was considered worth knowing, region by region, including maps, narratives, and descriptions of natural resources (Cañizares-Esguerra 2006, 7–10). Cataloguing these new ethnobotanical resources, as was also the case with the United Kingdom's possessions, was part and parcel to their entire empire-building enterprise (Gascoigne 2009). Finally, the quest for the healing secrets of the humid tropics in the early years of oceanic discovery, as was true of other forms of knowledge, was fueled by the European Renaissance mentality and by the intellectual turn of the Enlightenment (Disney 2007, 284). Commercial desires to exploit new-found natural wonders were certainly in evidence, but at its core these endeavors were symbolic of a scientific era "marked by a thirst for wondering, wandering, and collecting" (Finucci 2008, 524). As Hendrik Van Reede noted in his magisterial *Hortus Malabaracus* (1678–1693), illuminating these many hundreds of new plants and their virtues "would please many students of botany in Europe" (Váczy 1980). Chivalric duty and pure curiosity, in nearly every instance, trumped avarice as motivation for botanical knowledge gathering.

Although less documented than these imperial efforts to marshal the healing virtues of tropical nature, there were likely thousands of efforts by individual Europeans to discover botanical cures for what ailed them. Most became forgotten footnotes in the quest for green gold. Such is the case of seventeenth-century Italian Duke Vincenzo Gonzaga, instructive in the very triviality of his purpose. For at the relatively young age of 47, the Duke determined that he had a sexual performance problem. Perhaps others in his day would not have been as concerned, but the Duke was notorious for his libidinous recklessness and extramarital relations. To shore up his flagging reputation with the ladies, he sent the pharmacist Evangelista Marcobruno on a secret two-year odyssey throughout Europe and the New World to track down "a Viagra-like remedy." The druggist learned from a missionary in Barcelona that the famed "gusano," an aphrodisiac insect with a hairy-colored body and the head of a fly, was to be found in Peru, in the distant Americas. And so Marcobruno went packing to the New World in an attempt to rescue his master's diminished libido. There was already considerable pedigree to the notion that exotic biological offerings from the East and West had the required tonic effects; ginger was believed to bolster the emission of sperm, aloe strengthened the organ of concern, and vanilla and cinnamon were rumored to perk it up. From the East Indies, according to sixteenth-century Italian herbalist Pietro Andrea Mattioli, there existed a marvelous herb that could stir a man "to make love twelve times" or more. Likewise the African surnag, and the American maca, velvet bean, catuaba, and many others meant to stimulate, engorge, and otherwise enhance the sexual experience of men and sometimes women. The critical feature of these barely known locales was that they were tropical, for "hot was the operative word in the business of erotica." In any case, the druggist's frivolous mission was, in the end, a complete failure. The Duke died in 1612, his fantasies of acquiring a New World cure for his deflating sex drive unrealized, and Marcobruno, on his return voyage to Europe, was captured and enslaved in North Africa (Finucci 2008, 532).

The Duke's unsuccessful search for an aphrodisiac, however impractical, highlighted another feature of the early quest for green gold; short-term visitors lacking biological training frequently made wildly erroneous claims regarding the presence of legendary spices and herbal medicines. Thus Columbus reported on his first Atlantic crossing as having sighted medicinal aloe (*Aloe vera*), which he knew from Marco Polo's text to grow in Asia. What he actually saw was a native agave (*Agave bahamana*), of which he transported large quantities on his return trip to Spain. It ultimately proved useless. He also collected several local fruits in Cuba, such as *Canella winterana*, which he incorrectly identified as true cinnamon, a hugely valuable tree bark native to Indonesia. And from Columbus's second voyage, Nicoló Scyllacio noted, "The place abounds in rhubarb [*Rheum rhabarbarum*]," which at the time was a much sought-after Asian medicinal imported to Europe since at least the fourteenth century (cited in Quinn 1990, 74). This sighting was again a mistake, as the highly valued purgative was endemic to China. Ironically, Columbus filled his holds with piles of these worthless botanicals, but he failed to collect others he encountered that would in time change the course of history, especially maize, cassava, and tobacco (Flint 1992, 71–73; Morison and Obregón 1964, 48 and 51; Shaw 1ʗ_ _, 15–16).

The Admiral can perhaps be forgiven his sloppy science as he fully believed that he had arrived in the Asian heartland of exotic spices and medicines. He found exactly what he expected to find. Less than a decade later, during Pedro Álvarez Cabral's serendipitous encounter with Brazil, his scribe likewise catalogued the presence of rice, figs, wheat, grapes, and the European chestnut, none of which were in fact present in the New World (Filgueiras and Peixoto 2002). Some years later, in 1541, Francisco de Orellana made his infamous journey down the Amazon River, with the primary objective of discovering the fabled "land of cinnamon," rumored to be in those parts. Cinnamon was a valuable medicinal flavoring, imported from Asia since the time of the Egyptian pharaohs, and recognized as an antiseptic, digestive aid, and respiratory stimulant. The original source of the aromatic bark was known to be somewhere in the mysterious East Indies, but there was every reason to believe that God had likewise blessed the Amazonian Eden with the valuable tree. But God was not so generous, it appears, as two-and-a-half centuries later Alexander von Humboldt and his field collaborator Aimé Bonpland discovered that the "Orinoco cinnamon," the object of Orellana's desire, was in the same plant family as true cinnamon (Lauraceae) but was in a quite different genus (*Aniba*). The fabled cinnamon aroma and flavor of the original was present, but the compounds were different and quite inferior to true cinnamon (Gottlieb and Borin 2009). These erroneous identifications can be credited to biogeographical confusion, understandable for the time, and to the wishful thinking of entrepreneurial discoverers. But underlying these flawed expectations was a historically constructed idea several millennia in the making that mysterious medical cures were harbored in the foliage, bark, and roots of these warm and wet equatorial forests. Fueled with these beliefs, the great green quest began. And the first to be identified was guaiacum (*Guaiacum officinale*).

Widely known also as lignum vitae (tree of life), or by its Dutch name *pokhout*, guaiacum rapidly entered the European market as the primary treatment for syphilis. Native to the Antilles and south Florida, guaiacum replaced mercury treatments, which often had nasty side effects (Huguet-Termes 2001; Shaw 1992, 17; Wear 2004). In 1574, Nicolás Monardes, who had considerable dealings in the Americas, provided a narrative of how syphilis was contracted, and how the virtues of the guaiacum tree were first transmitted to Europeans:

There was a Spaniard that suffered great pain of the pox [syphilis], which he had taken by the company of an Indian woman, but his servant, being one of the physicians of that country, gave unto him the water of guaiacan [guaiacum], wherein not only his grievous pains were taken away that he did suffer, but he was healed very well. (Frampton 1580, fol. 11)

A scant sixteen years after the Columbian landfall, guaiacum was being exported to Europe to treat syphilis. The new remedy exploded in popularity following the publication of German Ulrich von Hutten's influential "Of the Wood called Guaiacum" in 1519, wherein the author claimed to have been healed of the horrible French pox by guaiacum. His rendition of its discovery went as follows: "A certayne noble man of Spayne . . . was greatly troubled with that infyrmitie: And after the people of that lande hadde taught hym that medycine [guaiacum] so he brought it to Spain" (Estes 2000, 113; Von Hutten 1536, fol. 10). The species was quickly adapted to the humoral system, and found ready acceptance, in spite of its absence from the *De materia medica*. But new diseases, according to the Galenical dictate, required new medicines. Guaiacum was enormously profitable for well over a century, and served as a model for other colonial bioprospecting efforts. Nearly two centuries passed before it was recognized that the plant was quite useless in the battle against syphilis. This medical mistake might have been identified earlier had anyone bothered to notice that the author Von Hutten died of advanced syphilis in 1523, a mere four years after the publication of his book. The message here is that the common refrain "the species has been used for centuries, so it must be effective" may well be an illusion. Lack of evidence to the contrary is no deterrent to the true believer. As for the tree species itself, after several centuries of overexploitation for a disease it could not cure, guaiacum is now endangered in its natural habitat (IUCN 2016).

Enormously popular and lucrative in its day, guaiacum represented a starting point in the great colonial search for green gold. But the means and motives of its discovery, as well as the preconceptions that encouraged the search for this and other plant-based miracle cures, mirrored the scientific pursuit that was to follow.

Ethnobotanical Axioms

The search for medicinal plants in the tropical realm was underpinned by several preconceived notions under which colonial physicians and men of science operated. Most found their inspiration in scripture and in the writings of the ancient philosophers. First among these was the biblical doctrine that the healing properties of nature were originally gifts of God to his green creations in the Garden of Eden. As seventeenth-century poet Guillaume de Salluste Du Bartas informed in his epic poem *The Divine Weeks*:

God, not content t'have given these plants of ours,
Precious perfumes, fruites plenty, precious flowers,
Infused phisike [physic = medicine] in their leaves and mores,
To cure our sickness and to salve our sores.
—DU BARTAS 1979 [1604], 191

In order to guide humankind in its search for his healing leaves, God had blessed each botanical remedy with a mark of its medicinal virtue. This principle, known as the Doctrine of the Signatures (or like cures like), holds that medicinal plants retain an identifying sign, or simili-tude. The healing properties of plants could thus be gleaned from their various attributes, such as shape, size, aroma, or color. The red juice from bloodroot (*Sanguinaria canadensis*), for instance, indicates that the plant should be applied to ailments of the blood. The phallic-shaped stem of various species of liana, such as escada-de-macaco (*Bauhinia* spp.), is believed to serve as aphrodisiac when soaked in white wine. And the rounded reproductive structures of the tiny herb quebra pedra (*Phyllanthus amarus*), whose Portuguese name means "stone-breaker," suggests this plant's ability to rid the patient of kidney stones (Bennett 2007; Voeks 1997, 64). Seventeenth-century botanist William Coles was a proponent of the concept, noting that "Wall-nuts have the perfect signature of the Head . . The Kernel hath the very figure of the Brain, and therefore it is very profitable for the Brain" (Coles 1657, 3). This concept dates at least two millennia to ancient India and China. But by the time the idea of botanical signatures had reached Christian Europe, it had been attributed to the hand of God and to man's expulsion from the Paradise. Eden was once brimming with healing herbs, it was under-stood, and residence in the Garden is what kept Adam and Eve free of illness. But following their expulsion, they were prone to diseases of the flesh that could only be cured with prayer and those leaves and roots appointed by the Almighty (Prest 1981, 63–64). The signatures were intended to assist people in their quest for God's many healing gifts.

The Doctrine of the Signatures was not universally accepted, particularly as Enlightenment skepticism and empiricism challenged religious dogma. Even in the time of the ancients its usefulness was questioned. Dioscorides was a fan of the signature concept, but Galen was not (Court 1985). Indeed, many colonial era naturalists and doctors found it impossible to identify which part of the plant was the actual signature; was it the heart-shaped leaves, or the latex-rich stem? English naturalist John Ray saw little value in the antiquated theory or "the foolishness of the chymists who chatter and boast so loudly of the signatures of plants" (Ray 1660, cited in Raven 1950, 98). English "feverologist" Robert Talbor was of a similar mind. "Why Eye-bright was specifical to heal the distempers of the Eyes, because its flowers they say resembles a Birds eye," and "Beans for the Reins [kidneys] and Testicles, from the similitude they have to those parts." "What rational man," he argued, "would be satisfied with such reasons?" (Talbor 1672, 8–10).

The second principle provided important insights into the initial discovery of health-restoring plants. It was widely held that knowledge of Eden's healing properties was not universally distributed among God's creatures. Rather, it was said to be blessed upon those entities perceived to be closest to nature. This followed from the view that within the hierarchy of the Great Chain of Being—God, Angels, Man, Animals, Plants, and Minerals—each link had an equal claim to existence, but each was also ' "unequal in dignity" (Lovejoy 1964, 186). Understanding of the various virtues of nature, in turn, was seen to diminish in quality and quantity along a gradient from the wild beasts that graze and browse on the vegetation, to the primitive folk who toil in close association with the forest and its bestiary, and finally to "civilized" people. According to Englishman Robert Talbor in1674, this progression in knowledge of nature, from animals to enlightened men, was another consequence of the expulsion of Adam and Eve from Eden. "Now the first man Adam had a perfect knowledge of the virtues of all plants, minerals, or animals . . . But since the Fall, Soul and Body have deviated from their first Perfection." Having lost their "primitive purity," humans were now consigned to following the observations of "the irrational creatures, as Birds, Beasts, and Fishes." Talbor offered the example of the herb celandine (likely *Chelidonium majus*), the ophthalmic virtue of which "was learnt from the swallow, who hath been often observed to squeeze the juice of the herb with her bill upon the blind eyes of her young, by which means they gain their sight" (Talbor 1672, 1–4). From this observation, the tradition of using celandine began in ancient Greece, and continued in John Gerard's famous 1597 *Herball, or Generall Historie of Plantes*: "the juice of the herbe is good to sharpen the sight, for it cleanseth and consumeth away slimie things that cleave about the ball of the eye and hinder the sight" (M. Woodward 1927, 247). Similar observations were provided by the German botanists Johann von Spix and Karl von Martius in their early nineteenth-century collecting journey through Brazil. Of "the Indians and Negroes" who harvest *ipecacuanha* (*Carapichea ipecacuanha*, the source of the powerful emetic ipecac), "We are assured that the savages had learnt the use of the *ipecacuanha* from the irara, a kind of martin, which is accustomed, they say, when it has drunk too much of the impure or brackish water of several streams and pools, to chew the leaves and the root, and thereby excite vomiting" (Von Spix and Von Martius 1928 [1824], 221). And from Malaysia, German physician Engelbert Kaempfer reported in the early eighteenth century that locals believed that the mongoose, if bitten by a poisonous snake, would dart into the forest, seek out mungo root (unknown species), and chew it as an antidote. The native people, Kaempfer recorded, discovered the value of the plant by observing the mongoose, and Kaempfer in turn learned it from the local people (Kaempfer 1996 [1712], 96–97).

How did these associations come about? According to Du Bartas, it was because animals in Eden had been paired by God with their natural medicines:

Yet each of them can naturally find
What simples cure the sicknesse of their kind;
Feeling no sooner their disease begin,
But they as soone have readie medicine.
The ram for phisike takes strong-senting rue;
The tortoise, slow, cold hemloke doth renew.
—DU BARTAS 1979 [1604], 374

This ancient belief—that medicinal plant knowledge is transmitted from wildlife to humans—has gained traction in recent years with the observation that numerous animals self-medicate with plants, and that local people observe and learn from animals. In one case, a Tanzanian healer discovered the use of an effective medicine against a dysentery-like illness by observing local wildlife. The village was suffering something of an epidemic of the illness, and their healer noticed that a porcupine, suffering from a similar ailment, gnawed on the roots of a local plant. Villagers previously had thought that the plant was poisonous and so had avoided using it. But armed with these new observations, they experimented with the root, found it effective, and adopted it into their pharmacopoeia, and to this day the plant is employed in the village for its curative properties (Huffman 2001).

Considerable evidence for self-medication in animals has been forthcoming in recent years, especially among our closest simian relatives (Masi et al. 2012). Studies of chimpanzees in Tanzania and Uganda, for example, demonstrate that when chimps are sick with heavy parasite loads, they chew the bitter pith of *Vernonia amygdalina*, a local shrub in the daisy family (Asteraceae). The chimps carefully peel away the bark before sucking out the bitter juice from the pith. Infected chimps are observed to dramatically reduce their parasite load in a short time, whereas those that fail to ingest the plant tissue continue to be infested. The species is likewise employed by numerous African human groups to treat various illnesses, including parasite infestation. The secondary compounds that give rise to its bitter taste (glucosides and sesquiterpene lactones) have been shown in the lab to inhibit the life cycle of the parasites (Huffman 2001). And it's not just primates. Domesticated goats in Uganda self-medicate for internal parasites as well. When plagued by a heavy worm infestation, they browse on particular parasite-killing plants. The African goatherds, who spend much of their lives in the field tending their livestock, are well aware of which species of plants the goats browse when they are ill. And many admit that they learned what plants to use when they suffered from parasites by observing goat behavior (Gradé et al. 2009).

Regarding this particular Edenic spin on people-plant relations, colonial observers were clearly on to something. But for men and women of early modern era science, this perceived connection between nature and people, especially technologically primitive people, carried considerable philosophical baggage. Europeans were convinced of their own superiority over native tropical people, and this link between plants and "savages" was taken as a clear signal of their brutish instinct rather than inherent intelligence (Schiebinger 2004, 82). Indigenous people, in the words of English physician Thomas Trapham, were "animal people," with no more entitlement to intellectual rights than the monkeys in the trees (Trapham 1694, 117). French merchant Jean-Baptiste Tavernier, whose historical claim to fame is having originally purchased the Hope Diamond, commented widely on the native peoples he encountered in his late seventeenth-century travels. The most "hideous" of these, he reported, were the Cafres (Khoi) of the Cape Province, who "live almost like beasts." But "brutal as they are" he confessed, "these Cafres . . . have nevertheless a special knowledge of simples, and know how to apply them." Ship captains and others frequented the Cafre doctors, and in each case, "were totally healed" (Tavernier 2001 [1676], 395). Likewise Edward Long, in his 1774 *History of Jamaica*, summed up the link between ethnobotanical knowledge and native peoples, stating that "brutes are botanists by instinct" (Long 1774, 381). Clearly, few doubted that indigenous people had a deep understanding of nature's medicinal virtues, and many believed that this knowledge had the potential to be medicinally useful and possibly highly profitable. But central to this narrative is the notion that the discovery process of drug plants by rural indigenous people was purely by blunder, instinct, and intimate association with the other beasts in the jungle, rather than by skill and thoughtful experimentation. Such an argument served to systematically dehumanize native people as well as negate any need for compensating those who chose to share their medicinal knowledge, whether voluntarily or through violence. Botanist brutes in this view had no claim to intellectual property.

A third widespread principle of the time regarded the geographic provenance of medicinal plant cures. In the vast and ever expanding known world, where were the best places to search? Hippocratic theory held that for each illness there must be a cure. That was a clue. But attached to this pre-Christian aphorism was the near-universal belief that God in his great beneficence had placed the cures for man's many maladies in the lands from which these illnesses originated. And because so many diseases were believed to have originated in the mysterious equatorial latitudes, the pursuit of drug plants was focused naturally on the hot and humid tropics. The theory was commented upon in the late 1500s by Englishman John Frampton. Referring to the general belief that syphilis "came from these parts of the Indies," he informed his readers that "Our Lord God would from whence the evil of the pox came, from thence should come the remedy for them" (Frampton 1580, fol. 11). This idea was developed further by the blind botanist Georg Rumphius from his post in Indonesia: "the Creator . . . provided each . . . country with its own medicaments . . . all countries have their own and singular illnesses which are to be cured by its native remedies" (cited in Beekman 1981, 12). Likewise Dutch physician Jacobus Bontius, in his influential 1631 treatise on tropical medicine, reported that "where the diseases above spoken of are endemical, there, the bountiful hand of nature has planted herbs whose virtues are adapted to counteract them" (Bontius 1769, 24). Thomas Trapham noted similarly from Jamaica that "the overflowing bounty of the great healer of us all, who hath given a balm for every Sore, and that not to be far sought and dear bought, but neer at hand" (Trapham 1679, 93–94). And English intellectual Samuel Hartlib connected this geographical concept to specific healing plants, "where any Endemical or National disease reigneth, there God hath also planted a specifique for it." It followed that, "in the West Indies, (from whence the great Pox first came . . .) there grow the specifiques for this disease, as Gujacum, Salsaperilla, Sassafras, and the Savages, do easily cure these distempers" (Hartlib 1655, 81).[3] God may have cursed his flock with newfound tropical diseases, but he also blessed them with endemic plants for their treatment.

This geographical connection was bolstered by ancient climatic determinist theories. As noted earlier (chapter 3), the Greek philosophers had long ago reasoned that human traits and intellectual capacities were under the influence of temperature and precipitation, and these age-old theories dovetailed nicely into medieval notions of medical cause and effect. The barren northern regions of the Earth, the Frigid Zones, were believed to be too cold to foster intellectual development, whereas the excessive heat and solar intensity of the equatorial regions, the Torrid Zones, permanently dulled the intellect and fanned the libido. For outsiders not accustomed to these enervating conditions, the hot and humid air blocked the pores and thickened the blood and mucus, while the noxious air (miasma) caused all sorts of awful ailments. The elevated incidence of infection and malaise in the tropics was counter-balanced, it was reasoned, by the many medicinal spices that grew in the equatorial zone—cinnamon, clove, nutmeg, pepper, ginger, and many others. German East India Company (VOC) physician Engelbert Kaempfer, who had traveled to Russia, Persia, Arabia, and many other locales in East Asia in the late seventeenth century, pondered these questions. As a proponent of the various theories on disease etiology of the era that connected the Sun, the Moon, and the stars with human health, he suggested that there existed a latitudinal gradient in the effectiveness of herbal remedies. "The boiling blaze of the tropic sun," he argued, "intensifies and sharpens the salutary powers of nature's resources." This held true as well for botanical poisons, such as the famous Makassar poison, which could achieve its deadly virtues only in the equatorial latitudes. "In cooler regions," he offered, "solar influence is weakened" (Kaempfer 1996 [1712], 96). Thus, not only were the lower latitudes overrun with diseases and their botanical remedies, but the healing power of tropical plants was overall more powerful than those inhabiting chillier climes.

Whether accepted or not, these archaic principles served more as general guidelines than as canons. An explosion of exotic drug plants was arriving daily from all points of the tropical compass, and European physicians became increasingly experimental, regardless of the provenance of the disease or the resemblance of a nut or a leaf to a body part. This was particularly true of diseases that failed to respond to the traditional humoral torments administered by physicians, particularly purging, vomiting, sweating, and bleeding. One example was treatment of the "green disease" or the "disease of virgin's," which affected young European women. Symptoms included very pale or even green-tinged skin, fatigue, tension, and a suppressed menstrual cycle. Reputed causes varied over time, but included among others unrequited love, constipation, and even excessive masturbation. The recommended treatment was often marriage. As related by Thomas Trapham, he had "never yet found a better Medicine against the Green Sickness than the Bermudas Berries." The berries (unknown species) had just begun to arrive from the Caribbean, and were quickly being incorporated into European pharmacopoeias. He goes on to record that "I'll tell you a remarkable Observation lately made by me upon a young Virgin, who dyed of this Disease, she was extremely far gone into the Green Sickness, lost her Strength and Spirits, had taken great quantities of Powder of Steel, been blooded in the Foot twice, and at length dying" (Trapham 1694, 7).[4] Although the origin of the disease had no connection to the Americas, Trapham pleaded that the girl would have been saved had she been administered these "marvelous berries" out of the tropics. In another example, John Peachi advocated the use of banella (species unknown) from the West Indies for the treatment of melancholy. He describes a fifty-year-old minister who had serious problems with melancholy and who had failed to respond to the standard treatments: "He had been let Blood several times in the Jugular Veins with a Lancet, and in the Hemorroide Veins with Leeches" and, "He had been Vomitted and Purged very often." Only after giving him banella for a month was he cured (Peachi 1694a, 4 and 5).

"The Woods Are Their Apothecaries"

As explorers, settlers, and priests quickly discerned, the ethnobotany of tropical medicinal plants, including their identities, habitats, modes of preparation, and the relevant illnesses they treated, was held either as collective or as specialized knowledge by the native populations. To learn the healing properties of these mysterious and unknown floras, Europeans needed the help of locals. As French missionary Raymond Breton reported from Guadeloupe, "One must have a great leisure to learn from the savages the names and virtues of the plants, the trees, and other things of these lands. They have soundly great knowledge and experience of the rare virtues of several things of which we don't know the name in Europe" (Breton 1978 [1647], 50). John Woodward prepared a concise toolkit for collecting and transporting plants and animals from the tropics, including descriptions of the "pagan" people's customs and virtues, and that travelers should take note of "their Physick, Surgery, and the Simples they use; [and] their Poysons" (J. Woodward 1696, 10). There was no question in the minds of Europeans that India and China and other Asian civilizations represented cornucopias of possible remedies, given the already lengthy international trade in their healing spices. But the "backwater" societies of the other equatorial zones also held secrets that could revolutionize medical practice. According to Robert Boyle, "Nor should we onely expect some improvements to the Therapeutical part of Physick, from the writings of so ingenious a People as the Chineses' but rather should also 'take notice of the Observations and Experiments' even of the 'Indians and other barbarous Nations'" (cited in Gascoigne 2009, 554). British merchant Henry de Ponthieu noted that the medicinal knowledge held by enslaved Africans in the Caribbean was constantly infused by new arrivals, and that British subjects in the Americas would do well to "interrogate" their slaves for ethnobotanical information (letter dated 1778, from de Ponthieu to Banks, cited in Drayton 2000, 93). In colonial Brazil, according to Friar Vicente do Salvador, the indigenous shamans were clearly "barbarous," but they were at the same time privy to "herbs and other medicines" that were sorely needed in the colony (Vicente do Salvador 1931 [1627?], 62). Similarly from West Africa, Willem Bosman reported, "The green Herbs, the principal Remedy in use amongst the Negroes, are of such wonderful Efficacy, that 'tis much to be deplored that no European Physician has yet applied himself to the Discovery of their Nature and Virtue." "I have several Times observed the Negroes cure such great and dangerous Wounds with them," he commented, "that I have stood amazed thereat" (Bosman 1721, 225). Englishmen Robert Knox succinctly summarized the view of many visitors to the tropics from his post in Sri Lanka (Ceylon), noting, "The woods are their apothecaries" (Knox 1681, 19). But would native people be willing to reveal their exotic remedies to strangers?

Descriptions of colonial-era ethnobotanical interactions between indigenous peoples and Europeans often followed a similar storyline. The visitor describes a malady from which he or a comrade suffers, sometimes acquired as a result of contact, hostile or otherwise, with the local population. He then briefly records the situation under which a local woman or man provided the plant remedy. In many cases, the indigenous informant is depicted as humble and willing to share information of free will. Regardless, he or she is nearly always anonymous. In almost no case is the informant dignified in the narrative with a name, other than "the Indian" or "the mulatto" or "the Negro." Nor are indigenous healers given credit for the intellectual merit of their knowledge. Thus, for example, Nicolás Monardes reported on an herbal treatment in the Indies that was identified by "an Indian" who was "a servant to a Spaniard called John Infant." The nameless indigenous informant revealed the botanical remedy, in this case for wounds, but to this day, as Monardes reports, they call the herb "John Infant" (Monardes 1580, fol. 10). Similarly John Peachi, in his description of the efficacy of "Mexico seeds" (species unknown) says that he "observed the Indians make most use of [it] against Worms" (Peachi 1695, 3). Alexander von Humboldt, who was plagued by chiggers on his Orinoco expedition, notes that he was cured with the leaf of "uzao," shown to him by "an Indian" (Von Humboldt and Bonpland 1827, 5: 245). From Java, Thomas Horsfield described the properties of the mysterious "Antshar," a powerful tree poison that was revealed by "an old Javanese . . . celebrated for his superior skill in preparing the poison" (Horsfield 1823, 83). And from Malabar, Van Reede reported that the natives were "happy" to disclose the names and the curative virtues of plants (cited in Váczy 1980, 44).[5]

But these examples may have been the exception, akin to currently investigating the "open access" collective medicinal plant knowledge in a rural community. Shamans and other specialized healers, on the other hand, were a different story. Then as now, specialized healers zealously guarded their botanical remedies as personal intellectual property. Reporting from Barbados in 1759, for example, William Hillary reviewed the local treatments for yaws, reporting that the "Negro doctors" discovered a treatment using the caustic juices of certain plants which "they keep as a secret from the white people, but preserve among themselves by tradition" (Hillary 1759, 341). Among the Amerindians in Dutch Suriname, Swedish botanist Daniel Rolander similarly reported that they are "too jealous of their [medicinal plant] secrets to reveal them to anyone unless he has inspired their trust, and can provide them valuable information in return" (Van Andel 2015; Rolander 2008 [1755], 1487).

4.1 Charles de Rochefort on St. Kitts Island (1681) collecting useful species. In : *Histoire Naturelle et Morale des Iles Antilles de l'Amerique*. Pg. 53. RB 330382. The Huntington Library, San Marino, California. Digital Image by Robert Voeks.

Jacques Bouton reported from Martinique in 1640 his amazement at the stellar health of the indigenous people and on the "beautiful" botanical secrets they possess, "but it is impossible to get them out of them" (Bouton 1640, 45). Charles Rochefort similarly noted that the Caribbean people "are extremely jealous of their secrets in medicine, especially their women . . . they have yet to want to communicate. . . . their sovereign remedies" (Rochefort 1681, 53) (Fig. 4.1). Some tried financial incentives, such as Nicolas Bourgeois who offered money for medicinal plant remedies in Guadeloupe. But "the cleverest" among them, he reported, guarded the secret "of medicinal herbs that we do not know nearly as well as they do" (Bourgeois 1788, 482). In Hispaniola, it was necessary to "gain their confidence, as I was able to do with some of them," in order to "pull from them the secret" (cited in Weaver 2002, 440). Others resorted to friendly persuasion. Michel Descourtilz, working in Haiti in 1799 in the midst of the great revolution, uncovered several medicinal recipes from an anonymous mulatta woman. She was very reluctant to disclose any of her secrets in the beginning, but he "softened" her temperament over time by giving her some of his paint-ings of plants, "which she coveted" (cited in Schiebinger 2004, 81).

In the case of Brazil's indigenous Tupinambá, browbeating replaced persuasion as a means of extracting ethnobotanical information. The Jesuit community maintained the principal botanical pharmacies (*bóti-cas*) for the colony, and they discerned early that their Mediterranean-based pharmacopoeia was no match for the legion of new diseases they were encountering. They were eager to exploit indigenous knowledge of the healing flora (Walker 2013), but they were also keenly aware that the most botanically savvy individuals among the natives, the shamans (*pajés*), were also instruments of the devil who spread their idolatry by means of their barbarous medical practices (Leite 1938, 87). In a 1549 letter, Padre Nóbrega boasted about his strategy for acquiring ethnobo-tanical knowledge from a famous Tupinambá shaman who claimed to be a god. After being verbally demoralized by Nóbrega and the other good brothers, the *pajé* begged to be baptized into the Christian faith "and now is one of the converted" (Cartas do Brasil 1886 [1549–1560], 67; Leite 1938, 1: 21–34 and 2: 14–22). Only after being shamed and bullied were the shamans encouraged to divulge their medicinal plant secrets.

In addition to infectious and parasitic diseases, European explorers and military personnel were plagued by the various plant-derived poi-sons employed by indigenous peoples. Toxic plants, which were used by enslaved Africans to poison their masters, caused considerable anxiety among slave owners in Brazil (Voeks 1997, 47). Similarly in Suriname, Daniel Rolander reported that "wicked people," meaning enslaved Af-ricans and Indians, serve *Mimosa* sp. poison in food to their masters (Rolander 2008 [1755], 1296). Of even greater concern, however, were the poisoned arrows and darts employed by indigenous combatants. Of these, Europeans were wholly ignorant. In Suriname, esc d slaves (ma-roons) cultivated *Canna indica* for its rock-hard seeds. These ere soaked in secret poisonous liquids and used as rifle shot against the Dutch mili-tary (Van Andel et al. 2013). Similarly in sub-Saharan Africa, there was the continuous fear of arrows poisoned with "potent plants, snakes, fermented urine and scorpions," in particular the dreaded kombe vine (*Strophanthus hispidus*) (Osseo-Asare 2008, 274). Native peoples were be-lieved to maintain antidotes to their poisons, and when the identities of these could not be ascertained peacefully, violence was often employed. Sir Hans Sloane in 1707 relates the story of the identification of *con-tra yerva* (likely *Dorstenia contrajerva*) in Guatemala, which was used to counteract the effects of arrow poisons. A Spanish doctor, having been "wounded by one of their poison'd Arrows, to find out his Cure, they took one of their Indian Prisoners, and tying him to a Post threatned to wound him with one of their own venomous Arrows, if immediately he did not declare their Cure for that Disease, upon which the Indian immediately chaw'd some of this Contra Yerva, and put it into the wound, and it healed" (Sloane 1707 and 1725, lv). Although Sloane's story seems to be missing some relevant details, the important message is that physical pain and threat of death were considered acceptable methods for acquiring ethnobotanical knowledge of poisons.

The most protracted hunt for a poison antidote involved the infamous *upoh* (or *upas*, or Makassar poison) tree of Southeast Asia. First reported in Europe in 1320 by Franciscan Friar Odoric of Pordenone, the upoh (*Antiaris toxicaria*) was believed to be the most deadly and insidious toxin in the world. Sixteenth-century Portuguese historian Diogo do Couto spread the bizarre legend that the *upoh* "kills those who stand in its shade on the western side, but standing on the eastern side serves as an antidote" (cited in Carey 2003, 525). Even the generally sober-minded German botanist Engelbert Kaempfer perpetuated his own set of "*upoh*-isms" in the early 1700s, reporting that "When the tree [*upoh*] has been located, it will instantly suffocate its attackers with an outburst of fumes, unless the tree is wounded from a distance and from the direction from which the wind is blowing." So dangerous was it to collect, Kaempfer noted, that the chore was delegated only to those "condemned to death for the crime of witchcraft" (Kaempfer 1996 [1712], 97).

The tree in fact exhibited none of these fictitious features, but it did generate considerable concern among Dutch colonists and soldiers in Indonesia. As they consolidated their seventeenth-century foothold on the islands, the poison-tipped projectiles of indigenous warriors proved a formidable and greatly feared weapon. But years of searching for the ethnobotanical antidote yielded only a single revolting remedy provided by the local population. According to German soldier Johann Saar, the only known cure for a wounded soldier was for him to immediately consume "his own excrement, as fresh as it goes from him" (cited in Carey 2003, 528). Perhaps this was dark humor on the part of the local people, but as noted earlier, the notion that human urine and feces contained medicinal virtues was not completely foreign to Europeans of the period. In any case, Kaempfer recorded that Dutch soldiers indeed carried a supply of their own ordure at all times. "Armed with this excremental antidote," he relates, "they joined battle with the dark people of Celebes." Although Kaempfer never records the true antidote, he says that the natives cherished a botanical cure more effective than excrement, the identity of which "could be extracted only by torture" (Kaempfer 1996 [1712], 99). Physical violence again was the key to unlocking ethnomedical secrets.

The darkest pages in the history of medicinal plant inquiry are reserved for the exploits of Gonzalo Pizarro, brother of the conqueror of the Inca Empire. In the mid-1500s, he set out to discover Amazonia's legendary "land of cinnamon." Cinnamon was a fabulously valuable medicinal spice, one with a history reaching back some 5000 years, and the Spanish were keen to find a New World source (R. Lee and Balick 2005). Pizarro was guided by local Indians to a small grove of *canelo* (*Aniba canellila*), which to the Spaniards appeared "of the most perfect kind." Unsatisfied with the quantities he was being shown, Pizarro pressed the local Indians further. They replied that being hemmed in by other tribes, they were only familiar with trees in this small territory. Furious at their answers, Pizarro had these innocents tortured, and when this failed, others were "stretched . . . on barbeques . . . until the fire penetrated and consumed their bodies." Others he ordered thrown to the dogs, who "tore them to pieces with their teeth and devoured them" (Cieza de León 1918 [1518–1554], 59 and 60).

Benefit Sharing

The question of fair compensation to local informants for sharing their medicinal plant knowledge seldom arose during European settlement and colonization. Nature's healing properties were treated as open access resources; whoever was sufficiently charming, clever, or cruel was usually able to ascertain their prize. There are a few documented examples of reasonable financial compensation being given to the local healer, however, and curiously each involves enslaved Africans in the Americas. For instance, on November 24, 1749, the South Carolina Commons House of Assembly considered the question of Caesar, an enslaved African owned by a Mr. John Norman. Testimony showed that Caesar was in possession of an herbal concoction that had cured several people of deadly snakebite, and that he was willing to divulge his secret formula "for a reasonable reward." A committee was appointed by the Assembly to review the claims, and some days later their investigation confirmed the preliminary reports. When Caesar was asked what he considered to be fair compensation, he responded "his freedom and a moderate competence for life." After further deliberation, the Assembly agreed; for the price of his mysterious herbal cure—actually the juice of common plantain (*Plantago major*) or horehound (*Marrubium vulgare*), both of which were ancient European medicinal plants, and "a leaf of good tobacco moistened with rum"—Caesar was granted his freedom as well as a lifetime annual allowance of one hundred pounds (Chappell 1975, 183 and 184). Other similar manumissions in exchange for medicinal recipes occurred in colonial America, including Papan, who was freed in 1729 by the Virginia legislature for revealing his treatment for venereal disease, and Sampson, who was offered his freedom in 1755 as well as a 50 pound annuity for disclosing his mysterious remedy for rattlesnake bite (Fett 2002, 64). In all of these cases, the question remains as it does today of whether these were novel discoveries, or had these men learned the cures from others?

The issue of botanical knowledge priority and just compensation appears again in the case of Graman Quassi (or Kwasimukámba, or Quacy). Born in West Africa in 1690 and later enslaved and transported to Suriname, Quassi was freed from servitude for his service hunting down African maroons for the Dutch military and planters. For these heroic actions, he was presented with a gold breastplate inscribed with the words "Quassie, faithful to the whites." Later on, as a reward for his many contributions to the colony, he was sent to The Hague and met with Willem V, the Prince of Orange, who presented him with, among other gifts, a fine gold-laced jacket, a golden medallion, and a gold-tipped cane (Price 1979) (Fig. 4.2).

Quassi went on to become a plantation master and slave owner himself. But he was also a famed herbalist and sorcerer who employed medicinal plants and magical amulets to cure what ailed the black and white population. The amulets allowed the Black Rangers, a corps of Africans who fought against escaped slaves, to "fight like bulldogs," according to Dutch soldier John Stedman. Stedman, however, had mixed feelings on Quassi's reputed medical abilities. On the one hand, he disparaged him as an indolent "blockhead" and his amulets as "trash" that has "filled his pockets with no inconsiderable profits." At the same time, he considered him "one of the most Extraordinary Black men in Surinam, or Perhaps in the World" (Stedman 1988 [1790], 581). Quassi's principal fame came from his discovery of the fever-fighting properties of a native treelet, later named by Linneaus *Quassia amara* in his honor. The bitter wood of this species compared favorably with Jesuit's bark (*Cinchona* spp.) for its ability to break a fever, and over time, according to Stedman, made Quassi a rich man. He may have sold the recipe for a "considerable sum" to the Swedish botanist Daniel Rolander (W. Lewis 1791, 529; Schiebinger 2004, 213),[6] who took it back to Europe.

The celebrated Graman Quacy.

London, Published Dec. 1st 1793, by J. Johnson, St Pauls Church Yard.

4.2   John Stedman. 1796. "The Celebrated Graman Quacy." *Narrative, of a Five Years' Expedition, Against the Revolted Negroes of Surinam, in Guiana, on the Wild Coast of South America, from the Year 1772 to 1777.* Image painted by William Blake. RB 23614. The Huntington Library, San Marino, California. Digital Image by Robert Voeks.

But did Quassi actually discover the drug plant? William Lewis, an English chemist, stated in 1791 that the species had long been used by the "natives" of Suriname, and Rolander recorded in his diary that Quassi had learned the usefulness of the plant from local Indians. Then as now, the issue of prior art and equitable sharing of medicinal drug plant benefits is contingent and complicated. In any case, the bitter brew made popular by an ex-slave is still widely used in Suriname, and known popularly as Quassi Bitters. Quassi lived into his 80s, addressed in personal correspondence as Master Phillipus of Quassi, Professor of Herbology in Suriname (Van Andel et al. 2012).

Intellectual ownership of botanical nature assumes many guises. Although this knowledge and skill often involves a novel medicinal application of a plant species, the concept also encompasses other unknown ethnobotanical attributes of the species, such as mode of preparation, plant part used, life history features, or harvest strategy. The story of the vanilla orchid and a young enslaved African boy, Edmond Albius, is especially instructive. Although the genus *Vanilla* naturally inhabits the Old and New World tropics, the species from which the vanilla flavor is extracted, *Vanilla planifolia*, is endemic to the tropical forests of eastern Mexico. Domesticated in pre-Columbian times as a flavoring and medicine, vanilla entered into the Aztec elite's luxurious chocolate concoction, an intoxicating brew thickened with maize, heated with chili peppers, tinted with blood-red achiote, and flavored with vanilla and other still unknown flowers (Norton 2008, 18). Its medicinal qualities were attested to by physician Francisco Hernandez, who recorded that a decoction of vanilla "beans" was used in sixteenth-century Mexico to regulate urine flow, aid in abortion, strengthen the stomach, and diminish flatulence. Centuries later, vanilla-flavored hot chocolate exploded in popularity, becoming "almost a vice" in Chiapas and Guatemala (Bruman 1948, 366). As an increasingly lucrative forest product, vanilla's many secrets were zealously guarded by its local tenders, the Totonac Indians, including even how to properly cure the beans. According to English explorer William Dampier, "The Indians have some secret that I know not. I have often askt the Spaniards how they were cured, but I never could meet with any could tell me" (Dampier 1703, 235). But curing was not the primary problem that vexed would-be vanilla cultivators. For although the orchids grew vigorously after being transported to various locations in Africa and the islands of the Indian Ocean, they failed entirely to produce the precious beans. The problem was that the requisite pollinators, which included several New World bee genera, failed to make the journey with the orchids. Consequently, wherever it grew outside of Mexico, the widely appreciated vanilla orchid was quite sterile (Bory et al. 2008). Enter the young Edmond Albius (Fig. 4.3).

French planter Ferréol Bellier-Beaumont had been cultivating a few vanilla vines on his Reunion Island plantation for some twenty years. His attempts to coax the much sought-after beans, like all such efforts outside of native Mexico, were unsuccessful. But one day in 1841 he noticed with astonishment the unmistakable swelling of ovaries in two vanilla orchid flowers. By some miracle, the long-desired vanilla beans were beginning to grow. He was even more surprised when his twelve-year-old slave Edmond informed him that it was he who had effected the pollination. Bellier-Beaumont was suspicious of his young protégé, and "demanded" to know how he could have done this. Edmond obediently showed how he pushed a small bamboo stick through the organ that prevents self-fertilization (rostellum), allowing the pollen to reach the sticky female organ (stigma). This simple hand-pollination procedure, discovered serendipitously by an enslaved African boy, quickly spread throughout Reunion, as young Edmond was sent from plantation to plantation to teach the art of "orchid marriage." In a few short years, this young slave's discovery revolutionized the world's source of the aromatic flavoring, relegating its eastern Mexico home to little more than a backwater producer of vanilla.

Edmond Albius's ethnobotanical innovation changed the economic geography of the western Indian Ocean, and created fortunes for plantation owners on Reunion and Madagascar. But the young man, after freely revealing his procedure, sank into obscurity. Years later, after slavery had been abolished in France's colonies, Edmond Albius ran into trouble with the law, and found himself serving five years in prison for his supposed offenses. His previous owner, Bellier-Beaumont, appealed to the French courts for his release on the grounds that he was a hero for his botanical discovery, and that France owed him his freedom and a reward for his contribution. The courts found in favor of the claim, and Edmond Albius was ultimately freed from incarceration, but without the requested financial reward (Ecott 2004).[7]

In spite of these rare records of financial compensation, medicinal plant exploration in the Americas and Africa during the colonial era was marked mostly by dehumanizing treatment of indigenous informants and blatant exploitation of their intellectual property. Native peoples were considered culturally and intellectually inferiors, and in most instances were treated as such. But the European perception of the lands and peoples of tropical Asia was quite different. Having fired the imaginations of Europeans for many centuries with their mysterious and highly

A. Roussin del. et lith. d'ap. nature.   1863.   Imp. A. Roussin. Ile de la Réunion.

EDMOND ALBIUS
Inventeur de la fécondation artificielle du vanillier.

4.3   Antoine Louis Roussin. 1863. *Album de l'Ile de la Réunion.* Vol. 3. Edmond Albius holding a vanilla plant. Public Domain.

sought-after medicaments, the Chinese and Indians were treated more as teachers than as apprentices, at least in the area of medical theory and treatments. Jesuit priests working in China, for example, expended considerable effort studying Chinese healing practices and comparing them to their own (Barnes 2005, 48). And in colonial India, even "governors and clerics relied on Hindu healers rather than European physicians" (Pearson 1996, 27). In 1679, Thomas Trapham sang the praises of Chinese medicine, noting from his post in Jamaica "our agreeable China [root] . . . this root among the wise Chineses, from whence it took its name" (Trapham 1679, 92). Indeed, while mining for medicinal plant treasures in Africa and the New World was usually carried out through deceit and even physical violence, in the East it consisted more of rearranging and reorganizing medical knowledge that had long ago been systematized.

The classic *Hortus Malabaricus* was a prime example. Compiled and published between 1678 and 1693 by Hendrik Adriaan Van Rheede tot Draakenstein (hereafter Van Reede), then Dutch Governor of Cochin (in Malabar), it outlines modes of preparation and application of medicinal plants based on pre-Ayurvedic knowledge as related by some of the great hereditary physicians of the region. Information presented in the *Hortus* was drawn from palm leaf manuscripts by Itty Achuden, a famous local healer. And unlike the largely anonymous contributors to our knowledge of the African and American healing floras, Van Reede took pains to give priority to the major Indian contributors to the volumes. Itty Achuden notes in the first volume that the medicinal powers of the plants he describes were culled directly from his family's book of Ayurvedic treatments, which had already been in use for generations. In addition to specialists, a board of 15–16 indigenous physicians and botanists from various parts of Malabar were engaged to examine the materials prepared and to present their opinions about the curative properties of the plants collected. The resulting twelve volumes, thirty years in the making, represented a supremely collaborative effort between European and Indian botanists, and medical practitioners (Ram 2005).[8] Their painstaking efforts and rigorous methods ultimately preserved an ancient body of ethnobotanical wisdom that otherwise would have been lost to the world (Manilal 1980).

The Age of Biopiracy

Individual efforts by Europeans to uncover the healing properties of tropical nature were nearly always exploitive of local knowledge. But these were not particularly conspiratorial plots, and with few exceptions, the consequences of their deeds did not lead to the fame and fortune so often associated with pilfering of intellectual property. Rather, it was the carefully orchestrated theft of enormously profitable species, some new to science, others known and traded for centuries, that came to characterize and stigmatize the colonial Age of Biopiracy. Many involved the same exotic spices that had been enriching the dining tables and apothecaries of Europe for over a thousand years. Spices traveled over land and sea from lands distant and exotic, and the perceived provenance of these botanical treasures was at times as fanciful and fabricated as the people who tended them. Many people accepted the word of French chronicler Jean de Joinville, who in the fourteenth century related how fisherman on the Nile River cast their nets in the morning only to later find them filled with precious healing spices—cinnamon, rhubarb, ginger, and aloes. They were thought to have entered the river from the Garden of Eden, at the time believed to be situated in Central Africa, perhaps as long ago as Noah's great flood (Prest 1981, 30).

Throughout the Middle Ages, exotic and astronomically expensive spices represented items of conspicuous consumption for Europeans of wealth and position. To serve highly spiced foods to guests was considered the highest form of snobbery. Sumptuous feasts often included hundreds of pounds of black pepper, cinnamon, cloves, nutmeg, grains of paradise (*Aframomum melegueta*), and many other imported and hugely expensive luxuries. But spices were not just flavor enhancers. They were also revered for their medicinal attributes, particularly in their ability to harmonize the bodily humors. Their often piquant and peppery properties were considered vital to establishing equilibrium with the usually moist and cool state of meat and fish. Mace and nutmeg were considered hot and dry. Ginger was hot but also moist. Sugar, which in medieval times was also considered a spice, was believed especially useful, since as neither hot nor cool it served to stabilize the body humors (P. Freedman 2008).

From the sixteenth through the nineteenth centuries, control of these profitable products, including ginger, cinnamon, clove, nutmeg, rhubarb, aloes, dragon's blood, black pepper, and later cinchona, waxed and waned as colonial states jockeyed for power and market share. As much as gold, silver, and slaves, botanical resources in the distant humid tropics were crucial to the colonial expansion enterprise. In London, demand for exotic medicines and spices rose dramatically in the sixteenth and seventeenth centuries, and exports rose twenty-fold by 1770, to 585,000 kilos (about 1.3 million pounds) per year (Wallis 2012). And changing perception of the efficacy of plant cures meant that old imports quickly fell into disuse, only to be replaced by newer and even more mysterious medicaments. Such was the case with guaiacum, which over time found intense competition from Asian china root (*Smilax* sp.), and later from American sarsaparilla (one or more species of New World *Smilax*) (Winterbottom 2014). Thus, of the 93 medicines carried by Columbus in his medicine chest on his second voyage, most had fallen out of favor as treatments by the eighteenth century (Rutten 2000, 31–34). Dramatic price fluctuations were common in Spain and Portugal, as the English and Dutch naval forces interrupted their access to medicinal sources in Malacca, Goa, Ceylon, and elsewhere (Almeida 1975).

But outside of outright military attacks on the medicinal spice lands of Asia and the Americas, how could these lucrative monopolies be broken? The answer was obvious—ascertain their legitimate source, secret their seeds or seedlings away, and transplant them to their far-flung colonies. For a few species, such intentional introduction would prove unnecessary, as the process had already begun spontaneously. Asian ginger, for example, an "incorrigible colonist," had been imported to Europe since before the Christian era and had the habit of sprouting from its rhizomes while in transit. Marco Polo noticed it festooning the decks of Chinese merchant ships in the fourteenth century, and by the fifteenth century, it was widely dispersed in Asia and Oceania (Keay 2007, 19). But other prized plant species were passionately guarded by their colonial custodians, and would require considerable stealth and imperial resource. The geopolitical stakes of these larcenous actions were immense, as the Dutch at least considered unauthorized shipment of their monopolized floral treasures as "an act of war" (Dalby 2000; Dean 1995, 86). Two endemic species stand out in the annals of biopiracy—nutmeg and cinchona.[9]

The Nutmeg Conspiracy

It is difficult to fully appreciate the historical and geographical significance of nutmeg. What today is not much more than a fragrant yellow powder used to liven up holiday spice cake and top off eggnog was for many centuries an article of tremendous global economic importance. Nutmeg and its sister spice mace were on the minds of the early Iberian mariners, Columbus, Cabral, da Gama, and Magellan, as they sailed into the cerulean unknown. Once Europeans had "discovered" and colonized the nut-bearing islands (discovery being relative, as Chinese, Indian, and Arab traders had been trading with the islands for centuries), they were quite willing to apply torture and murder to enforce their botanical claims. As for the indigenous peoples who husbanded the fragrant groves, slaughter and slavery were their fate. Of all the medicinal spices that would one after another captivate Europeans with their rarity and "unfathomable remote origin," none was more sought after by foreign powers and would cause more misery for native peoples than nutmeg (Keay 2007, 4).

The tree that bears nutmeg, *Myristica fragrans* (Family Myristicaceae), is native to only a tiny smattering of volcanic islands in far eastern Indonesia—the Bandas, better known as the Spice Islands. So fragrant were the evergreen nutmeg trees that sailors claimed to detect the aromatic isles miles before they caught sight of them. The tree is not tall, growing to only 10–12 meters (about 30 or 40 feet) in height, with alternate leathery leaves. It is usually dioecious, with male and female flowers situated on different individuals (Weil 1965). The fruit is about the size and color of an apricot. The dried seed (endosperm and embryo) is about two and a half centimeters long (one inch), two centimeters wide, light brown in color and rock hard. It is contained in a succulent fruit and surrounded by a fleshy, crimson-colored seed appendage (aril) known commercially as mace (Fig. 4.4).

Nutmeg has been an article of international commerce since ancient times, having navigated its way to China and India by the seventh century CE. Nutmeg and mace entered into the Roman pharmacopoeia by the time of Theophrastus (373–288 BCE) and Pliny (23–79 CE). Theophrastus reckoned it grew in distant Arabia, while Pliny pinpointed its origin as Syria. Neither was remotely accurate (J. I. Miller 1969, 58–60). The earliest pharmaceutical reference in the Arab world was in ninth-century Baghdad, where it seems to have been used mostly for digestive ailments (Weil 1965). By the end of the twelfth century, nutmeg and mace had reached as far as northern Europe (Lloyd 1911, 84), and its medicinal benefits were well known to the English by the fifteenth century (Stannard 1964). In 1696 Robert Boyle recorded that the "oyl of nutmeg" was good for colic, and others attested to its efficacy in treating colds, dysentery (bloody flux), and flatulence (Boyle 1696, 73; Milton 1999, 3). More importantly, it was the plant product of choice to treat the awful outbreaks of plague that ravaged Britain in the early seventeenth century. Nutmeg was also considered a powerful aphrodisiac for men (and still is by some) and was used to induce abortion (abortifacient). English women seeking to terminate their pregnancies were referred to as "nutmeg ladies" (Emboden 1979, 47). There was also suggestion of nutmeg's use as a tonic and to induce a state of drunkenness, even hallucinations. It was employed on slave ships to relieve weariness and for its hallucinogenic properties, and experienced renewed interest as a recreational drug plant during the "hippy culture" (Lennep et al. 2015,

A

B

4.4 A. Nutmeg plant (*Myristica fragrans*) with fruit, flower, leaves, and stem. Van Reede, 1683. Wellcome Library, London, Creative Commons, 43188. B. Fresh nutmeg seeds and arils (mace). Windsor Forest, Portland parish, Jamaica. Photo credit: Ina Vandebroek.

46; Rutten 2000, 46). In the 1960s, regular marijuana smokers, including prisoners and college students, reported substituting nutmeg when cannabis was unavailable, although the experience usually made them sick (Weil 1965). In spite of widespread reports, there is only anecdotal evidence to support these psychotropic effects (Gils and Cox 1994).

The Banda archipelago is so tiny and isolated that it seldom constitutes more than a faint dot on a small-scale map. But its lack of size was dwarfed by the enormity of its importance to foreign traders. For a thousand years or more before the European arrival, Indian, Chinese, Malay, and Arab traders had bartered with the Bandanese people for the precious propagules. It was not until 1512 that the islands were discovered for Europe by Portuguese Captain Antonio de Abreu. He was greeted warmly by the Bandanese, who were keen to introduce new clients to their forest products. They happily loaded the holds of their ships with their botanical treasures—nutmeg, mace, and some clove—which were later sold for roughly one thousand times what they had paid the local Bandanese. But the Portuguese for some reason neglected the Banda Islands in favor of other exotic ports, and the trading vacuum was taken up nearly a century later by rival representatives of the British and the Dutch East Indian Company (VOC) by 1601. The VOC, backed up by a sizeable armada of Dutch warships, were the more successful at creating a global monopoly of the important resource, and the more ruthless in its protection.

Nutmeg was at the time the most lucrative forest product on the planet. And the Dutch were determined to control the entire value chain, from cultivation to sales. But the British were serious competitors in the islands, and the loyalty of the local indigenous people, according to some, was not to be trusted. So the Dutch hatched a preemptive plan. Inventing a story that their competitors were conspiring to attack them, their ruthless governor, General Jan Pieterszoon Coen, rounded up the British agents from their chief holding on the island of Run. After submitting them to horrific tortures by fire and water (today known as waterboarding), Coen had them beheaded and quartered, their remains displayed on cane stakes as an example to any who questioned their authority over the prized plants (Anonymous 1665). The Amboyna Massacre, as it came to be known, was later rationalized by the Dutch as permissible given the norms of the day. As for the indigenous Bandanese who gathered and traded the precious kernels, the Governor General believed them to be in league with the British and, in any case, too untrustworthy for so lucrative a trade item. So most of their leaders, the *orang kaya*, were marched in chains to a circular enclosure and systematically beheaded and chopped into quarters by Japanese samurai on the VOC payroll. Then, moving from island to island, King Coen (as he was known by the British) had the entire Bandanese population herded into ships, men, women, and children, and sold into slavery in distant Java. Of the original fifteen thousand souls, probably one thousand survived within the archipelago. The few who were allowed to remain in the nutmeg-producing areas were kept only to instruct the newcomers, especially Dutch settlers and indigenous laborers from New Guinea, Timor, and Borneo, on how to care for the trees and properly harvest the precious nuts and arils. Finally, to better control production and avert any chance of piracy of the precious plants, the Dutch set about destroying thousands of nutmeg trees on the islands outside of their control. On the island of Run, every nutmeg tree was chopped down, and the vegetation was burned down to the roots (Hanna 1978, 46–90; Keay 2007, 238–242; Milton 1999, 343 and 360).

Given the resolve of the Dutch VOC to maintain their stranglehold on nutmeg, the obvious answer for those who lusted after the lucrative resource was to naturalize the species in a distant location. But this simple solution had long been thwarted by the Dutch, who treated the nuts with lime before shipment, thus rendering them sterile (Zumbroich 2005). Somehow, someone needed to spirit away some fresh seeds or seedlings from the islands. It would not be easy, but the French had a plan. And in a detailed series of letters, dated 1721 to 1772, they outlined a scheme to smuggle young nutmeg trees from the Bandas and transplant them on French soil. Two heavily armed ships were dispatched on a covert mission to secure the seedlings. The sailors carried drawings of the leaves with them so as to avoid confusion. When they reached a poorly guarded stretch of island, they were to use buckets to retrieve the valuable seedlings in local soil and carry them to the waiting ship for the journey to France's Reunion and Mauritius Islands. For various reasons, however, this and several subsequent attempts were completely unsuccessful. Biopiracy was not an easy enterprise.

Three decades later, still with no nutmeg trees to show for their efforts, the French offered a reward of twenty thousand pieces of silver to whoever could secure a batch of the coveted trees. Thus began a new series of botanical expeditions, this time under the direction of Pierre Poivre (Peter Pepper in English). Earlier a missionary assigned to Cochinchina (China and Vietnam), Poivre established his credentials as something of a botanical pirate, having secretly collected valuable seeds and seedlings from the Dutch gardens at the Cape of Good Hope, and successfully transporting seedlings of black pepper, cinnamon, and varnish trees from Vietnam to French gardens. Knowing France's longstanding aspirations, and having spent time imprisoned by the Dutch in Batavia (Jakarta, Indonesia), Poivre engineered a bold plan to smuggle the long sought-after nutmeg trees from the Bandas and naturalize them on the French island possessions of Mauritius and Reunion. In 1772, with the Philippines as a base of operations, he directed two cutters to slip into the Bandas and secure the seeds. This they accomplished with the full complicity of the remaining Bandanese, who quite sensibly despised the Dutch. Unfortunately, his ships arrived after the prime seed production time, and in his own account to the Secret Committee of the French East India Company, Poivre admitted that he was able to secure only twenty-five nutmeg plants. By the time of his arrival in Pondicherry, India, this number had shrunk to nineteen, and finally after months of waiting in India, he was successful in bringing only five living nutmeg trees to the island of Mauritius. These five later died under suspicious circumstances. Undeterred, Poivre again secured government support for his scheme. Although aware that the Dutch had murdered several Europeans for attempting to smuggle nutmeg seeds, he returned to the Spice Islands and was able to abscond with eleven more nutmeg plants, as well as several cacao and breadfruit trees. Once again, however, these attempts were a failure.

Years passed, and Pierre Poivre took a young wife and settled into a quiet life in Lyon, France. He spent his time writing and publishing his thoughts on botany and agriculture. For these contributions, he was awarded the noble title of the Order of Saint Michael and, most importantly, in 1766 the post of Administrator of Mauritius and Reunion. His return to the Indian Ocean afforded him one final opportunity to succeed in the dangerous spice wars. Once settled, Poivre sent clandestine collecting expeditions to the Bandas. This time, however, with the help of a disenchanted Dutchman, he was able to extract a huge cache of the priceless plants—450 nutmeg plants, 100 germinating nutmeg seeds, and five crates of approximately 10,000 nutmeg nuts arranged in layers separated by beds of local Banda soil. Nutmeg nurseries were quickly established in Mauritius, and young trees were soon transferred to neighboring Madagascar and Zanzibar. In time, seeds and seedlings were transferred to Cayenne in distant South America and in Grenada in the Caribbean, which is now the world's largest nutmeg producer. Pierre Poivre, the acclaimed French biopirate, had permanently ended the Dutch monopoly of nutmeg and globalized a once-endemic plant resource (Keay 2007, 1–17; Ly-Tio-Fane 1958, document 1, 4, 8, and 14; Maverick 1941).

Stories often take ironic turns, and ethnobotanical stories are no exception. The thousand-year lure of nutmeg that had helped drive the Age of Exploration and pushed European powers to the brink of war, was fueled as much by its exotic provenance as by its healing and flavoring virtues. As the French dispersed nutmeg trees across the tropical latitudes, the mysterious allure of the once-famous nut ebbed. No longer rare or hard to come by, nutmeg prices plummeted as the laws of supply and demand kicked in. Soon, new and more piquant products from the Americas would attract the attention of European physicians and apothecaries, further softening demand. For all their Machiavellian efforts, the French came to possess a plant product of very modest export value. For the long-suffering indigenous people of the Banda Islands, however, this was a fortuitous outcome to the story. For as the Dutch, French, and English lost interest in their trees and their islands, the people of the Bandas were able to return to some semblance of their former lives. Nutmeg lost its medicinal lure for foreigners, but it continues to be used by local islanders in their traditional healing. They rub nutmeg oil on their stomachs to relieve pain, and on their heads to relieve headache. Combined with other ingredients, nutmeg is used to treat rheumatism, body aches, and diarrhea. Its rind is used in a popular local candy, *pala gula*, which is believed to relieve sores in the mouth. It is also used as a sedative, and it is placed as an amulet around the neck of sick infants with the hope that God will heal the child (Gils and Cox 1994).

The Fever Tree

According to legend, in 1638 the wife of the Viceroy of Peru, the fourth Count of Chinchón, lay gravely ill in her bed in Lima, racked with malarial fever and chills. As the frequent bleedings by her physician seemed only to make her weaker, he suggested to the Count that they try a native remedy, the powdered bark of a rainforest tree known locally as *quina-quina*. In desperation, the Count agreed, and the bitter-tasting powder was served to the Countess in a glass of wine. Choking down the concoction, the Countess's symptoms soon abated and she experienced a miraculous recovery. In gratitude, she later procured a quantity of the wonder drug, and dispersed it widely among the local population. The Countess later returned to her native Spain, where she lived a long and happy life, forever singing the praises of the exotic Peruvian fever tree. Her physician also returned to Spain, where he became a wealthy man selling the powdered bark in malaria-ravaged Seville.

This influential narrative, which circulated for several centuries, revealed the discovery of the most famous drug plant of all time, the source of quinine, as well as the generous nature of those who revealed it. The story was, however, entirely fabricated. The Countess it seems was quite healthy and in good spirits when she was supposedly at death's door, and she in fact never returned to Spain. The first documented appearance of the powdered bark of the fever tree was in Rome, not Seville, where it was introduced by Jesuit priests in the early 1630s, not by the Countess' physician (Jarcho 1993, 1–4).[10] But unlike other tales from the New World, the veracity of this and other cinchona stories would prove crucial to British hegemonic ambitions. Of all the plant-derived drugs that would appear in Europe from the sixteenth through the nineteenth centuries, none would go on to save more lives than the humble Peruvian fever tree. And none would better reveal the depth to which foreign powers would sink to deprive forest-dwelling peoples of their intellectual and genetic heritage.

The genus to which the fever tree belongs was named in memory of the late Countess of Chinchón by Linnaeus, who was enamored with the poignant story of her recovery. However, he unwittingly omitted the "h" in her name, so that ever after it was known as *Cinchona*. There are currently about 25 species recognized in the genus *Cinchona* (Family Rubiaceae). It is endemic to the highlands of Central and South America, extending from southern Costa Rica through the Andes and into Bolivia. Over the years, it became known variously as cinchona, the Countess' powder, fever-wood, and *quina*. But most knew it as Jesuit's bark. The name was likely taken because the Jesuits in the Americas held a virtual monopoly on apothecaries and medicinal drug plants. They introduced the medicinal bark to Europe early in the seventeenth century, and thereafter it was referred to as Jesuit's bark or powder (Brockway 1979, 103–139). Cinchona's primary medicinal value comes from the presence of the bitter-tasting alkaloid quinine, which occurs in the bark of several species. Taken orally, quinine acts to dramatically inhibit the presence of the protozoan parasite, *Plasmodium falciparum*, the cause of malaria. But unlike so many other exciting new ethnobotanical discoveries of the era that ultimately proved ineffective, cinchona truly was a miracle drug plant.

Jesuit's bark found an eager market in seventeenth and eighteenth century Europe, where fever and death from malaria (known as ague) were common. Following its initial success in Rome in the 1650s, cinchona's use spread rapidly throughout much of Europe. By the mid-1700s, it had become one of Spain's most important and lucrative New

World products. Only Britain was slow to adopt the miraculous fever medicine. Given its bitter hatred of the papacy, the term "Jesuit's bark" carried too much religious baggage for Protestants to accept. But the plant truly was a wonder drug plant, and reason eventually prevailed. Feverologist Robert Talbor surreptitiously slipped the banned South American powder into his successful treatment of British Lord Protector Oliver Cromwell, avoiding persecution by simply referring to the bitter brew as a foreign substance. French surgeon Nicolas de Blégny appears to have been the first to reveal in writing that the miraculous drug was derived from "the Bark of an Indian Tree, of the bigness of a Cherry-Tree." But knowledge of global biogeography was still a bit muddled, and he was uncertain whether "Indian" referred to the East or West Indies, that is, India or Peru (Blégny 1682, 2; Talbor 1672; Wallis 2012).

Scientists eventually understood that the source of the cure was in the South American Andes, and they were eager to learn the details of its ecology and medicinal qualities. The Spanish crown in particular was concerned with identifying the variation in effectiveness of the product and adulteration of the product, as the King frequently distributed the bark as gifts to foreign dignitaries. He could ill afford to have a foreign ambassador drop dead due to the poor quality of a royal gift (Crawford 2007, 195). William Arrot, a Scottish physician working in Peru, provided the earliest accurate description of the plant from the field. In a letter dated 1737, he stated that there were four types of cinchona, each with different qualities, and that the bark should be used when it was fresh, as it became "insipid and useless" with age. It can be found only in close proximity to the city of Loxa, between the Marañon and Amazon Rivers. Arrot's most important notations, however, involved the harvest of the valuable bark and the fate of the indigenous bark collectors. He reported that "large quantities of it are cut yearly" by the indigenous laborers, with "a great many of the fine large bark trees having been entirely cut down." Unsustainable harvest practice appeared to threaten the future of the fever tree. Arrot also commented, however, that the bark grew back nicely in ten to twenty years, meaning that if debarking were properly managed, the cinchona supply could be assured. But the fate of the forests, he suggested, was equally tied to the fate of the indigenous people who knew the trees best. And they were not faring well. Because of their cruel treatment at the hands of the Spanish, he predicted, "in a very few years their race in that country [Peru] will be quite extinct" (Gray 1737, 82 and 83).

Various stories circulated concerning the original discovery of the fever tree. The tale most often repeated was originally circulated by French geographer Charles Marie de La Condamine. Having returned from Ecuador via the Amazon River from an expedition to test Isaac Newton's idea that the Earth bulged around the Equator (he found that it did), he provided the first scientific descriptions of a number of tropical forest products, including rubber, curare, and cinchona. In a letter to the French Royal Academy of Sciences, he recorded an ancient tradition "of which he could not vouch for the truth" that the virtues of "*quinquina*" had been instinctively learned by South American mountain lions, which chewed the bark of the tree to relieve what ailed them. The lion's unusual actions were observed by local Indians, who then tried it successfully on their own fevers. Later, the Spanish Jesuits learned the treatment from the natives, and the circle was completed when the Countess of Chinchón was saved by the miraculous bark (Condamine 1738, 233; Schiebinger 2004, 81–82). Although giving priority to native people for the discovery, La Condamine's notations also fit neatly into European ideas of the Great Chain of Being and the primitive nature of indigenous knowledge (see chapter 3). Another legend told of how an Indian, who was consumed with fever and near death, crawled to a pool for a last sip of water. He found it bitter tasting, but after he drank, he experienced a remarkable recovery. He showed the healing pool to others, who eventually realized that the cure was due to a dead cinchona tree that had fallen into the water. They located other cinchonas, and so began the harvest of the medical bark (Hooker 1839, 7). In both cases, discovery was by serendipity rather than cerebral investigation, therefore negating any later claims of intellectual property and benefit sharing.

The issue of who originally uncovered the healing properties of cinchona seems like a politically benign question. But over time the subject became central to British designs on the precious botanical resource. Did priority go to the Andean natives, or did the Spanish themselves learn the secret? For good reason, the British opted to believe the latter. For by dismissing indigenous discovery and use of the tree, the native people could not be considered the rightful custodians of the plant's intellectual property. But not everyone agreed. William Arrot stated clearly that "its qualities and use were known by the Indians before any Spanish came among them, and it was by them applied in the cure of intermitting fevers, which are frequent over all that wet unhealthy country" (Gray 1737, 86). Spanish botanist José Celestino Mutis, who spent decades studying the flora of the Andes and months traveling with Humboldt, was of similar opinion. According to Mutis, the peoples of the Andes had long prepared a healthful fermented beverage out of cinchona bark (cited in Zimmerer 2006, 350). Others disputed these claims. Humboldt was of the opinion that the native peoples were not the discoverers of the drug plant, and that they "would rather die than have recourse to cinchona bark." He also considered it extremely improbable that "the discovery of the medicinal power of the cinchona belongs to the primitive nations of America" (Von Humboldt 1821, 22–2? English botanist Richard Spruce, who would be pivotal in the cinchona story, also wrote that the Indians did not use it for fevers because they considered it a heating drug (in the humoral system), which one would never employ to treat a fever (Brockway 1979, 111). And Sir Clements Markham noted that the local Indians "attached little importance" to cinchona. Indeed, he argued, their very ignorance of the plant explained why the medicinal virtues of cinchona were not understood earlier by the Spanish (Markham 1880, 6).

Who discovered the healing powers of cinchona? On the one hand, it is highly improbable that the Spanish independently discovered the properties of cinchona bark after a mere century in the highlands of Peru. Ethnobotanical studies of diaspora peoples (see chapter 7) suggest that the medicinal properties of forest trees are learned more slowly than herbs, shrubs, and other more common plant life forms (Hoffman 2013; Voeks 2013) . The medicinal use of cinchona bark had been learned most likely long before the arrival of Europeans by indigenous inhabitants during their thousands of years of occupation. Discovery (prior art in patent terms) must go to the native Andean people. However, perhaps the most credible explanation for the knowledge transfer of cinchona's properties was provided in 1663 by Gaspar Caldera de Heredia. He relates how Indians traveling from Quito to work in the mines were forced to cross a freezing mountain torrent. In order to alleviate the intense cold and shivering, they prepared a warm beverage with powdered cinchona bark. Jesuit priests, Caldera de Heredia suggests, seeing that this concoction supplied instant relief for their suffering, tested the powder on malarial fevers, which manifested similar symptoms. If this account is factual, it seems to give priority of discovery jointly to the local indigenous people and to the Jesuits. The priests, for their part, simply employed the ancient law of analogy, that is, the belief that a drug plant used to treat the shivering caused by cold and wet could likewise by used to treat the shivering produced by fever (Jarcho 1993, 4–11). Whatever the true explanation, it is clear that Britain had every reason to discredit indigenous knowledge and use of cinchona, which they ultimately did.[11]

The pretext for foreign appropriation of cinchona required more than simply priority of discovery. And this the British conveniently

4.5   Charles Laplante. "The Gathering and Drying of Cinchona Bark in a Peruvian Forest."
1867. Wellcome Library, London, Creative Commons, 20956i.

marshaled through a destructive environmental narrative. The Andean
Indians, they argued, could not be trusted to protect a botanical resource
of such critical importance to all of humanity. Arrot's early report of destructive harvest of cinchona bark was backed up by other eyewitnesses,
who noted that South American bark collectors (*cascarilleros*) took no
conservation precautions whatsoever during harvest, simply felling the
trees and stripping away their bark. Humboldt similarly commented
that the harvest practices he witnessed were unsustainable. "The tree is
felled in its first flowering season," he observed, and as a result, "The older
and thicker stems are becoming more and more scarce" (Von Humboldt
1850 [1807], 591) (Fig. 4.5). Humboldt was a keen observer of details, and
his observations cannot be discounted out of hand. But it is difficult to
gauge today whether Spanish colonial harvest practices in fact threatened
the resource at the landscape level. Humboldt spent at most only à few
months occasionally observing bark cutters, and long-term patterns of
tree demography cannot be discerned that easily. Moreover, there was
evidence that harvest techniques were less harmful in the long run than
Humboldt and others envisioned. For instance, the destructive exploitation of cinchona trees that had taken place in earlier generations was
apparently already being corrected by the Jesuits, who ordered that five
new individual seedlings should be planted in the form of a cross for

each tree felled to strip its bark (Gramiccia 1988, 10–11). Whether this policy was or was not an effective species reforestation technique is not recorded.

More important, however, were the results of a dissertation on cinchona written in 1839 by William Dawson Hooker. His opinions contrasted with everything that had been written by other outsiders. In this thesis, he argued that cinchona bark removal was "more" destructive than felling the trees, because insects actively attacked the debarked individuals. When the tree was cut down, as was done by indigenous harvesters, it actively grew back from basal sprouts (coppiced). And after six short years "the Cinchonas [were] again fit for cutting" (Hooker 1839, 14–15). The problem of overharvest, it would seem, was solved; cutting of trees as carried out by indigenous harvesters was the most sustainable harvest strategy. Unfortunately, young Hooker died of yellow fever only a year later at the age of 26, and his controversial findings were likely never noticed by those who needed to see them. Most importantly, this included his father, Sir William Jackson Hooker, who would soon oversee an international plot to secret the cinchona seeds out of the Americas and transplant them in Asia.

Whatever the merits of these prior art and sustainability arguments, the paternalistic attitudes of European observers fostered an elitist and top-down environmental attitude regarding native people and their natural resources. In the views of outsiders, inhabitants of the equatorial world were culturally backward, saddled with inefficient and corrupt governing bodies, and intellectually incapable of sustainably managing their own biological resources. This is a recurring them in ethnobotany that still plays out today (see chapter 9). Then as now, tropical biodiversity was conceived as global biodiversity, and critical to the greater good of humankind. Local stakeholders were assumed to be unable to sustainably manage their own resources, and so it was the duty of agents of the British Empire to step in and right things. "Filching from another country its commercial advantages," wrote British botanist Sir Joseph Banks in a 1787 letter to George Yonge, was a natural right if it fulfilled the goal of "the production of nature usefull for the support of mankind that are present confined to one or the other of them" (cited in Chambers 2000, 90). Biological endemism was the enemy of progress. Therefore, in order to protect this globally significant resource, Sir William Jackson Hooker drafted a bold plan to pirate cinchona seeds out of South America to Britain's colonies in Ceylon (Sri Lanka) and India. These actions were perfectly justified, as he confided to a colleague, because "given the profoundly destructive method of exploitation by the natives, the smuggling of precious cinchona seeds out of Bolivia represented a humanitarian act" (cited in Jackson 2008, 38).

A humanitarian act, perhaps. But the real motivation for naturalizing cinchona in distant India was Britain's ongoing imperial ambitions. They had serious designs on West Africa, yet every expedition, save those with adequate supplies of quinine, had succumbed to the effects of tropical fevers. More importantly, British India was witnessing massive malarial mortality, perhaps one million lives lost per year, and it was costing upwards of £100,000 per year to import cinchona bark from South America. If they could cultivate their own cheap source of quinine, British army officers and their families could operate without threat of bouts of malaria. Further, by making it safe for British wives to accompany their husbands to the fever-ridden tropics, they could enforce the cultural divide between themselves and East Indians, and thus sustain their sense of racial and cultural superiority. In so many ways, the future of Britain's colonial enterprise depended on an inexpensive and readily accessible supply of life-saving quinine (Honigsbaum 2002, 87–90).

The "cinchona project," as it was conspired by Kew Gardens in London and the Secretary of State for India, would furnish funds and logistics to facilitate the theft of cinchona seeds. These in turn would be transplanted in India. The project was to be carried out by Clements Markham and especially Richard Spruce, who was already busy collecting plant specimens in the Amazon watershed. Their letters made it clear that they would avoid any "jealousy" on part of the local governments by working undercover. Bribes and threats would be necessary, as the Bolivian government had granted a monopoly on their cinchona resource, and had banned any export of the seeds. Spruce used US $400 to secure some property in the Andes, moved into a secluded hut, and began collecting ripe seed pods and seedlings of *Cinchona succirubra*, one of the quinine-bearing species. It was hugely difficult work that pushed his health to the extreme, but he managed to transport almost 100,000 dried seeds and 637 plants over the Andes to the Ecuadorian port of Guayaquil, where he smuggled them onto a ship bound for England. Spruce's huge cache was later shipped to India's Nilgiri Hills, where it was successfully planted and became the nucleus for British cinchona production. This phase of the grand scheme was a success. As for Markham, he directed his efforts at Bolivia and Peru, where he collected seeds of *Cinchona calisaya*. His foray into the mountains and jungles was equally challenging, and he met considerable resistance from Peruvians, who were incensed that he was attempting to pilfer their national treasure. He was ultimately able to slip out of the country with a load of seeds, but by the time he finally reached India, the seeds were all dead. His major accomplishment, it seems, was to whip up nationalist resentment and hostility among locals towards European scientists. This enmity would come back to haunt one of the last actors in the drama, the British entrepreneur Charles Ledger (Brockway 1979, 114; Honigsbaum 2002).

Unlike his well-connected competitors, Markham and Spruce, Charles Ledger was an independent trader trying to make ends meet in Peru. Born in London in 1818, he had traveled to South America to seek his fortune. He settled into Tacna, Peru, and shortly began buying and selling cinchona bark and alpaca wool. He became proficient at identifying the various qualities of cinchona bark, and by 1842 had set up his own business. Ledger was a field man. He learned to travel and live like the indigenous Andean peoples, and for this he gained considerable local respect. In 1841, he made the acquaintance of a Bolivian Indian, Manuel Incra Mamani, with whom he struck up a lifetime friendship. It is not clear whether Mamani was of the Quechua or Aymara people, the dominant cultures in the region. Regardless, he possessed a profound knowledge of cinchona trees. He was able to identify 29 different "varieties" of the tree from a distance, each of which corresponded to a different quality of quinine. Mamani was also deeply religious, and would stop and pray at the cross-shaped groves of cinchona, created by the Jesuit-instructed plantings.

The idea of smuggling cinchona seeds out of Bolivia came to Ledger early, fifteen years before the British government thought to sponsor Markham and Spruce. He sent a letter to the British government about his "secret intentions" to help establish cinchona plantations in India, Jamaica, or wherever else was deemed appropriate. His request was ignored, it seems, but not forgotten. Fifteen years later, as Markham was trying to smuggle out the cinchona seeds, he carried a copy of Ledger's original letter with him. Clearly, the British government thought it was a good idea, but that Ledger was not the right man for the job.

Charles Ledger put the plan out of his mind for several years, concentrating on exporting alpacas to Australia. But Mamani had not abandoned the idea, and in 1865 arrived at Ledger's door with bags of the precious seeds. He had spent several years fulfilling his employer's bidding, and had collected from the highest-yielding variety known, the *Calisayo Rojo*, which would later be named for Ledger, *Cinchona ledgeriana* (now a synonym for *Cinchona calisaya*). Given the vicissitudes of the Bolivian highland climate, it had taken Mamani three full years of effort to finally secure the seeds, a task that only a highly knowledgeable local could ever have accomplished. For his efforts, Ledger paid him US

$500, and asked him to procure more. Ledger forwarded the sacks of seeds to his brother George in London, hoping for a big payday. But by this time, the whole situation had changed. Hooker had died, Markham was in India, and no one was really interested in these new seeds of unknown provenance. Ledger's brother ended up almost giving the seeds away—one pound to a Dutch trader for £20, and 14 pounds to an Indian merchant for £50. The single pound of seeds was transported to Dutch Java, however, where in a few years it was producing the highest quality quinine on the planet. In a short time, Dutch Indonesia became the world's greatest supplier of high quality quinine, all from the single sack of seeds collected by an indigenous Peruvian.

Following Ledger's request, Mamani returned to harvest more of the high-quality seeds. This time, unfortunately, the local chief of police learned that he was smuggling seeds for a foreigner, and threw him in prison. For twenty days and nights he was beaten and starved, but still Mamani stubbornly refused to divulge Ledger's name. Finally, barely alive, he was released to his son Santiago. He died of his wounds a few days later. In the years to come, Ledger never admitted any guilt for the death of his long-time friend and collaborator, but he did provide some financial support for his family into the future. Ledger in turn emigrated to Australia, but after further commercial failures with his alpaca venture, exhausted his meager resources and became a penniless ward of the state (Brockway 1979, 120–128; Gramiccia 1988).

The names Ledger, Mamani, Markham, and Spruce are forever linked to either a great act of selfless humanitarianism, or a monumental act of biopiracy, depending on your perspective. As quinine became readily accessible, literally millions of lives were saved, and continue to be so. Dutch Indonesia became the world's primary source for quinine, and Britain was able to secure its position in India and Africa. The real losers were Bolivia and Peru, which were deprived of their once endemic botanical resource, and the indigenous peoples of the Andes, whose knowledge of the healing properties of a local rainforest tree is now the world's knowledge.

The great colonial quest for healing plants in the tropical realm was fueled by ancient ethnobotanical principles and by the very real need to address a rash of newly encountered diseases. In the twentieth century, a new array of infectious maladies as well as diseases of affluence threatened modern society, and a new generation of bioprospectors began to plumb the pharmacological potential of tropical forests. This search was at the very heart of the emerging jungle medicine narrative, as it linked critical public health problems with the fate of tropical forests.

FIVE

Weeds in the Garden

Up until the 1970s, the news that no parent could bear to receive was that their child had been diagnosed with childhood leukemia (acute lymphoblastic leukemia, or ALL). Striking especially children between two and five years of age, ALL nearly always carried a death sentence. The disease is caused by a massive proliferation of white blood cells in the bone marrow, eventually crowding out the normal cells and spreading to other organs. Fortunately, a miracle cure appeared in the 1960s. It was developed from a tropical plant compound (the alkaloid vincristine), one that over time and in combination with other drugs increased ALL survivorship in children to nearly 90%. Today childhood leukemia is considered a curable disease. The plant species, the Madagascar periwinkle (*Catharanthus roseus*), went on to become the posterchild for tropical rainforest preservation efforts. And its circuitous path of discovery and drug development is crucial to the jungle medicine narrative.

The drug was discovered serendipitously by Canadian researchers Robert Noble and Charles Beer. Noble had learned of the possible effectiveness of a Jamaican plant against diabetes from C. D. Johnston, an African-American surgeon in Jamaica who had studied in Canada. At Noble's request, Johnston sent him a supply of the plant for testing. Researchers at the time were eager to identify an ingestible substitute for insulin injections, and this Caribbean herbal remedy was said to be administered as a tea. Unfortunately, initial tests as a diabetes treatment were not encouraging. So Noble took a more direct route, and injected rats with an extract of the plant. The rats promptly died, and soon it was determined that the cause of death was a collapse in their white blood cells. With this evidence, it was for Noble and his colleagues a logical leap to test the extract on leukemia. He first presented his team's findings at a scientific conference in 1958, but because of delays, he was unable to read his paper until after midnight. Few were still in attendance, but there was a group from the Eli Lilly drug company who had read the abstract and was keenly interested in hearing his results. There are twists and turns to the longer narrative, but in the end, the unsuccessful search for an improved diabetes treatment had several unexpected outcomes: it led to the development of a totally new class of cancer-fighting drugs (stathmokinetic agents, or spindle poisons); the American pharmaceutical giant Eli Lilly owned an immensely profitable drug patent; and the world had access to a potent new weapon against a once incurable disease (Duffin 2000; Foster 2010; Hudson et al. 2012; Piu 2009).

There was another malady in the 1970s, an environmental epidemic that was quietly burning through the heart of the world's equatorial forests. Scientists and the environmental community were just awakening to the realization that the tropical rainforests of Asia, Africa, Oceania, and the Americas were disappearing, and quickly. Earth's most biodiverse ecosystems were being chopped down and burned at an alarming rate—for monoculture plantations and cattle ranches, and to feed the families of poverty-stricken rural farmers. Various international efforts were marshaled with greater or lesser success to address this growing environmental crisis. But among the ethical and environmental justifications for forest protection that emerged, none resonated among an AIDS- and cancer-fearful public more than the discovery of vincristine from the Madagascar periwinkle (also known as the rosy periwinkle), the miracle drug of the decade. If one single plant species could save the lives of thousands of infants per year, what other medical wonders awaited discovery? "Save the rainforests," it was reasoned "and we save ourselves." In 1989, the US-based NGO called Rainforest Alliance launched the "Periwinkle Project," "highlighting [the] medicinal value of threatened tropical flora and collaborating with pharmaceutical companies, scientists, conservation groups and local people" to advocate an alternative to deforestation (http://www.rainforest-alliance.org/). It was a masterful narrative with elements that appealed to all—miracle cures, mysterious shamans, obscene profits, and nature preservation. The jungle medicine narrative emerged as the iconic savior of tropical Eden.

There were flaws in the periwinkle story from the start, which no one seemed to notice, or chose to notice. First, there was the thorny question of intellectual property. The periwinkle is native to the island nation of Madagascar, and the utility of the plant to treat diabetes may well have originated there. Therefore, because this traditional use inspired the experimentation and ultimate discovery of vincristine, the life-saving ALL drug, the people of Madagascar had a legitimate if modest claim to benefit sharing of drug profits. But as it often does, geography and time complicated the issue of intellectual property. The medicinal use of periwinkle to stabilize blood glucose, usually by chewing the leaves, was not limited to Madagascar, but rather had over time become widely dispersed (or perhaps independently discovered) in tropical Asia, Africa, and the Caribbean, disseminated centuries ago by seafarers (Osseo-Asare 2014). In this scenario, perhaps other regions and people should have shared in the pharmaceutical profits. And then there is the question of the Jamaican physician who provided the original lead to the Canadian research team. If anyone had a claim to some economic benefit sharing, surely it was C. D. Johnson. In truth, none of this was considered at the time. In the 1970s, just as had been the case during the great colonial quest for drug plants, indigenous medicinal knowledge was considered a public good, freely available for whoever was clever enough to secure it (Duffin 2000, 163; Frisvold and Day-Rubenstein 2008; Harper 2005).

Intellectual property is a contentious issue, with advocates on both sides of the debate. But even if financial rewards are not shared, perhaps as in the case of cinchona and malaria a century and a half ago, the public benefit of developing a life-saving drug outweighs financial concerns. Perhaps. But in the case of modern drug discoveries, including vincristine, access and costs are beyond the reach of the rural poor in developing world countries. In the case of Madagascar, it is one of the poorest countries in the world, and over 70% of its population is made up of small-scale farmers with little access to Western medicine. Thus, considering the exorbitant price of chemotherapy treatment and the difficulty of rural people in securing medical care, the likelihood that the people who originally discovered the various medicinal applications of their native periwinkle have benefitted from the discovery and development of vincristine seems remote at best. For the poor boys and girls of rural Madagascar, childhood leukemia is still likely to be a death sentence.

In terms of the jungle medicine narrative being crafted and publicized by scientists and environmentalists, there was an even larger problem. The green savior of the rainforest was not a majestic old-growth forest tree, or a serpentining liana, or even a rare epiphytic orchid clinging to a towering branch. It did not even inhabit old-growth forests. It was a shrubby garden ornamental, and in many places a roadside weed. As a posterchild for rainforest preservation, the Madagascar periwinkle was something of a red herring in the rainforest.

Madagascar periwinkle normally grows as a woody herb or small shrub, with opposite shiny leaves, and is distinguished (like other species in the Apocynaceae family) by bleeding abundant white latex from its leaves and stem. Its most appealing characteristic, however, is its attractive 5-petaled, white to lavender flowers. These attracted the attention of foreign travelers, and led to its first written description from Madagascar in 1759. Seeds were shipped by a French diplomat to a botanist in the Netherlands in the 1750s and on to the Royal Gardens in Paris. By the 1790s, the Madagascar periwinkle was established throughout Europe. It was also likely taken to continental Africa and perhaps Asia much earlier by mariners, who appreciated its various medicinal properties. From tropical gardens it readily escaped cultivation, and in a short time the once-endemic Madagascar periwinkle was firmly established as a pantropical weed, one so ubiquitously distributed that its actual origin was for many years in dispute. In the National Herbarium in the Netherlands, for example, the earliest dried plant collections (vouchers) date not from native Madagascar, but rather from Brazil, Java, New Guinea, Cuba, and elsewhere. Many botanists assumed it was native to the Greater Antilles. But whatever its provenance, the Madagascar periwinkle is a competitor. It packs a potent punch of secondary compounds, numbering 130 different alkaloids, numerous glucosides, phenols, steroids, flavonoids, and anthocyanins. Almost no hungry insect or mammal can penetrate its formidable chemical defenses. And it is anthropogenic almost everywhere it occurs, thriving in disturbed nature and direct sunlight. In some locations, it is classified as an invasive species (Foster 2010; Harper 2005; Osseo-Asare 2014; Van der Heijden et al. 2004). Ironically, the iconic protector of the rainforest grows best where the trees have been cut and burned (Fig. 5.1). But surely this species is an exception to the rule. The towering and hyper-diverse old-growth rainforests must be teeming with potential miracle cures. The whole jungle medicine narrative is balanced on this single assumption.

Disturbance Pharmacopoeias

My first opportunity to search for medicinal plants in a tropical forest was in 1989, in the Atlantic Coastal Forests of eastern Brazil. Dwarfed in size and notoriety by the legendary Amazonian Forests to the north and west, the Atlantic Forests were by the 1980s increasingly drawing scientists

5.1 A. Weedy Madagascar periwinkle (*Catharanthus roseus*) growing in the street in Maputo, Mozambique. Photo credit: Robert Voeks. B. Vincristine sulfate. Photo credit: Naveen Qureshi.

due to their high number of endemic species and their astounding biological diversity. A clue to the species richness was revealed by a single tree census of a one-hectare (2.5 acres) old-growth forest plot, which yielded an astonishing 450 tree species (Anonymous 1993). Comparing this to the estimated 1000 native trees species in the United States,

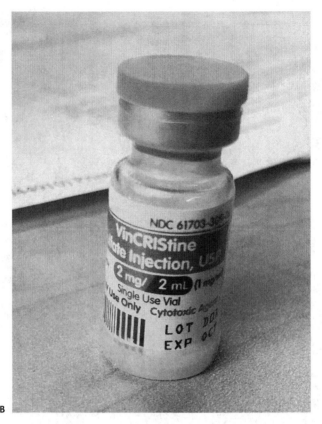

B

5.1   *continued*

which covers nearly one billion hectares (2.4 billion acres), the biological significance of this narrow corridor of rainforest becomes pretty apparent. I persuaded a man who came recommended as a local expert on the local healing flora to go out into the forest with me to hunt for medicinal plants. He agreed, but was insistent from the outset that most of the medicinal plants he knew of grew in backyard gardens, along trails, and in messy second-growth forests, not in old-growth forest. He was a Caboclo, a mix of indigenous, Portuguese, and African ancestry. His nickname was Indio (Indian).

I set out on foot with Indio from his *favela* (poor Brazilian neighborhood) in Ilhéus, Bahia. Nothing systematic, we just wandered and talked, first along muddy streets and through weedy empty lots, then through some cattle pastures and second-growth forest. We passed through the edge of a cocoa farm that used the *cabruca* management technique, which left most of the mature rainforest intact to shade the cocoa treelets. And finally we encountered some fine patches of old-growth Atlantic Forest. Along the entire walk, Indio provided a steady stream of names and uses of the plants we encountered. By the time we reached the real forest (*mata mesmo*) we had stopped and discussed twenty-five or so medicinal plants. But I admit that I was a bit disappointed. Most of the plants he pointed out were just weeds, most in the sunflower (Asteraceae), nightshade (Solanaceae), legume (Fabaceae), and pepper (Piperaceae) families. Surely these low-growing herbs and shrubs and climbers were not representative of the much-touted cornucopia of future pharmaceuticals that was at the time receiving so much Western press. I began to seriously question Indio's purported expertise.

When we finally walked into a patch of old-growth forest, Indio proceeded to point out the names and uses of understory and canopy trees. There was *jocára* (*Euterpe edulis*), which provided delicious palm heart, *piassava* (*Attalea funifera*), the source of thatch and broom fiber, *jangada* (*Apeiba albiflora*), an excellent source of timber, and *baba-de-boi* (*Cordia rufescens*), locally appreciated for its sweet fruit. There were a few large *jaca* (jackfruit—*Artocarpus heterophyllus*), somewhat surprising since the species is native to India. And there was *escada-de-macaco* (monkey's ladder, *Bauhinia* sp.), a prominent liana resembling a small wooden staircase that danged from a tall tree. Indio said this was "man's medicine," and that a piece of the vine should be soaked in white wine and drank to increase his "fortitude" (Fig. 5.2). But just as he had warned earlier, old-growth forest was not the ideal place to look for medicinal species.

I resolved to explore the question a bit more systematically in the future. So a few years later, working in southern Bahia near the town of Una, I set up four one-hectare plots (100 m × 100 m). Two plots were in old-growth Atlantic Coastal Forest (Estação Experimental Lemos Maia) and two were nearby in an area of second-growth forest. This site had been cut and burned seventeen years earlier, and planted in rubber trees (*Hevea brasiliensis*). The rubber tree scheme had failed immediately, and the abandoned space was now a tangle of small trees, shrubs, herbs, and lots of vines. For ease of sampling, each plot was further divided into 10 m × 10 m subplots. I then employed three local men, all reported to be knowledgeable about the local flora, although only one was considered a healer (*curandeiro*) per se. I walked through each subplot slowly with the men individually. He would point out plant names and utilities, and I made notations and permanent voucher collections.[1] My working

5.2 Escada-de-macaco (*Bauhinia* sp.) growing in Reserva Pau Brasil, southern Bahia, Brazil. Photo credit: Robert Voeks.

hypothesis was that the old-growth forest plots would be inhabited by significantly more medicinal plant species than the second-growth plots. But I was wrong. The results did not remotely support my hypothesis. In terms of overall ethnobotanical value, the old-growth plots were inhabited by a total of 66 useful species, including fruit, fiber, fuelwood, timber, palm heart, and medicine. The second-growth plots covered a similar array of uses, although there were fewer useful species overall. Only in the medicinal category was this not the case. The old-growth plots included four and seven species, respectively, for a total of nine medicinal plants. The second-growth plots included 14 and 15 medicinal species, respectively, for a total of 19 species. In addition, the floristic similarity of medicinal plants in the old- and second-growth plots was about equal (Joccard S = 57.1% for the old-growth plots, and S = 52.6% for the second-growth plots), suggesting that simply increasing the number of plots would not likely have changed the outcome. And so Indio was correct. In this region at least, old-growth forest was not the preferred collection habitat for medicinal plant species.

I later published the results (Voeks 1996), but at least one reviewer took me to task for my choice of field collaborators and region of study. The Atlantic forests were entirely too disturbed to be representative of "good" tropical forest, he or she contended, and my field collaborators were not sufficiently "indigenous" to be representative of forest-dwelling people. If I wanted to examine this question, I was informed, I should have done so in Amazonia. I never was able to explore this question elsewhere in Brazil, but a couple of years later I was able to pursue the topic again in Brunei Darussalam on the island of Borneo in Southeast Asia. In this case, almost all of the old-growth forests were intact, and there were numerous indigenous groups carrying out traditional hill rice swidden cultivation. I was unable to replicate the Brazilian study—at the time there was not much recent second-growth forest to census—but I did spend considerable time collecting medicinal plants, including with two indigenous Dusun men at the time in their late 60s (Fig. 5.3). In each case, we explored the disturbed habitats near their homes, including paths, swiddens, pastures, and small patches of second-growth forest, as well as relatively undisturbed dipterocarp rainforest. Although most of the old-growth forest appeared primary to me, Kilat and Umar both were able to point out spaces in the forest that had been cleared and planted in the distant past, usually evident by the presence of a few tall betel nut palms (*Areca catechu*) and often concentrations of native fruit trees, such as rambutan (*Nephelium* sp.) and wild durian (*Durio* sp.).

A

5.3　A. Kilat bin Kilah from Bukit Udal, Brunei. B. Umar Putel from Bukit Sawat, Brunei. Photo
credits: Robert Voeks.

B

5.3   *continued*

Medicinal Plant Life Forms (%) in Borneo and Brazil

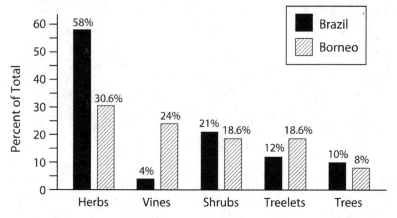

5.4   Medicinal plant life forms in eastern Brazil and northern Borneo. Image: Kelly Donovan.

The results of this informal survey were revealing. Between the two men, they were able to identify, name, and provide the medicinal properties of 77 plant species. A few of these maintained more than one material value, such as timber-medicine, or insecticide-medicine, but this was rare. What stood out was that most of the medicinal species (nearly 80%) were collected in humanized habitats, and most were herbs, vines, or shrubs. Both men could identify the names and uses of numerous old-growth trees (well over 100 species) used for a variety of purposes, including timber, crafts, fish poisons (piscicides), fruit, nuts, or in the case of palms, thatch. But just as was the case in far off Bahia, Brazil, trees were not often appreciated for their therapeutic value, in this case, only 4%. Combined with the data that I had collected in Brazil (Fig. 5.4), it is evident that at least in these two instances, tropical healing habitats are strongly associated with spaces created and occupied by people (Voeks 1996 and 1997; Voeks and Nyawa 2001).

Numerous studies, several more extensive and sophisticated than mine, have subsequently examined the same or a similar question. And with very few exceptions,[2] they all came to similar conclusions—tropical pharmacopoeias are dominated numerically by disturbance (ruderal) species, weeds, and cultivated food plants. These studies point to the overwhelming importance of "secondary forest," "cultivated plants," "invasives," "successional habitats," "fallows," "accessible plants," "herbaceous," "managed landscapes," "homegardens," and especially "weeds."

Lots and lots of weeds. The plant families that predominate the healing floras are fairly consistent from region to region, even continent to continent, including sunflowers (Asteraceae), mints (Lamiaceae), nightshades (Solanaceae), parsleys (Apiaceae), and spurges (Euphorbiaceae), all for the most part herbaceous and often weedy (Moerman et al. 1999). And this is not a localized phenomenon. The anthropogenic nature of medicinal plant pharmacopoeias is geographically widespread, having been reported in Bolivia, Brazil, Costa Rica, Ecuador, Mexico, and Nicaragua from the Americas; Benin, Gabon, Togo, Kenya, Liberia, Madagascar and Ethiopia in sub-Saharan Africa; Brunei, India, Indonesia, Nepal, and Thailand in Asia; and the Solomon Islands in Oceania.[3] Compared to humanized tropical spaces, old-growth forests represent a relatively minor source of medicinal plants to traditional communities. But how is this possible?

Weeds may be the pariah of the plant kingdom today, but this hasn't always been the case. Although they often possess a suite of distinct ecological and life history traits—they can germinate in diverse environments, they can self-fertilize (autogamy), they have a short life span, they are sun lovers (heliophytes), and they have small, often easily dispersed seeds. But for most people they are defined simply as plants growing where they are not wanted. Most importantly, however, "they grow entirely or predominantly in situations markedly disturbec_ _, man" (Baker 1974, 1). Weeds go where we go. But weeds have not alw'ys been so maligned. They have a lengthy legacy of contributions to traditional cuisines and healing practices. Indeed, much of Dioscorides' *De materia medica* is a celebration of the glories of healing weeds and foodstuffs. In medieval Europe, the great medicinal herbal texts of the time reveled in the various healing virtues of weeds. Robert Talbor sang the praises of lowly medicinal weeds in his seventeenth-century physic garden "for from these weeds (that start up in every corner of our Physick Garden) I have . . . sucked honey, as well as from the flowers of learning and ingenuity" (Talbor 1672, 13). A century ago, even the US Department of Agriculture encouraged farmers to "make these pests sources of profit" by collecting and curing weeds for the pharmaceutical trade (Henkel 1904, 7).

Weeds have always had a split personality, seesawing between good and evil, friend and foe. For ranchers, some weeds are noxious pests whereas others are important forage sources for livestock (L. L. Santos et al. 2014). For people growing food crops, weeds have always been associated with hard labor, even in the proverbial time of Genesis. When Adam and Eve were banished from the bounty of the Garden of Eden, their hardscrabble subsistence depended on taming the wilderness and conquering the weeds. In the Gospel according to Matthew, weeds became ominous symbols in his Parable of the Weeds. "The field is the world; the good seed are the children of the kingdom. The weeds are the people of the evil one, and the enemy who sows them is the devil" (Matthew 13: 37–39). But weeds also occupy contested space, at least in the minds of Westerners, for although they are loathed by those who grow food, they are coveted by those who make medicine.

The division between food and medicine is often fuzzy, as many weeds, in the past and the present, take on both roles. Indeed, among indigenous Mesoamerican farmers, there isn't even a proper name for weeds, at least with a negative connotation. Because most disturbance-loving plants have some food or medicinal value, they are simply referred to as herbs (Vibrans 2016). In the Lao PDR, weeds grow in the rice paddy fields, levees, and slopes. Many of these are purposely spared for later collection as tasty boiled vegetables (*sup phak*), or medicine, or both (Kosaka 2013). Similarly in the Republic of Benin, dozens of weedy species such as bitter melon (*Momordica charantia*) double as vegetables and medicine. Not only is the fruit of this weedy climber a bitter addition to soups and stir fries, but it is employed throughout the tropical world to lower serum glucose in sufferers of diabetes (Joseph and Jini 2013). And the quintessential edible and medical "weed," marijuana (*Cannabis sativa* and *C. indica*), has served dual purpose (and much more) since at least the end of the last Pleistocene glacial retreat, some 10,000 years ago. Its seeds are rich in protein and remarkably high in omega-3 and omega-6 fatty acids. In fourth-century Rome, *Cannabis* seeds were made into a confection to be taken at symposia, according to physician Galen, to help the participants "feel warm and elated." Its use as a medicine can be traced back at least 5,000 years in China, and nearly 4,000 years in Egypt and Mesopotamia (Clarke and Merlin 2013, 200–207, 241–248). Ongoing research is just beginning to verify many of the therapeutic effects of cannabinoids (Hill 2015; Mechoulam 2012). *Cannabis* is truly a wonder weed.

Weeds and other ruderal species would seem to represent ideal medicinal plants, at least from a collector's perspective. They are readily accessible, easy to harvest, and locally abundant. If you are sick and in need of a botanical remedy, "Plants that take several days to find and collect are not very beneficial" (Stepp and Moerman 2001, 21–22). The question of accessibility is a testable hypothesis, and several researchers have explored this question by applying the "ecological apparency hypothesis."[4] The idea, simply stated, is that species that are particularly apparent or "visible," like weeds, food crops, and other common plants, are more likely to be discovered and used as medicine by people than rare or otherwise inaccessible species (Lucena and Albuquerque 2005). The plant's "visibility" is enhanced as well by whatever properties make it able to be experienced by the senses (organoleptic)—fragrant and/or showy flowers, unusual textures, and pleasing or bitter flavors. Researchers have investigated these questions by various means. In one case, two contrasting environments were compared, one with high species diversity (upland Amazonian rainforest) and the other with much lower diversity (highland Andean *puna*). Although the Amazon rainforest provided much more raw material for plant medicine (a higher number of species), the species-poorer Andean vegetation provided many more medicinal plants species for traditional healers (Leonti 2011; Vandebroek et al. 2004b). Thus, because *puna* habitats are less species rich, individual elements in their flora are relatively more apparent than the often rare species inhabiting tropical rainforest. Another study considered greater or lesser accessibility by examining distance and travel time to collection. If a species grows relatively close to home, one of the dimensions of apparency, then it is more likely to have medicinal applications than those that grow far away. The results indeed showed that medicinal species tend to grow significantly closer to home (Thomas et al. 2009).

Other researchers investigating the question compared the use value of a species (UV—measured in number of uses per species) and the importance value index (IVI—measured in terms of species density, abundance, and basal area). If the apparency hypothesis is correct, then we would expect to find more uses associated with ecological importance, since these are more visible species. The results to date have been equivocal; some support the idea, others do not. In a Bolivian study, there was a positive association between apparency and plants that provide construction material, but a negative association for medicine and food. In this case, the more visible the species, the less likely it was to serve as a local medicine. Unfortunately, this study only included old-growth forest trees to the exclusion of the likely myriad medicinal herbs, shrubs, climbers, and crops (Guèze et al. 2014). However, in a similar study carried out in the northeast of Brazil, the results were nearly the same. Plants that served for construction, technology, and fuel were relatively apparent, whereas medicinal plants were not (Lucena and Albuquerque 2005). Human exploration and discovery in nature appears to be more nuanced than simple visibility.

Several characteristics of ruderal species help explain the prominence of disturbed habitats as sources of medicinal species. One of these is the apparent concentration of biologically active compounds (allelochemicals). A brief review of plant defense theory suggests why this is the case. Returning to Levin's (1976) global alkaloid census (chapter 1), recall that he discerned a latitudinal cline in secondary metabolites; the closer one gets to the Equator, the higher percentage of plant species possess alkaloids. But he also discerned that plants exhibiting annual life cycles, such as weeds and many cultivated plants, are twice as likely to employ alkaloid defense systems as are longer-lived species (perennials). In North America, for example, annual species averaged 33% alkaloid presence, compared to only 17% in trees. Because most weeds are annuals, they have an inherently higher likelihood of deploying an alkaloid defense system to stave off plant eaters. In addition, many weedy plants defend their short-lived territory by developing compounds to fend off other plant competitors. When these compounds are deposited in the soil around the plant (allelopathy), other species are unable to encroach on their space. Finally, it has been suggested (apparency theory and resource availability hypothesis) that slower-growing plants, like rainforest trees, tend to invest more heavily in plant defenses, whereas faster-growing plants, like weeds and other pioneer plants, tend to invest less. Thus, it makes sense for a tree to invest in metabolically expensive structural attributes, such as tannins, bark, and thorns, since the organism will be around for a long time and will very likely be discovered by many herbivores. Weeds and other ruderal plants, however, are fast growing and have short lives, and so they are better off investing in metabolically inexpensive chemical defenses. These cheap compounds, including akaloids, phenolics, and glycosides, by coincidence, are also of the appropriate molecular weight (<900 daltons) to be incorporated into medicinal formulas (Coley and Barone 1996; Endara and Coley 2011; Stepp 2004; Voeks 2004).

Other ideas have been put forth to explain the over-reliance of people on medicinal weeds. One of these, the "diversification hypothesis," suggests that non-native, usually weedy plants may contain classes of secondary compounds that by chance are missing from one or another local flora. In such cases, rural healers incorporate recently arrived species into their medical repertoire in order to diversify the local medicinal stock and thus fill therapeutic vacancies that native plants fail to satisfy (Alencar et al. 2010). This hypothesis is supported by the global biogeography of weedy taxa. Not only are traditional pharmacopoeias top-heavy with weedy species, but these taxa often trace their origin to cosmopolitan, temperate zone (especially Eurasian) plant families. Because these species often have huge geographical ranges, much greater than the average tropical species, they are likely to contain a broader range of ecologically relevant information in their genes, including bioactive compounds. More compounds translate to greater medicinal versatility.

Finally, it is important to remember that the European colonial period led to the reformulation of the useful floras of the world. Weeds were distributed to all points of the geographical compass, as were many of their Old World cultural uses. It is no coincidence, for instance, that Eurasian weeds are over-represented in New World pharmacopoeias to treat respiratory illnesses. Since many of these viral "crowd illnesses" were introduced during the colonial period by Europeans, who used many of the same species in their country of origin, it is safe to assume that Old World diseases, the weedy vagrants used to treat them, and the knowledge of how to prepare and administer them arrived and became prominent in newly colonized regions of the world (Leonti et al. 2013; Molares and Ladio 2009). If as historian Alfred Crosby notes, "the sun never sets on the empire of the dandelion," this maxim is doubly true for medicinal weeds (Crosby 1993, 7). Consider the humble castor bean plant.

The Palma Christi

The castor bean (*Ricinus communis, Family Euphorbiaceae*) has one of the longest recorded histories of any medicinal plant, reaching from the dawn of Western civilization up to the very present. It is, in addition, one of the world's most invasive weeds. It may be native to Egypt, since early Greek writers referred to it as "the Egyptian oil," but it was widely naturalized in the Mediterranean, North Africa, and Mesopotamia before recorded history. In the fourth century BCE, Herodotus saw it being cultivated in Egypt, but also said it grew spontaneously in Greece. Its earliest recorded use may be in the *Ebers Papyrus*, or *Egyptian Medical Texts*, dated around 1500 BCE (but derived from documents perhaps 1000 years older). All of the remedies outlined in the *Papyrus* were magical spells and incantations, which was the core of ancient Egyptian medicine. The single exception seems to have been the castor bean, which was prescribed for various maladies—from headaches and infected sores to hair restoration. Most importantly, it was prescribed as a powerful laxative: "a few of its seeds are chewed with beer by a person who is constipated, it will expel the faeces from the body of that person" (Dawson 1929). In ancient Assyria, the castor bean was recorded on clay cuneiform tables as being used for eye ailments and, among other drug plants, was recommended for "when a man is full of guraŝtu [perhaps scabies or itch]" (Dawson 1929, 53; Manniche 1989; Scarpa and Guerci 1982; Thompson 1924, 10). Dioscorides noted

A

5.5   A. Castor bean (*Ricinus communis*) male flowers and developing fruit, Fullerton, California. Photo credit: Robert Voeks. B. Tick-shaped castor beans. Windsor Forest, Jamaica. Photo credit: Ina Vandebroek.

that the oil was used in lamps, but he also prescribed the seeds to treat cancer and "to provoke vomiting," as well as the leaves to cure styes and other disorders of the eye (Riddle 1985). Pliny the Elder (first century CE) tells us that the species was labeled "ricinus" because its seeds resembled the creature of the same name—a tick (Fig. 5.5). He also recorded that the leaves were prepared with vinegar and applied to skin infections and to swollen breasts (Dawson 1929, 59–60).

Castor bean is a woody shrub, growing into a small tree in favorable locations. Its crimson and spiny fruits catch the eye, but its formidable, hand-like (palmate) leaves are its most salient feature. It is believed that

B

5.5 *continued*

castor bean is the large-leaved plant referred to in the Bible that provided shade to Jonah "And the LORD God prepared a gourd [or leafy shrub, or ivy, depending on translation], and made it to come up over Jonah, that it might be a shadow over his head, to deliver him from his grief. So Jonah was exceeding glad of the gourd" (Duke 2010, 374–6; Jon. 4:6, King James Version; Zohary 1982, 193). Perhaps it was this Biblical reference, or its universal therapeutic quality, that gave rise to its widespread common name—the "Hand of Christ" (Palma Christi). Whatever the answer, its barbed and explosive fruits aided the useful weed in its seeming goal of world domination, while its distinctive leaves allowe'´ ˀven the most untrained traveler to the tropics to readily identify it.

Within two centuries of the Columbian landfall, the geographical shadow cast by castor bean's leaves was indeed considerable. Tomé Pires recorded in his famous *Suma Oriental* [1512–1515] that castor bean was being employed medicinally in India. "If they want to purge themselves, they take the crushed leaves, or the juice or the seeds of *figueira do inferno* [*Ricinus*], and they are well purged" (T. Pires 1967 [1512–1515], 69). Two centuries later, Frenchman Pierre Poivre reported that the oil was considered caustic in Europe, but in India they used it as a powerful purgative,

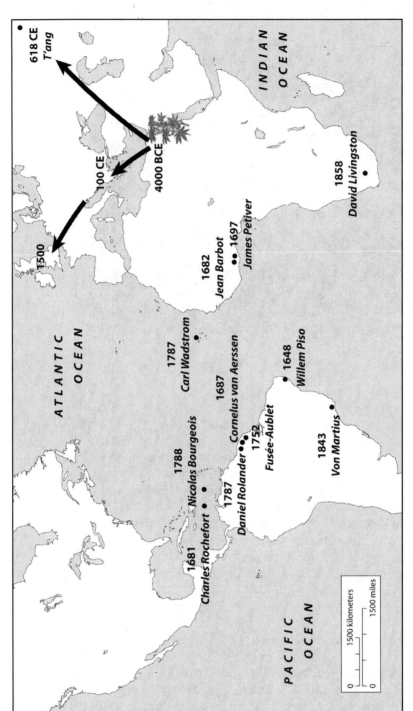

5.6　Early recorded presence of castor bean (Palma Christi—*Ricinus communis*) in West Africa and the Americas. Locations: Andel et al. 2012; Barbot in Hair et al. 1992; Bourgeois 1788; Fusée-Aublet 1752; Livingston 1858; Von Martius 1939 [1844]; Petiver 1697; Piso 1948 [1648]; Rochefort 1681; Rolander 2008 [1755]; Wadstrom 1795. Image: Kelly Donovan.

especially for nursing infants, who were given "every month, a spoonful of it, mixed in an equal quantity of their mother's milk" (Poivre 1770, 57). London pharmacist James Petiver received samples of the plant from John Smythe in West Africa, where it was likewise used medicinally (Petiver 1697). Jean Barbot also noted in early eighteenth-century West Africa that "the oil drawn from this plant is medicinal" and that it is grown in gardens on the island of Goree (cited in Hair et al. 1992, 74). Michel Adanson, for whom the iconic African baobab was named by Linnaeus, reported it growing wild in the hills above the Niger River (Adanson 1759, 118). By that time, castor bean had successfully colonized the Americas, as numerous physicians and travelers recorded the presence of the easily identifiable Palma Christi (Fig. 5.6). Charles de Rochefort reported from the Antilles that Palma Christi was present, and that the oil was used to heal certain ulcers and was "rubbed into the hair [of enslaved Africans] to protect themselves from lice" (Rochefort 1681, 107). It was also present in northern South America, as seeds of Palma Christi were reportedly sent from Dutch Suriname to the botanical garden in Amsterdam in 1687 (Van Andel et al. 2015). And in Brazil in the late 1600s, Dutch physician Willem Piso made accurate drawings of castor bean, calling it *Ricino Americano* (Piso 1948 [1648], 441–445). Among enslaved Africans in the Americas, castor bean took on special cultural significance. As a treatment for body lice and skin disorders by slave traders, it was symbolically connected with their status as the possessions of others. But given the historical legacy of the species' use in distant Africa, castor bean was embraced by many New World Africans as cultural markers of their homeland (Carney 2013). The weedy shrub is known today by West Africans and their Brazilian descendants by its Yoruba name, *ewe lara*, or its Bantu name, *mamona*, and continues to be employed on both sides of the Atlantic for similar ritual purposes (Voeks 1997).

In spite of its villainous standing as an invasive organism, the castor bean plant continues to be applied by traditional healers throughout the world. And it has found numerous uses in Western medicine. It is still used in an oil form (castor oil) to treat constipation, although it would be an exaggeration to call it "mild." It is mixed with other drugs as an antifungal agent and in cancer chemotherapy. It has widespread use in industry, for soaps and paints. It is used as a lubricant, especially for race cars, and there is considerable interest in developing castor oil into biodiesel, especially in Brazil.

Finally, there is a sinister side to the medicinal merit of this plant. As a powerful laxative, it has been used as an implement of torture, literally to purge prisoners to death. Worse still, the protein ricin, which can be extracted from the beans, is one of the most toxic natural substances known to humankind. The United States and the Soviet Union experimented with weaponizing ricin, although these schemes were later abandoned. Ricin's most infamous use was the assassination by poison of Bulgarian dissident Georgi Markov in London in 1978. It was later determined that a ricin pellet had been secretly injected into his leg by the tip of an umbrella, probably wielded by the Bulgarian secret police. Thankfully, given the difficulty of producing it in sufficient quantities, the potential for terrorists to manufacture and use ricin is minor (Schep et al. 2009). In any case, as with myriad other medicinal weeds, the tropical Sun never sets on the empire of the multi-talented castor bean.

The castor bean is only one example of the scores of weedy invaders that reached the far corners of our planet during the European Age of Exploration. Very few weeds were recorded by colonial observers simply because not many had the sorts of distinguishing morphological features that would have drawn their attention. One of the few in the tropical Americas was lion's ear (*Leonotis nepetifolia*), a weed native to Africa, but now naturalized throughout the Old and New World tropics. First recorded in the Americas in Dutch Suriname in 1680, the herb continued its African therapeutic uses among the enslaved African population as well as their descendants (Van Andel 2015). But most of these lowly species were what botanist Herbert Baker called "belly plants," noticed and observed best from the prone position. We know they were widely dispersed in the past, by one means or another, because humanized spaces today are overrun with them. Weeds today are vilified for being exotic, opportunistic, and a threat to native ecosystems (M. Davis et al. 2011; Larson 2011). But many if not most have a role in curing what ails people. Food crops are non-natives as well, but they obviously have not received such scorn because they were purposely conveyed around the world and, in most cases, they were welcomed by local people. And like their underappreciated cousins the weeds, most crop plants also possess medicinal value.

Food as Medicine

The medicinal benefits of weeds and other ruderals are complemented, now as in the distant past, by crop plants and spices. Indeed, the connection between eating and healing is likely as old as the process of plant domestication, or older. Food plants represent the flora with which we are most familiar, and because we consume them, their effects can generally be trusted. Random experimentation with wild plants, it seems, "may have had little to do with the acquisition of plant-based knowledge" (Johns 1990; Logan and Dixon 1994, 36). It is also likely that the earliest domesticated foods served a dual mission as food and medicine. For instance, figs (*Ficus carica*) are our earliest known domesticated crop, at over 11,000 years BP (before present). In the Bible, they are referred to more than once for their curative effects on cutaneous anthrax, a particularly loathsome skin disease. "And Isaiah said, take a lump of figs. And they took and laid it on the boil, and he recovered" (2 Kings 20:7, King James Version; Ben-Noun 2003; Duke 2010, 187). Some edible species might in fact have been appreciated originally for their healing rather than nutritive values. The eggplant (*Solanum melongena*) has been used as food and medicine in India and China for at least 2000 years. At present, there are some 77 different medicinal applications for eggplant in just India, China, and the Philippines (Meyer et al. 2014). The weedy and bitter-tasting ancestors of eggplant were probably first employed for their medicinal rather than culinary contribution (Daunay and Janick 2007; Weese and Bohs 2010). The same is true for carrots (*Daucus carota* var. *sativus*), whose wild ancestor's seeds were used medicinally in the Mediterranean well before the root was domesticated for the cooking pot. Dioscorides reported on both wild and domesticated carrots, noting the medicinal qualities of their seeds, leaves, and roots (Stolarczyk and Janick 2011). How important for the ancients were medicinal foods? In a remarkable recent discovery, Italian archaeologists located a two-thousand-year-old shipwreck off the Tuscany coast. In it were numerous tin boxes filled with tablets. DNA analysis showed that the tablets, now believed to be medicinal pills, were made up of powdered onions, carrots, leeks, cabbage, parsley, and garlic, all considered part of the Hippocratic arsenal of medicinal foods (Touwaide and Appetiti 2015, 27).

In the earliest human civilizations—Assyrian, Chinese, Indian, and Mediterranean—cultivated grains, roots, fruits, and spices were employed frequently in the curing arts. In the age of the pharaohs, lettuce, sesame, onions, leeks, cucumbers, plums, watermelon, and many other fruits and vegetables were included in the healer's arsenal (Manniche 1989).[5] According to the ancient *Greek Hippocratic Collection* (fifth century BCE), 77% of the plants reported for therapeutic purposes were food crops or otherwise edible, and most were cultivated in gardens (Touwaide and Appetiti 2015).[6] The connection between food and health promotion is particularly evident in regards to spices. The lavishly expensive Asian condiments that flowed into the Mediterranean along the ancient Silk Road (or roads) over two thousand years ago reveal the extent to which their healing attributes were appreciated. Black pepper (*Piper nigrum*) appears first, and was prescribed for a wide range of health problems, especially "diseases of women." These maladies included swelling of the womb, indurated cervix, hysteria, and even bad breath, and were treated by chewing pepper, or packing it into pills, potions, and pessaries. There is also mention of pepper use for eye pathologies, a therapy that may have been assimilated from India's ancient Ayurvedic traditions. Ginger (*Zingiber officinale*) was also nearly always associated with medicinal cuisine. In fifth-century CE Gaul, it was added to foods for its warming influence and for stomach ailments, similar to current traditional Chinese medicine. In ancient India, ginger was considered not a spice at all, but rather "the great cure" for a panacea of ailments, from gout and elephantiasis to stomach upset (Ravindran and Babu 2004, 4). And the practice of baking ginger with honey into a human-shaped object—the iconic gingerbread man—may have developed historically more for medical and magical ends than for dessert purposes.

Today, many of the health benefits of plant foods and spices are being rediscovered and confirmed, as the field of functional foods and nutraceuticals draws increasing attention. There are literally thousands of plant-derived compounds in wild and domesticated food plants, and many are linked with one or another health benefit, including antioxidants, antimalarials, anti-diabetics, and many others. The brilliant red, orange, and blue corn (maize) varieties developed in ancient Middle America, for instance, are loaded with anthocyanins and carotenoids, important immune system enhancers and protectors against age-related blindness (Ryu et al. 2013). Wild and cultivated mushrooms are widely reported to have anti-cancer, anti-tumor, anti-inflammatory, antibacterial, and many other disease preventive and fighting benefits (Rahi and Malik 2016). Chili peppers, native to tropical America and first recorded by Columbus, constituted food, medicine, and even weapons for Native Americans (Andrews 1995). Hispaniola's indigenous Taino battled Columbus's troops with gourd bombs "stuffed with ashes and ground hot peppers" (C. Mann 2011, 9). And of course capsaicin, the active compound in chilis, is still weaponized in the guise of pepper sprays. In addition to being a "thrill food," capsaicin is being used by physicians to treat post-surgery pain, inflammation, itching, and injury (Bode and Dong 2011). In Uganda, people living with HIV/AIDS are constantly battling infections. To combat these afflictions, many rely on an array of common fruits and vegetables—papaya (*Carica papaya*), avocado (*Persea americana*), guava (*Psidium guajava*), and many others—to maintain and strengthen their immune systems (Asiimwe et al. 2013). And consider chocolate, native to the rainforests of South America and according to its Latin genus label *Theobroma*, "the food of the Gods." For centuries it was consumed as a tonic and aphrodisiac, first by the Mexica (Aztec) king Montezuma, who said that a cocoa bean allowed a man "to walk a whole day without food," and later for amorous exploits by the likes of Giacomo Casanova and the Marquis de Sade (Norton 2008). Today it is known as both the "love drug" and as one of the richest flavonoid-containing foods known. Can it be a coincidence that the longest recorded lifespan, 122 years, was achieved by the Frenchwoman Jeanne Calme, who consumed nearly a kilogram of chocolate per week her entire life (Keen 2001; Jeune et al. 2010)?

Food crops were almost always domesticated in one location, such as oranges in Southeast Asia, bananas in Papua New Guinea, or watermelon in southern Africa. But whatever their provenance, it would not have been long before people began to disperse them. Sometimes the process was slow, as crops passed from one neighboring group to another (cultural diffusion), or as culture groups expanded their territory and carried their crops and culinary traditions with them (demic diffusion). Other efforts were rapid and far flung, such as crops moved by merchants along distant trade routes or by ocean voyagers. Such was the case with "canoe plants," which were introduced during the great Polynesian voyages of discovery across the Pacific Ocean to Tonga, Samoa, Hawaii, and even distant Easter Island (Rapa Nui). Moving quickly and over great distances beginning about four thousand years ago (the express-train hypothesis), these ancient mariners transported coconuts (*Cocos nucifera*), bananas (*Musa* ssp.), breadfruit (*Artocarpus altilis*), sweet potatoes (*Ipomoea batatas*), taro (*Colocasia esculenta*), and many others, mostly for food and medicine (Kayser et al. 2000; Whistler 2009, 1–7). Similarly around the Indian Ocean, food plants had traveled back and forth from India to Africa thousands of years before Europeans arrived. Cowpeas (*Vigna unguiculata*), sorghum (*Sorghum bicolor*), pearl millet (*Pennisetum glaucum*), and other African domesticates had reached India by 1500 BCE or much earlier. In the opposite direction, banana and plantain (*Musa* ssp.), coconuts and taro likely reached sub-Saharan Africa over two thousand years ago (Rangan et al. 2012). There was also considerable globalization of starchy foods, the "Trans-Eurasian Exchange," in which rice (*Oryza* sp.), wheat (*Triticum* sp.), barley (*Hordeum* sp.), and others had spread from one corner of Eurasia to the other several millennia before Europeans began their journeys of discovery (M. Jones et al. 2011). And as noted earlier, culinary and medicinal traditions probably travelled with these species, although in most cases this is difficult to document (Touwaide and Appetiti 2013).

Millennia later, beginning with Columbus's Caribbean encounter, a new epoch in food crop diffusion began: the Columbian Exchange (Crosby 1993). The scope of this biotic homogenizing process was astonishingly rapid and global, linking the distant and vaguely known shorelines of East and Southeast Asia, India, sub-Saharan Africa, and the Mediterranean with tropical America. In the New World, the arrival of European, African and Asian cultivars doubled or even tripled the number of indigenous cultivated food crops (see chapter 7). Some of these arrivals were the result of acclimation efforts carried out by European governments and religious orders. Missionaries were especially effective botanical conveyors, their efforts to disseminate recently discovered edible and medicinal plants matched only by their proselytizing zeal. The Jesuits in particular, for whom studying and exploring nature was "one way of worshipping God," were committed to learning and dispersing the world's exotic foods and medicines (Anagnostou 2007, 294). Stationed in missions throughout the world and in long-term intimate contact with indigenous people, the clergy was probably better placed to discover and distribute the world's useful species than any other group (Harris 2005).

Other spiritually motivated efforts to acclimate newfound foods and medicines were likewise initiated early by European medical schools and physic gardens. When it became apparent to men of science that the Garden of Eden did not physically exist in Asia, Africa, or the Americas, they hatched a new plan—to reconstitute the biblical Garden by collecting the most wondrous plants from around the world and transplanting them in botanical gardens. Because it was widely accepted that many of God's green beings were intended as remedies for what ailed mankind, a complete collection of His plants from all over the world would supply "a ready remedy for every injury and infection" (Prest 1981, 10). This floristic reconstitution of the purity of the Almighty was finally believed possible due to the discovery of the botanically rich forests and fields of the Americas. And these botanical gardens were not intended as mere symbols of Eden. They were meant to be scientifically informed resurrections of the actual birthplace of humanity. For Renaissance Christians, a walk through a botanical garden was meant to be a "journey through the mind of God" (Drayton 2000, 11).

Economic motives for mixing the world's useful flora complemented Christian-inspired efforts. So that farmers and merchants would have access to the cornucopia of novel foods, flavorings, beverages, and spices being discovered in distant lands, seeds and seedlings were dispatched to European centers and to remote tropical colonies, such as those in the Cape of Good Hope (1652), Calcutta (1787), Ceylon (1821), Java (1811), and the Jardim Botânico in Rio de Janeiro (1808) (Gunn 2003). The prestigious European botanical gardens, including Kew in London and the Jardin du Roi in Paris, were pivotal in this botanical enterprise, particularly when the intent was to breach lucrative monopolies. Agents of the garden circuits orchestrated various strategies, from patronage to outright theft, to effect the transfer of clove, cinnamon, vanilla, ginger, and of course, nutmeg and cinchona (Gramiccia 1988; Spary 2000). These transfers of biological resources, in the minds of Europeans, represented socially useful enterprises, if not Christian obligations. Carl Linnaeus summed up this view in 1756, noting that "Nature has arranged itself in such a way, that each country produces something especially useful; the task of economics is to collect [plants] from other places and cultivate such things that don't want to grow [at home] but can grow [there]" (cited in Koerner 1996, 151).

Considered in this light, the great global relocation and range expansion of useful crops and weeds that took place before and during the colonial era should be viewed not only as a grand experiment in biogeographical reorganization, but also as an accidental but highly effective effort to enrich the world's plant pharmacopoeias. Many of the indigenous healing traditions associated with exotic fruits, vegetables, spices, and weeds accompanied these species on their journeys, and in so doing augmented the medical options of distant healers. Numerous healing features of Indian and Chinese foods and cuisines, for instance, arrived and survived in Europe. Likewise, Mediterranean medicinal food traditions diffused with the Spanish and Portuguese diaspora to the New World, allowing continuity in their culinary and healing traditions. In many other instances, indigenous peoples redefined the healing properties of newly arrived foods to match their own cultural conditions. In West Africa, for instance, the leaves of South American papaya became a treatment for malaria (Titanji et al. 2008). In Suriname and Brazil, Asian bitter gourd (*Momordica charantia*) became a treatment for diabetes and hypertension (Van Andel and Carvalheiro 2013). Consequently, as European visitors were inadvertently infecting the Americas and Oceania with Old World contagions, newly introduced medicinal foods, spices, and weeds were being assimilated into indigenous healing ceremonies as shamans and herbalists worked feverishly to aid the afflicted. Although their efforts did little to repulse the worst of these alien microbes, such as smallpox and influenza, the ultimate impact of these long-distance plant introductions is plainly evident in present-day tropical pharmacopoeias. Today, nearly all tropical healing floras include a rich array of medicinal plants that were introduced over the past five centuries, testimony to the crucial role of humanized landscapes and species to tropical pharmacopoeias. In northern South America, for instance, over 200 of the plants employed as medicinals are non-natives, and 88 of these were originally introduced as food plants (Bennett and Prance 2000). Indeed, if you take the time to examine in detail the botanical composition of widely separated tropical plant pharmacopoeias, you may be surprised to discover how many include an identical core of medicinal foods, spices, and weeds.

Foods with medicinal values are most often cultivated in tropical homegardens (backyard gardens). Since the time when people first came to occupy permanent dwellings, homegardens have served as readily accessible sources of vegetables and fruits, medicines and spices, as well as areas for rest and contemplation. They constitute vibrant social spaces where horticultural history is made and transmitted, and where botanical resources contribute to economic and nutritional security. Industrial farming and fast food may have killed the food-medicine connection in much of the developed world,[7] but in the less developed world, the innumerable fruits, vegetables, and spices cultivated in homegardens continue to serve double duty as food and medicine. Tropical homegardens and swiddens are notoriously rich in crop varieties, and many of these enter into medicinal recipes (Coomes 2010; Sander and Vandebroek 2016; Srithi et al. 2012; WinklerPrins and de Souza 2005). Among the African Hausa, for instance, 49% of medicines taken for gastro-intestinal disorders are also consumed as foods. In the Oromia region of Ethiopia, 27 of the 36 species that are cultivated as medicinals serve for both food and medicine (Hunde et al. 2015). And in the Bolivian Amazon region, hundreds of useful species are cultivated or spared in homegardens and fallows, with most serving as either medicine or food, or as both, such as cassava, avocado, papaya, and many others (Thomas and van Damme 2010). As Nina Etkin commented, many plant species "move fluidly between the categories of food and medicine" (Etkin 2009, 71). For many traditional societies, now and in the past, food is medicine, and medicine is food.

A brief review of tropical plants that have successfully transitioned from the field to the pharmacy highlights the prevalence of cultivated, weedy, and otherwise managed elements of anthropogenic space. Most are weedy herbs, shrubs, climbers, garden crop plants, or common trees. Of the 100-plus plant species from which current pharmaceutical drugs are derived, 36 are considered weeds (Stepp 2004). The glaucoma alkaloid pilocarpine, derived from *Pilocarpus jaborandi*, is a common Brazilian herb or shrub. The digestive drug papain, derived from papaya fruits, is a widespread trash-heap food plant. The anti-Parkinson drug L-Dopa, derived from the *Mucuna pruriens* var. *utilis*, is a pantropical leguminous climber.

The antimalarial alkaloid quinine (see chapter 4) is derived from Andean *Cinchona* trees, which are mostly treefall gap species. The original source of the widely used contraceptive diosgenin is *Dioscorea alata*, an invasive weed of the southern United States (Soejarto and Farnsworth 1989). And the antimalarial compound artemisinin, discovered in the 1970s, is derived from the widely dispersed Chinese weed *Artemisia annua* (Callaway and Cyranoski 2015).

According to the jungle medicinal narrative, pharmacies and medicine cabinets are crammed with precious healing compounds that originated in the forest primeval, and scientists learned the identities and healing virtues of these species through consultation with forest shamans. The first element in this assumption is false; most drug plants in the past and at present occur in disturbed habitats, and many are derived from domesticated foods and weeds. Old-growth tropical forests maintain a wealth of globally significant ecosystem services, as well as locally significant economic and cultural values, but a "pharmaceutical factory" of medicinal plants is not one of them. The second element, the knowledgeable and mysterious shaman, represents the crucial nature-society intermediary in the story. Only with the direction provided by forest healers will tropical nature reveal its secrets. But in nearly every case, shamans and other healers are depicted as men. Women are almost always invisible elements in the story line. In the next chapter, I examine the issue of gender in the jungle medicine narrative.

SIX

Gender and Healing

The iconic image of the mystical male shaman guiding his Western scientist apprentice through the forest primeval in search of life-saving drug plants is a key element in the jungle medicine narrative (Fig. 6.1). Secretive, self-assured, and frequently masters of mind-altering drug plants, shamans seem to be the personification of the mysterious and enigmatic rainforest. The more convoluted their rituals and the more arcane their paraphernalia, the more noble and authentic they seem to be. But this romanticized representation in the Western world is a relatively recent phenomenon. During the protracted period of European colonization and settlement, shamans were the bitter adversaries of missionaries in their battle for the bodies and souls of indigenous "heathens." Not only did they hold sway over the spiritual lives and health concerns of their followers, they were also believed to harbor knowledge of the odious plant poisons that were deployed against their enemies, including European invaders (chapter 4). If they could only conquer the shamans, missionaries would be free to command the souls of the savages towards the one true Christian God. In Africa, shamans were derisively labeled as witches, witch doctors, and wizards, whose practices included black magic, fetish worship, and tribal fanaticism. European visitors reveled in chronicling their "snake gods," their magical fetishes, and especially their supposed ability to communicate with the spirits. As John Atkins recorded in 1737, the shamans "pretend Divination, give Answers to all Questions, . . . like our Fortune-tellers" (Atkins 1737, 104; T. Pires 1967 [1512–

6.1 Iban shaman during pig liver divination (*betenong ati babi*) ceremony, Sarawak, Malaysia. Photo credit: Robert Voeks.

1515], 43). The perception of shamans in colonial South America was perhaps less ominous, but they were still seen as obstacles to Spanish and Portuguese ambitions. In sixteenth-century Brazil, Jean de Léry reported that shamans blew sacred tobacco smoke on their followers who wished to "overcome all of [their] enemies" (De Léry cited in Purchas 1625, 1338.) Shamans were instruments of Satan, it was widely agreed, who spread their insidious influence through barbaric council with the devil and through their so-called healing ceremonies (Leite 1938, 21–22; Taussig 1987, 142–143). And because they employed macabre divination rituals and drug-induced vision quests to see the future and cure what ailed their followers, shamanic healing practices were for centuries roundly scorned by outside observers as too mired in the occult and witchcraft to be of relevance to science and medicine. Enter the future doyen of modern ethnobotany, Richard Evans Schultes.

Shamans

Often referred to as "the father of ethnobotany," Schultes was a Harvard-trained botanist who became fascinated early in his career with the use of psychotropic drugs in tribal societies. He carried out many years of ethnobotanical investigation in the American Midwest, Mexico, and especially South America. Much to the annoyance of the sober and politically conservative Schultes, his pioneering research with drug plants became a catalyst for the 1960s psychedelic generation. He carried out ground-breaking studies on the use and cultural significance of peyote cactus (*Lophophora williamsii*), "magic mushrooms" (*Psilocybe* spp. and *Panaeolus* spp.), ayahuasca (*Banisteriopsis caapi*), and several other intoxicants quite new to science. He spent many years in the Amazon rainforests, especially during World War II, during which time he amassed a huge inventory of medicinal and poisonous plants employed by forest shamans. As an academic, Schultes published much of his research in professional botanical journals with limited circulation. But popular journals, such as *Life* magazine and *Popular Mechanic*, and later his own books, especially *The Vine of the Soul*, and *The Healing Forest*, transported his tropical adventures and exotic discoveries into the living rooms of Western readers (Enright 2009, 210). For disillusioned young people seeking a getaway from the monotony of postindustrial existence, the mind-altering botanicals Schultes acquired from indigenous shamans and shared with the Western world provided their vehicle for escape. Amazonian ayahuasca, in particular, could be used to dig deeply into the human psyche, producing in many users profoundly spiritual experiences. Today, legions of foreign visitors embark on journeys of self-realization into the Amazon forest, seeking mystical enlightenment from the visionary vine with the aid of a local shaman. Thanks to Schultes, the Garden of Eden became once again situated squarely in the heart of Amazonia, complete with the biblical "Tree of Knowledge"—ayahuasca. The view that jungle shamans have mastered mysterious drug plants that allow them to interpret the past, predict the future, and indeed, reveal the metaphysical meaning of existence, feeds into many of our deepest visceral beliefs. For better or worse, Schultes helped resurrect the long-maligned rainforest shaman into the noble and mysterious healer he is today (Bussmann 2012; Madera 2009; Polari de Alverga 1999, 2; Prayag et al. 2015).

Shamans play numerous social roles in their communities. On the one hand, they are often ambivalent figures, just as likely to do harm as to heal. They are in many instances masters of the black arts, able to advance the cause of one person at the expense of another. Some can prepare "medicine" that will attract a potential lover, repel an insistent suitor, or bring calamity to the life of a competitor. At the same time, they are able to resolve real-world health problems by communicating with the spirit world. Shamans are highly specialized healers, having heard the call early in life to their vocation, and usually apprenticed under an elder shaman. They are often charismatic by nature. In Amazonia, where so much of this research has been carried out, shamans employ psychoactive "master plants" to open doors of perception into the causes of illness. Although fifteen to twenty psychoactive species are used in the region, including ayahuasca, tobacco is considered by Amazonian shamans to be "the father of all plants." In most instances, the modes of preparation of psychoactive species are carefully guarded secrets, contributing even further to the mystique of the shaman (B. Freedman 2012, 151).

Given the fascination of the Western world with these botanical intoxicants, as well as the influential role played by shamans in tribal societies, it is no surprise that so much scientific inquiry has been focused on these (usually) male specialists, almost to the exclusion of female healers. Indeed, considering simply the time constraints and cultural norms that allow male researchers greater access to male informants, the standard operating procedure in most ethnobotanical research until very recently was to select a few key older men for study (Atkinson 1992; Browner 1991; Kothari 2003; Pfeiffer and Butz 2005). These sorts of gender-imbalanced ethnobotanical inquiries not only lead to a skewed understanding of the great diversity of medicinal plants employed in traditional communities, but also reinforce the untested assumption that most of this knowledge is the purview of men. For instance, in an article that explored the medicinal plants used to cure snakebite in India, the authors concluded, "Women of this community have very little knowledge of medicinal plants." However, a quick look at their selected field informants showed that they consisted of 41 men and one woman (Panghal et al. 2010). Not surprisingly, the shamans and healers that are depicted in prehistory, as well as in popular books and the big screen, such as *The Teachings of Don Juan: A Yaqui Way of Knowledge* (Castaneda 1968) and *Medicine Man*, continue to be drawn from "a largely male crew" (Tomášková 2013, 14).

Shamanism represents a complex and contested role in hunter-gatherer and small-scale agricultural societies. Indeed, the term "shaman" covers a range of magical and religious practitioners, including shamans, shaman-healers, mediums, priests, and sorcerers, with various degrees of social and political power (Winkelman 2010). Although the core position is often occupied by men, there are in fact numerous instances in which women serve as shamans, or at least play pivotal roles in shamanistic rituals. In Siberia, where the original term "shaman" first appeared to Europeans around the early seventeenth century, female shamanism appears to have been common. According to Tomášková (2013, 151), there were "both male and female shamans, as well as those who resist a gender category, be they men dressed as women or women who looked too much like men." In Korea, over 95% of the shamans are women, and numerous North American indigenous nations included both male and female specialized healers (Atkinson 1992; Pfeiffer and Butz 2005).

By coincidence, the two groups with which I've had most interaction, practitioners of Candomblé in Brazil and the Dusun of Borneo (Brunei Darussalam), both have significant representation by women shamans. In the case of Candomblé, about half of the shaman-healers that I've worked with are women, and the most famous priests in the country are women (females are known as *mães-de-santo*, males as *pais-de-santo*). Like their male counterparts, they divine the source of illness or other individual problems via communication with spirits (*orixás*), they are able to associate symptoms with specific illness, defined in the broadest possible sense, and they prescribe culturally acceptable treatments. Many of these treatments include elements from a large pharmacopoeia of "sacred leaves," about which every *mãe-de-santo* is well versed (Voeks 1997). In the case of the Borneo Dusun, the egalitarian nature of their society dictates that both women and men can experience the call to become a shaman. For men, most assume the role of *dukun*, whose province is largely magical, currently focused on love charms and preventing the loss of a partner, and often including the use of saliva, uncooked rice, lime, and turmeric. Female shamans, or *belian*, assume the role of intermediary between humans and the spirit world. Their role is to orchestrate the range of Dusun rituals, or *temarok*, which address important events such as harvest festivals or community dilemmas. One of the most often performed is the crocodile ritual, the *temarok buayo*, in which milled or unmilled rice is arranged on the floor in the shape of a crocodile. Eggs are placed in the chest, symbolizing the heart, and bananas are arranged at the tips of the legs, symbolizing the claws. The ritual is carried out to the accompaniment of several gongs (*sangang*), as women circle the *buayo* throwing palm-leaf spears, and as the *belian* goes in and out of the trance state, alternating "between her own female identity and the male identify of her spiritual partner . . . At one moment she is the flirtatious, ostentatiously feminine companion and euphoric bride, at the next the stereotypical masculine hunter of beasts, even heads" (Kershaw 2000, 153) (Fig. 6.2).

Sex and Space

Just as tropical landscapes constitute mosaics of ecological space, from nearly pristine to overwhelmingly cultural, so too does the nature of human relations with plants and habitats vary over space and time. One of the most apparent of these variables is sex. Men and women may inhabit roughly the same territorial space, but their relationship with the various physical and biological entities is culturally negotiated by power, religion, historical precedent, and especially division of labor. Men and women are often subject to profound separation of work and space in subsistence-oriented communities. And these divisions are further reinforced because instruction by elders on how to perform in these different roles is often gendered as well—men teach boys how to become men; women teach girls how to become women (Grenier 1998; Müller-Schwartz 2006). Thus, in much of rural sub-Saharan Africa women's work includes fetching water, gathering fuelwood, cultivating garden greens, cooking, doing laundry, and caring for children. Consequently, the spaces with which they are most familiar are those that provide resources for these gendered duties, especially homegardens, swiddens, trails, and

A

6.2 A. Female Dusun *belian* orchestrates a crocodile ritual, the *temarok buayo*. B. Men and women play gongs (*sangang*) during the celebration. Photo credits: Robert Voeks.

B

6.2   *continued*

secondary forest. At the simplest level of labor separation, men are frequently engaged in hunting, fishing, livestock herding, and timber extraction, activities that take them to less disturbed habitats that are distant from their settlements. Women, on the contrary, are more likely to spend their time managing local resources in anthropogenic spaces relatively near their homes. There are, in truth, too many variations on this theme to make grand generalizations; women in some societies engage in hunting and fishing, and men tend gardens. In northwestern Amazonia, for instance, women are in charge of most domesticated crops and homegardens, whereas men's activities and knowledge are focused on the forest. However, men control the cultivation and use of specific garden plants, mostly mind-altering cultigens like coca, tobacco, and pineapples for fermenting into alcohol. Exceptions notwithstanding, in most rural societies, indigenous or diasporic, there are clearly delineated and culturally reinforced men's and women's spaces, as well as men's and women's plant species (Colfer and Minarchek 2013; Howard and Nabanoga 2007; Kevane 2012; Pfeiffer and Butz 2005; Reichell 1999).

Gendered divisions of work and space are rooted in religious and social conceptions of what it means to be a man or a woman, and thus what is the appropriate behavior for each sex. These norms determine not only the social position and activities of women and men, but also the biological arena in which day-to-day interactions with nature are negotiated. Women, for example, often have social or religious sanctions prohibiting their entering or working in specific habitats, and in many cases, these are forested habitats (Fadiman 2005; Howard and Nabanoga 2007; Momsen 2004, 143–152; Pfeiffer and Butz 2005). Sometimes women are excluded from sacred forests, while in other situations they are restricted from certain ecological spaces due to notions of feminine "impurity." In Madagascar, for instance, cultural taboos (*fady*) prohibit Antanosy women from going into the forest alone (Lyon and Hardesty 2012). In many other situations, women are prohibited from journeying far from the village due to fear of kidnapping and sexual assault. This was the explanation I received years ago when I was carrying out research with Dusun cultivators. At the time, I had hoped to duplicate walk-in-the-woods identification of medicinal plants with female collaborators. But I was informed that it was not appropriate for women to travel far from home, particularly in the company of a male outsider. Thus, although kidnapping had not been an issue in this region for generations, the cultural constraints have persisted.

Spatial divisions have deep cultural and spiritual support in small-scale hunter-gatherer or horticultural groups in which wild game is central to cosmology and identity. The Desana people of Colombia (see chapter 3) maintain a complex set of gendered taboos and rituals that regulate hunting of preferred game, such as tapirs, peccaries, monkeys, and various other forest wildlife. Game are symbolically labeled "females," regardless of their biological sex, and so hunters pass through elaborate rituals prior to the hunt in order to attract these female game in the forest, including wearing perfumes prepared by the women in the village, and sexual abstinence so as not to provoke jealousy. But although hunting by men is perceived as fundamental to Desana subsistence, fishing and horticulture in fact provide most of the community's daily calories and nutrition. Men and boys cut and burn the swidden sites, but otherwise, all gardening and garden species are considered part of female space. This is similar for the Runa people of Ecuador, where hunting in the forest and gardening are deeply gendered and infused with myth and ritual. In a "sensuous exchange of disciplined female work for male work," a husband trades the fresh game he has worked so hard to acquire for the sexual favors of his wife, as well as manioc drink (*asua*) from the garden she cares for "as if it were her own baby." In Runa mythology, many of these homegarden plants are believed to symbolize flawed men or women from the distant past, either lazy or promiscuous or both. Over time they were transformed into "plant people," and are carefully tended in the garden by women, who celebrate these medicinal and other culturally infused species through love songs, as if they were "moody male lovers" (Swanson 2009, 37 and 41). In hunter-horticultural societies such as the Desana and the Runa, deeply gendered and ritualized relations with green nature are the rule rather than the exception.

Property ownership is a complicated and fluid issue, with considerable implications for gendered knowledge of nature. In northeastern Thailand, for example, customary inheritance of agricultural land and gardens passes through women rather than men. As a result, particularly in anthropogenic habitats, women are the main gatherers, selectors, transplanters, and propagators of edible and medicinal plants (Cruz-Garcia and Price 2011). The situation in sub-Saharan Africa, however, is quite different. Men are much more likely to hold formal title to land, placing women in positions of considerable dependency. In Kenya, husbands and wives navigate the same garden space, but the men actually own the land. Nevertheless, it is the women who carry out most of the labor and management, and who are far more familiar with its various vegetative and ecological properties. At the garden perimeter, women have nested spaces representing "left over" habitat unwanted by the husband, such as ditches, or strips on the edges of orchards, or steep and inaccessible slopes that are inappropriate for commercially valuable species. Even within these less valued spaces, however, plant resources are partitioned further among the sexes, such that men may "own" the trees, as well as their monetary values, such as charcoal or timber, but women have priority in regards to the subsistence properties of the tree's renewable resources, such as fruit, sticks for fuelwood, leaves for fodder, or wild vegetables (Rocheleau and Edmunds 1997). Indeed, the separation of commercial- vs. subsistence-valued plants often represents the line that divides men's from women's botanical domain (Tragett 2012). If it has commercial value, it belongs to men; if it has subsistence value to the family, it belongs to women. This is the case with the Mandinka in the Gambia where men control peanuts, millet, sorghum, and maize in upland fields, which are all destined for commercial markets. Women on the other hand control rice and vegetable production in the low-lying swamp zones, both of which are destined for domestic consumption (Schroeder 1993). These are classic examples of male-dominated cash crops and female-dominated subsistence.

There are myriad permutations on the theme of gendered ecology in rural tropical landscapes. But in most instances, forests and trees are associated with male privilege and space, whereas anthropogenic habitats dominated by cultivated and ruderal plants around the home and settlement, especially in homegardens, are considered women's space and women's plants. Most of the latter species are vegetables, fruits, spices, ornamentals, and weeds, precisely the same subset of botanical nature that dominates the medicinal category of so many tropical pharmacopoeias. In Dominica, this cognitive separation of space and knowledge is learned early through "chore curricula." Young boys and girls learn about trees and domesticated crops early in their lives. But over time, girls learn more and more about medicinal plants, which is more the province of women, whereas boys continue to focus their knowledge on trees, which come to occupy a greater part of their adult livelihoods (Quinlan et al. 2016).

Tropical homegardens in particular are crowded with medicinal species. Many are planted and exchanged with friends and neighbors as gifts, while others arrive spontaneously through wind or animal dispersal (Ban and Coomes 2004). Although they serve multiple purposes, homegardens are usually women's domain, and they are always crowded with medicinal herbs, foods, and especially edible plants that serve double duty as medicinals. In a homegarden survey in Pernambuco, Brazil, for example, numerous fruit trees were cultivated for food, but nearly all were employed also for their medicinal properties (Albuquerque et al. 2005). Similarly in Bahia, Brazil, a medicinal plant census of a Afro-Brazilian community revealed that in spite of abundant nearby old-growth rainforest, nearly 84% of the medicinal species were collected from predominantly "women's space," that is, homegardens and associated trails, and many doubled as food and medicine (Santana et al. 2016). In a survey of 31 households in Oaxaca, Mexico, 100 food plants and 81 medicinal species were recorded. But many were also multipurpose plants, that is, food-medicine-ornamental, or spice-medicinal-ornamental-ritual (Aguilar-Støen et al. 2009). The situation in tropical Asia is similar. In northeastern India, most of the medicinal species, whether wild or domesticated, grow in homegardens (Barbhuiya et al. 2009). In northern Vietnam, the healing flora is dominated by herbaceous species that grow on the forest floor, along paths, and in homegardens (Hoang 2009). And in northeastern Thailand, fully 95% of edible species cultivated in homegardens are also employed in traditional medicine (Cruz-Garcia and Struik 2015). The intimate association between food and medicine, anthropogenic space and women, is a recurring and important theme in tropical traditional communities (see chapter 5).

Women Healers

Women carry out much of their daily lives in highly humanized habitats, the same cultural landscape elements that are so conducive to the ecology of medicinal plants. Whether growing in homegardens, or clearings near the home, or the many trails that crisscross occupied space, the species that inhabit anthropogenic habitats are highly visible and accessible, are considered safe due to long-term familiarity, and are often packed with a powerful array of bioactive compounds. But women's knowledge of medicinal plants runs deeper than simple habitat association. In many tropical (and extra-tropical) landscapes, one of the most important duties of rural women is addressing the day-to-day medical maladies that crop up in their families and neighbors. In rural Brazil, for instance, older women frequently represent the primary healthcare providers for the family and the community (Voeks and Leony 2004; Wayland and Walker 2014). In the town of Ilhéus, Brazil, where I lived many years ago, there was an elderly woman on my street who was considered a community healer. Her homegarden was overrun with dozens

6.3   Dona Maria in her backyard medicinal garden, Ilhéus, Bahia, Brazil. Photo credit: Robert Voeks.

of medicinal plants, especially herbs and climbers. Whenever anyone in the community had a minor ailment, they often turned to Dona Maria, who graciously shared her knowledge and her plants. She had a medicinal tea for nearly every problem, and was always happy to pass on cuttings of her plants to neighbors (Fig. 6.3). This was historically the case in North America as well, and certainly prevails in many other regions in the tropical developing world today (Ban and Coomes 2004; Begossi et al. 2002; Kothari 2003; Schrepfer 2005, 84–85; Voeks and Nyawa 2001; Wayland and Walker 2014). If a child or adult becomes ill, someone will call on an elderly woman herbalist for the remedy, which can usually be sourced from her homegarden. They may not deploy mystical powers and psychotropically charged visions to solve medical maladies, but women are in many cases master herbalists. Indeed, in the past decade, a number of researchers in various parts of the tropical world have asked the simple and testable question—who knows and uses more medicinal plant species—men or women?[1]

Considerable research on gendered medicinal plant knowledge has been carried out in Latin America.[2] But in most cases, the results are equivocal. One culture group that has been extensively studied is the Caicaras, small-scale peasant farmers and fisher folk who reside along the coast of southern Brazil. Caicara women and men were determined to have roughly equal knowledge of medicinal plants species. However, men's knowledge was more evenly distributed among them (they knew on average roughly the same species), whereas women's knowledge was more heterogeneous or idiosyncratic (women tended to know different species than other women). A second Caicaras study discerned no differences whatsoever between the medicinal knowledge of women and men, whereas a third showed that women knew more medicinal species than their male counterparts, a separation especially pronounced regarding non-native species. Moving away from the coast, a further study in the region showed that there was no quantitative difference in medicinal knowledge between men and women, but that most of men's species inhabited forests, whereas women's were cultivated in the garden or collected in nearby open spaces (Begossi et al. 2002; Hanazaki et al. 2000; Miranda et al. 2011; Poderoso et al. 2012). Similar findings were reported among the Amazonian indigenous Fulni-ô. Men knew slightly more medicinal species than women, but this was because men could collect in the forest as well as in anthropogenic spaces, whereas women were restricted to the latter (Albuquerque et al. 2011). Among Venezuelan Mestizos, men and women knew roughly the same number of medicinal plant species (Souto and Ticktin 2012).

Results from other regions of Latin America are less ambivalent. In Nicaragua, Garifuna women are much more knowledgeable about medicinal species than men. This seems to be a function of the life history of the local healing flora, which is mostly ruderal, and thus within the domain of women (Coe and Anderson 1996). Similarly in the northeast of Brazil, women maintain a significantly larger corpus of medicinal plant knowledge than men, a cognitive divide that increases with age. And the majority of healing species are associated with habitat disturbance (Voeks 2007a). In a narrower study restricted to palm species in the Brazilian Amazon, Araújo and Lopes (2012) discovered that women know significantly more medicinal uses than men. They attributed this to sexual division of labor, in which women dominate areas of food and health care, whereas men are occupied with species used in construction and other commercial activities. Among the Peruvian Asháninka, women in nearly all age groups know significantly more medicinal plants than men, and most of these are native herbs (Luziatelli et al. 2010). And among the indigenous Rarámuri of northwestern Mexico, women know and harvest significantly more medicinal plant species than men (Camou-Guerrero et al. 2008).

Studies of gender and medicinal plant knowledge from the Asian and African tropics have likewise yielded mixed results. In Indonesian Borneo, older women are reportedly more knowledgeable about medicinal plants than men. In this case, women's knowledge is confined to ruderal and garden areas, whereas men's knowledge is concentrated in old-growth forests (Caniago and Siebert 1998). Similarly, women in Karen and Lawa communities in northern Thailand know and use significantly more medicinal plants than men (Junsongduang et al. 2014). However, elsewhere in northern Thailand, male and female immigrants from Myanmar (Tai Yai people) exhibit no differences in medicinal plant knowledge (Khuankaew et al. 2013). And in Vanuatu, men "retain a richer body of ethnomedical knowledge than women" (McCarter and Gavin 2015, 257). Results from Africa suggest that medicinal plant knowledge is generally more gendered than in other tropical regions. In southwest Ethiopia, women are the principal cultivators of medicinal plants in homegardens (Hunde et al. 2015), whereas in Madagascar, there are no reported quantitative differences in the total known corpus of medicinal plants. However, women know more than men about plants near the village and in other anthropogenic spaces, but less than men about the medicinal properties of local forests (Lyon and Hardesty 2012). In eastern Tanzania, women know more about medicinal herbs, whereas men are more knowledgeable about medicinal trees (Luoga et al. 2000). In Benin and Togo, however, neither gendered spaces nor differences in knowledge of medicinal plants exist between the sexes (Rodenburg et al. 2012). In South Africa, men list more medicinal species than women. However, this study included only woody species, which are likely part of male habitat domain (Dovie et al. 2008). Finally, Ghanian men and women maintain roughly the same quantitative knowledge of medicinal plants. However, the species named by men and women are quite different. In this case, because men and women operate in different ecological spaces, and often experience different types of illness, women tend to know species near the home and that are used for birth and delivery, fever, stomachache, neurological ailments, and worms in children. Men on the other hand are more knowledgeable about timber and farm species, and are more likely to be familiar with plant medicines for impotency, hemorrhoids, cuts, and snakebites (Dudney et al. 2015).

Overall, the suggestion that women know more medicinal plants than men in tropical landscapes is not supported by recent quantitative research (Torres-Avilez et al. 2016). There are some instances when women know and use a larger repertoire of species than men, but this is far from universal. What these and other studies do demonstrate, however, is that men and women in most rural communities have quite different occupations and social roles, and that these take them to different geographic and biological spaces. Women tend to be more knowledgeable about the biological properties of anthropogenic nature, particularly where this involves health, food, and firewood. And most medicinal species inhabit these humanized spaces. Men are generally more knowledgeable regarding forest resources than women, including the medicinal species. But unlike women, who often do not spend much time in the forest due to cultural norms, men also live in the village, walk along the trails, toil in the swidden, and relax in the homegarden. Their forested habitats are not as rich with medicinal species as "women's space," but they know them much better than women. And men are reasonably well acquainted with the medicinal plants in anthropogenic space as well, even if caring for the health of the family is not one of their primary tasks.

The issue of whether women or men sustain a greater body of medicinal plant knowledge is a legitimate question, particularly because women healers and their knowledge of medicinal nature have for so long been ignored by science. The recognition that women's quantitative medicinal knowledge overall is about the same as th ⁺ of their male counterparts—sometimes greater, sometimes less, often jus⁻ different—undermines the rather clichéd role of the mysterious male ⁻haman in the jungle medicine narrative. Contrary to films and fictional depictions, the much-touted tropical plant pharmacopeia is as much a province of informal female herbalists as it is of specialized male shamans. But this revelation is not purely of academic interest. Medicinal plants are employed by a large proportion of people in the developing world, and women's ethnobotanical knowledge is of particular importance to children's health in tropical forests. In Bolivia, for example, McDade et al. (2007) found that indigenous women (but not men) who are more knowledgeable about the useful properties of the local flora are likely to have healthier children than women who are less knowledgeable. Ethnobotanical ignorance, it seems, comes at a cost. Male shamans may get all the headlines, but the health and well-being of the community in the end may well depend on the herbal skills of mothers and aunts and grandmothers.

Gendered relations with medicinal plants reveal not only the skewed power relations between the sexes in traditional societies, but also the increasing significance of nature-society relations among immigrant populations. The stereotyped healers portrayed in the jungle medicine narrative are drawn from indigenous societies that have lived and learned about their forests and fields for the proverbial "thousands of years." But the fact is that now as in the past, people have migrated locally, regionally, and globally—whether voluntarily or by force. And these immigrant groups (diaspora) have much to inform us about the process of continuity and change in plant pharmacopoeias. The case of the descendants of enslaved Africans in Paramaribo, Suriname, is instructive. Detailed ethnobotanical inventories have shown that women maintain a pharmacopeia of close to 200 plant species for use in genital steam baths. These are very much "women's plants." Most are ruderals or inhabit second-growth forests, but women also cultivate many of them in their homegardens for ease of access. Cleanliness is an essential element in Afro-Surinamese culture, and genital steam baths are seen as a means of purifying women and of making them more appealing to their husbands. Although both men and women consider these baths to be an aphrodisiac, in fact the primary purpose is to tighten the vagina so as to give more pleasure to their husbands. This "dry sex" practice can be uncomfortable and painful, and may increase the chances for passing STDs such as HIV. But women feel compelled to perform this risky ethnomedical ritual in order to please their husbands and curb the likelihood of infidelity. The practice of genital steam baths appears to have arrived during the course of the African slave trade, although the plant species and other paraphernalia used in Suriname are mostly different than they were in Africa. Today, this custom and its associated pharmacopoeia has continued its migration with Afro-Surinamese immigrants to the Netherlands, where dry sex using South American plant species is still being practiced. This example illustrates both the degree to which gendered relations with plants are subject to power relations in society, as well as the extent to which immigrants will go to maintain traditional ethnobotanical practices (Van Andel et al. 2008; Van Andel and Westers 2010). The next chapter examines the significance of diaspora knowledge and use of medicinal plants in the tropical realm.

SEVEN

Immigrant Ethnobotany

"They use it in *macumba* [black magic]." That was the response from my Brazilian botanist friend when I asked about an inconspicuous little sedge I noticed growing on the edge of a swampy area. He was clearly disinclined to talk further about this plant or its uses. Many Brazilians do not believe in *macumba* but they "respect it" sufficiently to not publically disparage it. But I pressed. He told me that its name was *dandá* or more properly *dandá-da-costa* (*Cyperus rotundus*), and that he believed it was native to Africa. In fact, as I discovered later, every plant with "*da costa*" attached as an epithet referred to species that were believed by Brazilians of African descent to be native to their homeland, "the coast" of West Africa. Most people in the northeast of Brazil trace their ancestry to sub-Saharan Africa, but this was my first encounter with an ethnobotanical tradition that seemed to have somehow survived the horrors of the slave trade. I later learned that the tubers of *dandá* were sold in open markets in the city of Salvador (Fig. 7.1) and were placed in a small cloth amulet, called a *patuá*, which was worn around the neck to ward off evil eye (apotropaic). The botanical components of the *patuá* included a piece of *dandá* rhizome carved into the shape of a tiny *figa*, a human fist with the thumb pushed between the second and third finger; this was sandwiched between small leaves of rue (*Ruta graveolens*) and *erva guiné* (*Petiveria alliacea*) and sewn into the small cloth bag. Christian prayers on small scraps of paper may or may not be included. Evil eye is taken seriously by many people in the northeast of Brazil, and employing some form of prophylaxis against its nefarious

7.1   *Dandá* (*Cyperus rotundus*) for sale at Feira de São Joaquim, Salvador, Bahia. Photo credit: Bruna Farias de Santana.

influence is common. But neither rue nor the concept of evil eye is specifically African and, indeed, rue has been used in Mediterranean Europe since Aristotle's time to counteract its perceived effects (Pollio et al. 2008). *Dandá's* other use, however, was more clearly of West African origin. In this case, a sliver of the fresh rhizome is placed in a person's mouth, similar to a plug of chewing tobacco. This practice seems to have arrived during the slave trade from Nigeria, where the aromatic tubers of *dandá* are (or were) employed by people who seek to influence others by their magically charged words. Defendants in court cases often concealed a piece of *dandá* in their mouth as a charm to secure acquittal, and this use survived relatively intact among the Afro-Brazilian population in the northeast of Brazil (Meek 1931, 304–305; Querino 1955, 90; Voeks 1997, 112).

Over the coming year, I thought about the insignificant little weed many times. How did it get here? Did enslaved Africans somehow smuggle it into Brazil? Or did it arrive serendipitously? And how did an arcane ethnobotanical tradition manage to survive the monumental adversities of the slave trade? As it turns out, *dandá* is just one example of the avalanche of plant species that arrived and colonized distant human-ized landscapes, and that were fortuitously recognized and readopted by immigrant people. Soon after the Columbian landfall, and in rare cases many centuries before, useful plants were intentionally and acci-dentally transported to nearly every point on the globe where the en-vironment allowed for their survival (see chapter 5). And these great waves of plant introduction were followed, immediately or in the com-ing centuries, by people who had known and used these species in their distant homelands. The European version of this ethnobiological saga was outlined masterfully by historian Alfred Crosby in his *Ecological Im-perialism* (Crosby 1993). Who can dispute the immense ecological and demographic impacts that resulted from the introduction of cattle and horses to North America, maize and cassava to sub-Saharan Africa, sugar and rice to South America and the Caribbean, and alien microbes almost everywhere? But these grand imperial acclimation efforts with mostly commercially valued biota tell only part of the story.

The jungle medicine narrative, indeed ethnobotanical inquiry in gen-eral, is built largely on a conceptual foundation of cultural antiquity and geographical continuity, not alien people and non-native plants. Reference is made often in the literature to the "thousands of years" over which biocultural wisdom had accumulated. And whether stated or implied, mature plant knowledge profiles are frequently conceived of as the trial-and-error consequence of long-term residence and gradual familiarity with the floristic environment. Particularly in the tropical realm, where biological diversity is extreme and individual plant fre-quency is low, the ability of people to recognize, label and categorize, and especially assimilate the material and spiritual values of individ-ual plant species is assumed to be a glacially slow process. Accordingly, native people are portrayed in the jungle medicine narrative as living repositories of ancient plant wisdom handed down as sacred oral text from generation to generation. Diaspora communities,[1] on the contrary, are more likely to be presented in environmental narratives as part of the problem rather than the solution. Their relatively recent arrival in a protean landscape is seen as inconsistent with the acquisition of signifi-cant ethnobotanical knowledge. And their alien worldviews and modes of subsistence are perceived as threats to the harmonious ecological bal-ance developed over time by indigenous societies. But this conceptual-ization of the people-plant interface implies a degree of sedentism in human and plant populations that is radically inconsistent with the geo-graphical and ecological record. Does ethnobotanical expertise depend on long-term residence in a particular region, that is, is it space-and-time contingent? And are mobility and migration by individuals, communities, or even entire culture groups inconsistent with retention and reformulation of ethnobotanical traditions? The answers to many of these questions are revealed by examining the consequences of the largest forced migration in Earth's history—the African slave trade.[2]

Candomblé Medicine

From the sixteenth to the mid-nineteenth century, nearly 12.5 million sub-Saharan Africans were abducted, sold into slavery, and transported to North, Middle, and South America. It was "the largest, long-distance coerced migration in history" (Eltis and Richardson 1997, 16). Brazil alone witnessed the arrival of nearly six million enslaved people between 1538 and 1851, more than any other colony or country. The Caribbean Islands—British, Dutch, French, and Spanish—likewise received huge numbers of African men and women, nearly equal in number to those arriving in Brazil. British North America, by contrast, witnessed the arrival of only 305 thousand Africans (Fig. 7.2) (Eltis and Richardson 2010).

Regarding the diffusion of plants and plant traditions, three features of the African slave trade stand out. First, over its entire duration, the vast majority of enslaved people, probably over 90%, were forcibly removed from one tropical landscape, especially rainforest, seasonal forest, savannas, and wetlands, and relocated to another, mostly in Latin America. The slave trade was in large measure a tropical-to-tropical enterprise. And for better or worse, arriving Africans found considerable that was familiar in the New World in terms of climate, soils, and vegetation structure (physiognomy). Second, given the tremendous volume of people transported across the Atlantic, especially from Senegal and the Gambia, the Bight of Benin, the Bight of Biafra, West Central Africa, and Mozambique, the American coastal frontier came to be dominated numerically by Africans and their descendants. The tropical Americas may have become in many regards an extension of European civilization, but the indelible imprint of African culture and lifeways is abundantly evident nearly everywhere. And third, the brutality of the infamous Middle Passage from Africa to the Americas, and the dehumanizing living conditions endured by chattel slaves, obviously mark one of the darkest pages in history. In colonial and later imperial Brazil, Africans suffered tremendous demographic losses due to overwork, malnourishment, disease,

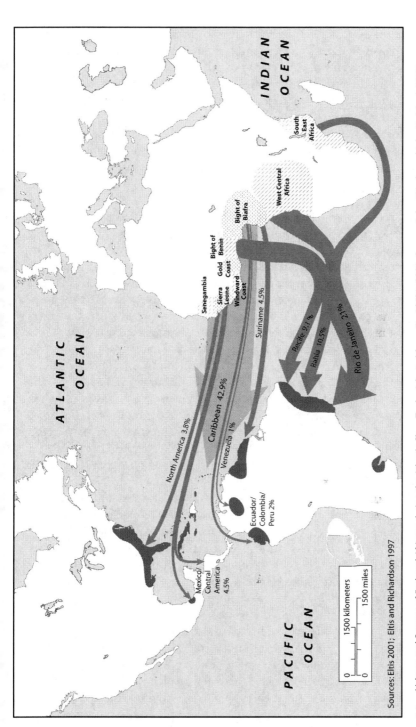

ATLANTIC OCEAN

PACIFIC OCEAN

INDIAN OCEAN

South East Africa

West Central Africa

Bight of Biafra

Senegambia

Bight of Benin

Gold Coast

Sierra Leone

Windward Coast

North America 3.8%

Caribbean 42.9%

Venezuela 1%

Suriname 4.5%

Recife 9.1%

Bahia 10.5%

Rio de Janeiro 21%

Ecuador/ Colombia/ Peru 2%

Mexico/ Central America 4.5%

0   1500 kilometers

0   1500 miles

Sources: Eltis 2001; Eltis and Richardson 1997

7.2   Volume and Sources of Enslaved Africans during the Atlantic Slave Trade. Adapted from: David Eltis 2001. Source: Robert Voeks and John Rashford. 2013. *African Ethnobotany in the Americas*. Image: Kelly Donovan.

suicide, and colonial policies that discouraged family units. The average slave survived a brief seven years in captivity, and his or her life expectancy at birth was a mere 23 years. Replacement level fertility was seldom achieved, creating a steady demand for new slave shipments. Thus, for instance, even after nearly three centuries of the slave trade, the population of Rio de Janeiro in 1823 was still 73% African born (Schwartz 1992, 41–42; Thornton 1992, 162–182). Because slave owners viewed the importation of "fresh hands" as economically preferable to the high cost of maintaining a demographically healthy population, Brazil's captive labor force was numerically dominated by people whose traditions were shaped by their African heritage. Each new shipload of enslaved people reintroduced African practices and beliefs, a cultural "rescue effect" that continuously reinvigorated ancestral traditions and values among the enslaved and free black population. In spite of monumental obstacles, Africans added mightily to what would emerge over time as Brazil's unique cultural identity—from language to foodways, and music to dance (Carney and Voeks 2003). And nowhere were the consequences of these contributions more strikingly evident than the emergence of the Candomblé belief system and its associated plant traditions.

Candomblé constitutes a set of beliefs and practices introduced in large measure by Yoruba slaves and freedmen, negotiated and blended over time with Catholicism and with other African-derived traditions. Engenho Velho, the first Candomblé temple (*terreiro*), was clearly in existence by 1830, and some accounts trace its origin to the mid-1700s. Candomblé is separated into several denominations, known as nations (*nações*), including Candomblé de Ketu, Candomblé de Angola, Candomblé de Jeje, Candomblé de Congo, and others (Carneiro 1948; Prandi 2002). Each nation, to some degree, maintains its own unique lexicon, chants, ceremonies, spiritual entities, and offerings. Each also sustains its own medicinal, spiritual, and magical crop plants and wild species. Religious activities are directed by the *mãe* or *pai-de-santo* (also known by their Yoruba labels *ialorixá* or *babalorixá*), which translates literally to mother- or father-of-saints. She (or less commonly he) represents the principal line of communication between the world of mortals (*aiê*) and the world of the deities (*orun*). In addition to taking responsibility for temple functions, administrative and spiritual, the *mãe-de-santo* serves as temple shaman-healer, divining the source of medical, magical, and spiritual problems and prescribing culturally relevant remedies. These remedies, more often than not, are complemented by an array of plant-based recipes drawn from a pharmacopoeia of sacred foods and medicinal leaves (Voeks 1993). The *mãe-de-santo* can treat everyday illness episodes, such as infections and body aches and diarrhea, but she is more often called upon to address spiritually and magically derived health problems. Spiritual maladies are believed to result from an imbalance between the afflicted and one or more of the spiritual entities, and are treated (in part) with a combination of plant species. Trained as well in sorcery, the *mãe-de-santo* can also negate the effects of black magic and, if necessary, draw on the occult arts for her own ends or for those of her clients (Brazeal 2007).

Candomblé adherents recognize the existence of a supreme god, Olórun, the unknowable creator of all things, but he is perceived as distant and unapproachable by humans. It is the *orixás* of the Yoruba pantheon, serving as the earthly ambassadors of the high god Olórun, who are directly linked to the daily lives of mortals. The principal spiritual entities include Xangô, Ogun, Oxalá, Oxóssi, Omolu, Ossâim, Iroko, Yemanjá, Oxum, Iansã, Nanã, Oxumarê, and Exú, although some *terreiros* sustain a much larger pantheon. Each *orixá* in turn is associated with a distinct province of the natural world—water, air, plants, animals, and earth—and it is from these primordial reservoirs that each harvests and imparts his or her *axé*, or vital energy. *Axé* is the fulcrum upon which the success of the Candomblé temple depends. Nurtured and properly tended, it grows like a sacred flame, imparting spiritual health and good fortune to its human attendants. *Axé* emanates from various sources, but especially from the blood of sacrificial animals, such as goats, chickens, and doves, and the "blood" of plants, that is, the white latex, the oil, and clear liquids emanating from their leaves. The fundamental goal of Candomblé worship is to cultivate and sustain a state of spiritual equilibrium between adherents and the *orixá* "owner of his or her head," with the ultimate goal of maximizing prosperity, good health, fertility (at least in years past), and a general good life.

Ossâim is the guardian of sacred leaves and medicine. He is the spirit most intimately associated with health and healing, and his domain ranges from forests to fields to homegardens, wherever curative plants are found. According to Candomblé legend, Ossâim was once the sole custodian of the Yoruba healing flora. But this knowledge was coveted by other deities who sought to share in his secrets. The following oral text, recorded in Brazil, Nigeria, and Cuba (Cabrera 1971, 100; Verger 1981, 122–124; Voeks 1995, 2013), describes how the *orixás* usurped Ossâim's herbal power, and so came to possess their own individualized plant pharmacopoeias:

There is a legend of rivalry between Ossâim, the orixá of medicine and leaves, and Iansã, the orixá of stars, winds, and storms. Everything began as a result of jealousy.

CHAPTER SEVEN

Iansã went to visit Ossâim. Ossâim is very reserved, quiet, silent. Iansã wanted to know what he was doing. When Ossâim has the opportunity, he explains things. But Iansã is always rushed, she wants everything done immediately. She is always asking questions, and she needs to know everything that's going on. When Iansã arrived at the house of Ossâim, he was busy working with his leaves. It happens that there are certain types of work with leaves that you can't talk about, you need to remain silent. Iansã started asking, "What are you doing? Why are you doing this? Why are you doing that?" And Ossâim remained silent. "Alright, if you don't want to tell me what you're doing, then I'll make you talk." That's when Iansã began to shake her skirt and make the wind blow. The house of Ossâim is full of leaves, with all of their healing properties, and when the wind began to blow, it carried the leaves in every direction. Ossâim began to shout "Ewe O, Ewe O!" [Yoruba for—"my leaves, my leaves!"]. Ossâim then asked the help of the orixás to collect the leaves, and the orixás went about gathering them. And it happens that every leaf that an orixá collected, every species, he or she became the owner of that leaf.

Scattered in every direction by the winds of Iansã, the sacred leaves drifted into the habitats of the other spirits. Oxum, goddess of freshwater, collected leaves near her rivers; Yemanjá, deity of the oceans, collected her leaves near the shoreline. Oxalá gathered white leaves, his color preference, while Exú, the most capricious of the spirits, adopted those leaves that burned, pricked, or otherwise brought pain to the user. In this way, Ossâim was able to retain the mysterious power of the plant kingdom, for he alone comprehends the inherent meaning of each leaf. But each deity nevertheless came to possess his or her personal healing flora.

The resultant leaf-deity correspondence represents a fundamental feature of Candomblé ethnomedicine. Each plant species, at least in principle, is an element in the personal pharmacopoeia of each individual *orixá* (Voeks 1995). For human devotees who "belong" to one or another deity, healing is mediated through recourse to the inherent spiritual energy of his or her guardian spirit's leaves. A leaf bath (*abô*) for one of Oxum's followers, for example, will usually include three or seven of Oxum's species. A leaf whipping (*sacudimento*) intended to clean away negative energy from an Ogun devotee will include several of his leaves. If a client suffers from an ailment associated with another deity, such as Xangô's notorious anxiety or Oxum's obsession with material wealth, then the *mãe-de-santo* will incorporate leaves from the appropriate *orixá*. The sacred leaves of Candomblé thus represent a symbolic connection between the world of mortals and spirits.

The knowledge of how to collect, prepare, and administer liturgical and medicinal plants represents a highly salient feature of African Yoruba religious traditions, as well as those that arrived and survived during the slave trade as Brazilian Candomblé. An esteemed *mãe-de-santo* is said to be one who "knows all the leaves" (Costa Lima 1963). Plants and their products are integral components in almost every Candomblé ritual, ceremony, and celebration. And because ethnobotanical knowledge is central to the practice and ultimately to the existence of Candomblé in Brazil, the primary challenge during its formative years was biogeographical. How were enslaved and later free Africans in Brazil able to maintain plant-based traditions, and develop new ones, in an alien floristic landscape?

Botanical Conversations in the Black Atlantic

Over the course of several years, I collected and identified roughly 140 species used by Candomblé adherents for medicine and magic (Voeks 1997, 170–192). The resultant species list reveals the pharmacopoeia to be floristically rich, with 117 genera distributed in 54 families. Ninety-six of the genera, or 82%, are represented by only a single species. Although it is tempting to view this taxonomic diversity as a reflection of the region's hyper species richness, such a conclusion misses the mark entirely. The Atlantic coastal rainforests are dominated floristically and structurally by trees and treelets, but arborescence does not at all characterize the Candomblé healing flora. Trees and treelets make up less than 14% and 11%, respectively, of the species. Many of these sacred plants (41%) are exotic rather than native, and several of the native trees are domesticated or otherwise managed. Less than 10% of the Candomblé flora is collected in old-growth rainforest, and several species[3] are more abundant in secondary than in old-growth forests. Roughly 43% of the plants identified by Candomblé *mães-de-santo* are cultivated or otherwise spared or encouraged, and many of these, such as achiote (*urucum—Bixa orellana*), soursop (*jaca-de-pobre—Annona muricata*), and southern elder (*sabugueiro—Sambucus australis*), are domesticated species that depend on humans for their existence. Others exist naturally outside of cultivation, usually in highly disturbed habitats, including castor bean (*mamoeira—Ricinus communis*), broadleaf plantain (*transagem—Plantago major*), and devil's trumpet (*corneta—Datura metel*), but are often purposely planted in homegardens as ornamentals or for medicinal use. Still others, such as Madagascar periwinkle (*bom dia—Catharanthus roseus*) and coffeeweed (*fedegoso—Senna occidentalis*), occur spontaneously as ruderals or garden weeds, but are spared because of their cultural value.

Many of these plants are not native to Brazil. Of the 122 species for which origins are known or suspected, 25% are of Old World origin. A few of these, such as kola nut (*obí-da-costa—Cola acuminata*), were purposely introduced from West Africa and naturalized for use in Candomblé worship. Kola use by West Africans reaches deep into their history, representing a trans-Saharan, camel caravan trade item from at least the twelfth century (Niane 1984, 615–634). Willem Bosman noted from eighteenth-century Guinea that Africans sucked the bitter juice out of kola nuts, sort of an "African Beetel" in his view (Bosman 1721, 462). Kola was prized by West Africans in the early nineteenth century, and this demand stretched across the Atlantic to the New World African slave and free population. Botanist Friedrich Welwitsch reported from Angola in the mid-1800s that the nuts represented a lucrative item of export to South America, being "much sought out by the slaves there imported from Africa" (Welwitsch 1955 [1862], 320). Kola export was also pursued by ex-slaves who had purchased their own freedom, such as José Francisco dos Santos. Repatriated to Whyda, Dahomey (current Republic of Benin), from Brazil, he carried on a successful business from the 1840s to the 1870s shipping slaves and kola nuts to Bahia (Verger 1952). Some of the seeds must have been planted during the latter part of the nineteenth century, for by the early twentieth century there were several recorded populations in the northeast of Brazil (Voeks 1997, 26–27). Today, as a symbol of its enduring trans-Atlantic cultural connection, kola is cultivated widely in Brazil specifically for use in Candomblé initiation, divination, and healing rituals.

But intentional transport and naturalization of useful plants by Africans was the exception rather than the rule. In spite of personal narratives of the enslaved secreting seeds of useful plants during their transatlantic journey, kidnapped Africans headed for the horrors of the Portuguese slave ships, commonly known as *tumbeiros*, or "tomb ships," were not likely conspiring for their future in the Americas. It is certainly plausible, that with millions of enslaved people moving from east to west, that there were some seeds of useful species carried and transplanted, if not by the enslaved, then perhaps by "black jacks," the free African seamen who manned the ships. But the fact is, at least in terms of the flora Africans would encounter in the tropical Americas, there was little necessity, for whoever survived the Atlantic crossing would have been greeted in Brazil and other colonies and possessions with a considerable list of Africa's most useful species, including those often used in medicine and magic. It would be no Garden of Eden. But as Africans were confronted with a bewildering array of new and unknown plants, they did what all newcomers do; they exercised the "principle of attachment," that is, they sought out and found solace in what seemed familiar (Jahoda 1999, 9–12; Pagden 1993, 47–48). And in the increasingly humanized spaces of the American tropics, they encountered an abundance of familiar plant life, including crops, spices, ornamentals, and weeds.

By most measures of botanical similarity, Africa and South America have little in common. So taxonomically different are the two continental floras and so impoverished is Africa's flora compared to that of South America, that Africa was early termed "the odd man out" (Richards 1973, 21–26). The combined effects of continental drift, 100 million years of geographical isolation, pronounced climatic oscillations, and taxonomic divergence add up to a relatively minor shared floristic ancestry between these distant biomes.[4] Indeed, in terms of "native" plant species, newly arrived Africans would have encountered scarce opportunities to reconstitute their culinary and healing traditions in Brazil. One of the few was beach morning glory (*salsa-da-praia*—*Ipomoea pes-caprae*), whose notoriously effective flotation device accounts for its arrival in South America probably well before the European colonization effort. Known in some Candomblé *terreiros* by the Yoruba term *aboro aibá*, this attractive and widely dispersed beach colonizer was probably the first species that enslaved Africans recognized as they were herded off of the slave ships (Fig. 7.3). Today, beach morning glory is used similarly in West Africa and Bahia in spiritual baths dedicated to Nanã, the elderly Yoruba female divinity of rain, soil, and mud (Verger 1995; Voeks 1997).

African forced immigrants encountered few familiar native species in Brazil, but there is considerable similarity between Africa and Brazil at other taxonomic ranks. At the level of plant genus, the rank most salient in terms of human cognition (Berlin 1992, 52–78), and certainly at the family level, eastern Brazil and western Africa would not have seemed so very different. Some of the most ubiquitous families in Brazil's Atlantic forests are common components of West Africa's flora, for example, myrtle (Myrtaceae), sapodilla (Sapotaceae), spurge (Euphorbiaceae), laurel (Lauraceae), melastomes (Melastomataceae), and mulberry (Moraceae), to mention only a few. There are, moreover, roughly 110 genera common to tropical Africa and South America (Renner 2004), and the species in these pan-Pacific genera provided a wealth of possible candidates for substitution.[5] A striking example is provided by the genus *Bauhinia*, a small leguminous tree. In Nigeria, Yoruba priests employed *abafé* (*Bauhinia thonningii*) in magical baths. In Bahia, lacking the same species, their Candomblé counterparts substituted the morphologically similar *Bauhinia ovata*, which is also known to this day by its Yoruba name *abafé*, and which is similarly employed in healing baths (Fig. 7.4).

Crop plants also play a prominent role in Candomblé liturgy and healing. Fortuitously, by the time of the peak of slave arrivals, beginning in the eighteenth century, newly arrived Africans would have encountered a wide array of their traditional food plants being cultivated in Brazil. Whether for profit, nostalgia, curiosity, or alleviation of food shortages, people have been transferring edible plants from one place to another since well before recorded history (see chapter 5). As a result, many of the most common food plants originating in tropical Asia, Africa, Oceania, and America had navigated the globe well before state-supported imperial enterprises. For example, China witnessed the arrival of the American peanut (groundnut, *Arachis hypogaea*) by 1540 and the sweet potato (*Ipomoea batatas*) by 1563 (Ho 1998; Pires 1998). New World maize was being cultivated in the Chinese interior in 1555, suggesting its arrival several decades earlier. Jesuit Michal Boym's mid-seventeenth-century illustrations of New World papaya (*Carica papaya*), cashew (*Anacardium occidentale*), pineapple (*Ananas comosus*), and avocado (*Persea americana*) being cultivated in Chinese gardens hint at the antiquity of East-West exchanges (Shaw 1992, 78–81). American guava (*Psidium guajava*) was growing in India by 1673 and in Kampuchea by 1676. Brazilian pineapple was cultivated in India by 1564, and was introduced by the sixteenth century to West Africa for comestible as well as medicinal reasons, as they were widely known to be good "for them that are troubled with the stone" (Achaya 1998; Knivet 1625 [1591], 1310). Michel Adanson working in present-day Senegal noted in 1750 that on dry ground you can see "guavas, acajous [cashew], two sorts of paw paws [papayas], with orange and citron trees of exquisite beauty . . . [and] the roots of manioc [cassava, *Manihot esculenta*]," all of which are East Asian or American domesticates (Adanson 1759, 165). In the early seventeenth century, Dominican missionary João dos Santos observed from present-day Mozambique that "there are lots of pineapples, excellent as the ones in Brazil" (J. dos Santos 1609, 9). American prickly pear (*Opuntia* sp.), papaya, guava, soursop, and cashew were all present in West Africa before 1650 (Alpern 2008). Cassava from South America

A

7.3   A. Albert Eckhout. 1641. "Homen Africano" (African Man), oil on canvas, from Pernambuco, Brazil, housed in the National Museum of Denmark. Public Domain. Note beach morning glory (*Ipomoea pes-caprae*) growing at the base of the palm. B. Beach morning glory on the beach near Salvador, Bahia, the first plant that arriving slaves would have encountered. Photo credit: Bruna Farias de Santana.

B

7.3   *continued*

was cultivated in the Congo Basin by 1558, and maize (*Zea mays*) was recorded as early as 1540 growing west of Senegal, on the Cape Verde Islands (Camargo 2005; McCann 2005). In the Caribbean, fruits of Asian origin were so thoroughly acclimated by the late sixteenth century that Jesuit missionary José de Acosta could report "whole woods and forests of orange trees" (Acosta 1970 [1604], 265). African sesame (*Sesamum indicum*) and okra (*Abelmoschus esculentus*) were being cultivated in Dutch Suriname by the late 1600s, and within a few decades, over forty Old World tropical food plants were being grown in Suriname, including the provision gardens of enslaved Africans (Van Andel 2015). Southeast Asian banana and plantain (*Musa* spp.) were so common in eighteenth-century Suriname that John Stedman described them as "the Bread of this Country" (Stedman 1988 [1790], 296–297). By the 1550s, Padre Nóbrega could boast from Salvador, Brazil, that several Mediterranean fruit species were acclimatizing well (Cartas do Brasil 1886 [1549–1560], 69). From present-day São Paulo, Brazil, Padre Joseph de Anchieta noted in 1585 that "from Africa there are lots of melons and beans that are better than those from Portugal" (Cartas Jesuíticas 1933). Clearly, by the end of the seventeenth century, wherever one journeyed in the equatorial latitudes, and certainly between tropical and subtropical Africa and the Americas, you were bound to encounter a wealth of familiar edible plants, many of which served double duty as food and medicine (Carney and Rosomoff 2011; Gallagher 2016) (Fig. 7.5).

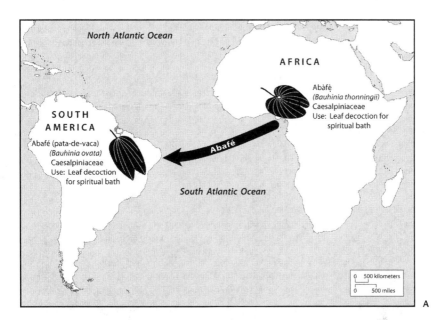

7.4   A. Abafé in West Africa (*Bauhinia thonningii*) and eastern Brazil (*B. ovata*). B. *Bauhinia ovata* growing in Bahia, Brazil. Source: Robert Voeks and John Rashford. 2013. *African Ethnobotany in the Americas.* Image: Kelly Donovan.

This ebb and flow of esculents across the vast Atlantic Ocean encouraged and facilitated cultural continuity and reformulation of foodways and ethnobotanical traditions among Brazil's black diaspora. Some species were purposeful transfers, such as *dendê* (African oil palm—*Elaeis guineensis*), introduced by the Portuguese to colonial Brazil early for commercial purposes. Captain William Dampier recorded its presence in Bahia, Brazil, in 1699, and noted that it had retained its Yoruba name: "Palm-Berries (called here Dendees) grow plentifully about Bahia. . . . These are the same kind of nuts or berries as they make the palm-oyl with on the coast of Guinea, where they abound" (Dampier 1703, 3: 71). Today *dendê* palms represent crucial elements in many Candomblé ceremonies and rituals, as well as overall Afro-Brazilian identity (Watkins and Voeks 2016). The shredded and woven inner fronds of the palm, known by the Yoruba term *mariuô*, are hung from the windows and main entrance of Candomblé temples as a line of defense against negative energies and *eguns*, ancestral spirits of the dead. The rich orange palm oil, sometimes referred to as "African blood," is used liberally in animal sacrifices and offerings (*ebó*) to *orixá* Exú. It is also used as cooking oil or an ingredient in many consecrated Afro-Brazilian dishes, including *omalá, acaçá, angu, vatapá, acarajé*, and others (Lody 1992, 10).

Some native African food plants arrived in the Americas as provisions on the slave ships and were transplanted, either by Africans or Europeans, to the kitchen gardens of the enslaved and freedmen (Carney and Rosomoff 2011). One of the signature species, now considered a sacred plant among followers of Afro-Brazilian religions, was okra, a native of West Africa.[6] Dutch physician Willem Piso reported from Brazil in the early 1600s that *quigombo* [origin of the term gumbo] had been brought from Africa by slaves, and that "the Africans taught the indigenous Americans how to use and prepare them" (Piso 1948 [1648], 441–445). Known today in Brazil as *quiabo*, a Bantu term, okra is a consecrated food for *orixás* Xangô, male warrior deity of lightning and thunder, and Iansã, hot-tempered, female *orixá* of wind and tempests. It is served freely to the public on December 5th as *caruru*, a stew consisting of okra, *dendê*, and shrimp (Lordelo and Marques 2010). *Caruru* is also associated with the celebration of Ibeji, the mythical Yoruba twins, and in many cities is served as a religious obligation. The ingredients of Afro-Brazilian *caruru* are mostly Old World in origin, and its preparation is associated with West African spiritual healing ceremonies. But *caruru* in fact originated as an indigenous South American stew made with cultivated amaranth (*Amaranthus* sp.), a New World grain, cassava (*Manihot esculenta*), and various other weedy greens. Several of these edible amaranth species,

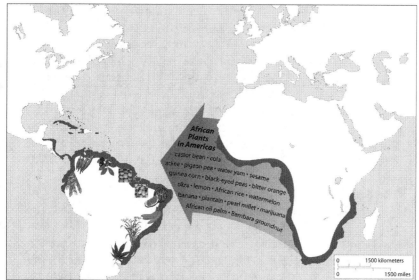

A

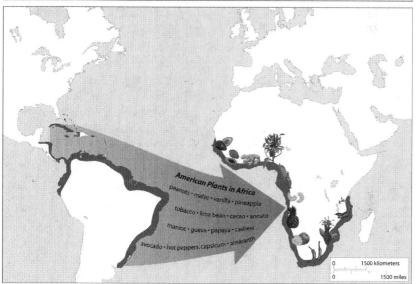

B

7.5 A. Early colonial diffusion of African crop plants to Latin America and the Caribbean. Sources: Acosta 1970 [1604]; Cardim 1939 [1584]; Carney and Rosomoff 2011; Crosby 1993; Edwards 1793; Labat 1724; Monardes 1580; Cartas do Brasil 1886 [1554]; Piso 1948 [1648]; Rochefort 1681; Sloane 1707; Sousa 1971 [1587]; Stedman 1988 [1790]; Trapham 1679; Rolander 2008 [1754–56]. B. Early diffusion of crop plants to sub-Saharan Africa. Sources: Achaya 1998; Adanson 1759; Alpern 2008; Atkins 1737; Bosman 1721; Dampier 1703; Garcia da Orta 1913 [1563]; Hair et al. 1992; Poivre 1770; Purseglove 1974; Santos 1609; Wadstrom 1795. Image: Kelly Donovan.

as well as the tradition of preparing them in *caruru*, were introduced to Africa early in the slave trade by the Portuguese (Camargo 2005). Over time, the amaranth component of this introduced South American cuisine was replaced in West Africa by native okra. Much later, having entered into West African "traditional" cuisine, this Africanized *caruru* was reintroduced to Brazil by the enslaved and their descendants. Today, an indigenous American *caruru* is enjoyed throughout Amazonia by native and mestizo peoples, whereas coastal communities with sizeable Afro-Brazilian populations prepare a highly Africanized version of *caruru* as a potent cultural marker of their distant homeland (Katz et al. 2012; T. Pires 1967 [1512–1515]; Voeks 1997).

The degree to which these edible plants constituted an ethnobotanical "conversation" between Africa and the Americas is nicely illustrated by the American peanut. Originally domesticated in present-day Argentina, peanuts were transplanted and naturalized by the Portuguese in West Africa by 1660, and possibly earlier (Alpern 2008). They spread rapidly from farmer to farmer, probably because of their similarity to the native Bambara groundnut (*Vigna subterranea*), a domesticated but less productive African groundnut (A. Smith 2002, 9–13). Over time, peanuts became an important staple crop and medicinal species in Africa, as evidenced by the frequency of their appearance in contemporary African pharmacopoeias (Dalziel 1948; Verger 1995). Later, peanuts became important provisions on the trans-Atlantic slave ships. As Hans Sloane witnessed from Jamaica in 1707, the conditions of a shipload of slaves arriving from Africa "was very nasty," but he was assured that the "Negros feed on Pindals, or Indian-Earth Nuts, a sort of Pea or Bean producing its pods underground" (Sloane 1707, lxiii). Peanuts were transferred to African-American homegardens along America's Atlantic seaboard. Among these was the Virginia variety, which had earlier journeyed to West Africa via Mexico and the Antilles, but was later (re)introduced in the seventeenth century to the Americas via the West Indies. As a result, having journeyed to and from West Africa, peanuts—as well as the aphrodisiac peanut beetle that infests them, *Ulomoides dermestoides*—over time came to be identified as an iconic food and sexual philter for Africans and their American descendants, one that from their perspective was native to both coasts of the Atlantic (Costa-Neto 1999; Voeks 2009).

The intentional transfer of culturally and medicinally valuable foods to the Americas was accompanied by the accidental and largely unreported appearance of weedy herbs and shrubs. Arriving as stowaways on the thousands of ships plying the tropical sea lanes, opportunistic plants, some of which became invasive species, silently colonized the in-

creasingly humanized landscapes of the New World. Merchants and settlers mostly failed to notice these weedy taxa, with Palma Christi (*Ricinus
communis*) being a notable exception (see chapter 5). Many arrived very
early in the colonial period, and these were tested by locals, who were
possibly instructed in their uses by recent immigrants who were familiar
with them in their homelands. But by whatever means, a steady stream
of newly arriving exotic species were incorporated into local pharmacopoeias (Pfeiffer and Voeks 2008.). One was the perennial herb *Bryophyllum pinnatum*, which doubles as a semi-invasive weed and as a valuable
medicinal and magical plant. Native to southern Africa but now cosmopolitan in distribution (Mort et al. 2001), it reproduces by sprouting
roots and tiny buds from the serrations of its leaf margins (viviparously),
hence its English names "everlife" or "neverdie." Less aggressive than
other invasive species, it represents an environmental problem only in
warm and wet climates, especially tropical islands. Most importantly,
everlife is used medicinally almost everywhere it occurs, its geography
mirroring its nearly universal utility (Kamboj and Saluja 2009). Its fleshy
and succulent leaves may well have been viewed as "signatures" for the
species' perceived cooling properties. In locations as distant as Ghana,
India, Peru, and Cuba, people press the thick leaves directly on their
forehead to alleviate the pain of migraine headaches (pers. comm. Moret
2006; Schultes and Raffauf 1990; Wezel and Bender 2003). My neighbor
Dona Maria in Ilhéus, Brazil, often placed the leaf of the plant on her
forehead for this purpose, wearing a scarf to keep it in place. Years later,
I encountered everlife being cultivated in a homegarden in interior Borneo (Brunei Darussalam). I asked the Dusun farmer about it, and he said
its name was *dingan-dingan*, meaning cool-cool, and he also used it to
relieve headache symptoms (Voeks and Nyawa 2001). In Brazil, arriving
West Africans must have readily recognized the plant, and to this day it
serves similar purposes on opposite sides of the Atlantic. In Brazil everlife is known as *folha-da-costa*, literally "leaf of the coast" of Africa, as
well as by its Yoruba name, *ewe dudu* (Verger 1976–77).

Whether intentional or serendipitous, the global movement of crops,
weeds, and even ornamentals that transpired during the early modern
era led to a wholesale reorganization and dramatic botanical enrichment
of the increasingly humanized tropical latitudes. Primary rainforest and
seasonal forest habitats that were distant from centers of human activities, then as now, sustained high numbers of endemic trees, lianas, epiphytes, and shrubs that were known in most cases only by Amerindian
peoples. But these were not the vegetation associations that drew the
ethnobotanical efforts of arriving Africans in Brazil. Rather, it was the

humanized spaces occupied and created by people—swiddens, fallows, trails, and homegardens—which over time were overflowing with many of the same useful plants, regardless of the continent of their provenance. The increasing globalization of tropical ethnofloras was noted or at least suspected early by colonial physicians and naturalists. From Goa, India, for instance, the Portuguese doctor Garcia de Orta concluded in the mid-sixteenth century, "the [botanical] medicines were never better known than at present, especially by the Portuguese . . . as a result of having transplanted them from one land to another" (Orta 1913 [1563], 121). Sir Hans Sloane, who both was well-traveled and had purchased or otherwise acquired botanical collections from Asia, Africa, Latin America, and elsewhere, also supported this view. In the Preface to his 1707 travel and natural history volume, he disputes the notion that the various tropical and temperate climates produce vastly different plant species. From his personal observations and those of others, he reported, "I find a great many plants common to Spain, Portugal and Jamaica and the East Indies and most of all Jamaica and Guinea [West Africa]" (Sloane 1707, preface). Likewise English pharmacist James Petiver writes of medicinal plants he received from John Smyth, who was Minister of the Royal African Company in Guinea (West Africa). Many of these are healing weeds, such as *attrummaphoe* (unknown), that he says grows and is used medicinally in Africa and India, *cuttofoe* (unknown), a West African species "which you so plentifully find in Jamaica," and *issong* (a legume), at the time known to inhabit Jamaica, West Africa, and India (Petiver 1697, 681–684). And the great Carl Linnaeus, whose formidable familiarity with the world's tropical plants came from his "apostles"[7] rather than personal experience, "tended to regard the tropical floras of the world as rather uniform" (Stearn 1988, 781). These comments from the specialists of the era may seem counterintuitive to us today, given our knowledge of the extreme biodiversity of moist tropical climes. But like all travelers, whether scientists or tourists, they were drawn to comment first on what they recognized, in this case the crops and weeds and other useful plants they had encountered elsewhere in their tropical travels.

Returning to the list of Candomblé spiritual and medicinal species, although many are of Old World origin, the majority are in fact originally from South America. Yet for religious practitioners, many are considered components of their African botanical legacy, and several have retained their West African names in Brazil. This curious circumstance—perceived African species that are in fact native to the Americas—resulted from the asynchronous transatlantic dispersal of plants and people. As discussed earlier, Europeans transferred many American species to Africa and Asia within a few short years of colonization, and most of these crops and weeds arrived during the first two centuries of exploration and colonization, that is, the sixteenth and seventeenth centuries. The bulk of African forced immigrants, however, did not cross the Atlantic until much later. Thus, of the nearly six million Africans brought to Brazil between 1538 and 1851, 80% (over 4.6 million) arrived after 1700 (Eltis and Richardson 2010). Consequently, during the first two centuries after American crops and weeds became established and naturalized in sub-Saharan Africa, African people must have come to perceive many as native species, which from a local perspective they were. And over time these American species were inculcated into "traditional African" cultivation regimes and ethnomedical practices. All of this would have occurred well before the height of the Atlantic slave traffic. Much later, during the peak of the slave trade and many generations after American plants had become established in Africa's humanized spaces, enslaved men and women arriving in the Americas would have encountered foods and medicinal species they considered to be their own. *Pai-de-santo* Vicente, one of my early Candomblé collaborators, was clear on this point. On several occasions, he noted that one or another spiritual species "*é nossa*" (it is ours), meaning originally from his African homeland. Consequently, whereas many of the species used in Candomblé worship are New World in origin, it is quite likely that they had become useful "African" species centuries before. New World maize, for example, was assimilated into Yoruba recipes in Nigeria for good luck, wealth, and healthy births (Verger 1995, 41–46). And pantropical weeds of American origin, such as *erva guiné* (*Petiveria alliacea*) and *alfavaquinha-de-cobra* (*Peperomia pellucida*), are now employed for similar medicinal and magical ends in West Africa, Brazil, and Cuba (Barros 1983; Moret 2013a and b; Voeks 1997).

Followers of the Yoruba-derived Candomblé belief system in Brazil developed a botanically rich and nuanced medical and magical system based on native and non-native species. Many of these plants had traversed and colonized the anthropogenic landscapes of the Atlantic world well before the height of the slave trade, encouraging continuity of species and associated medicinal and spiritual traditions. But this wasn't the end of the story. Many Candomblé plant species are in fact native to Brazil, are still endemic to South America, and lack a similar analog in Africa. These plants must have been completely new to African immigrants, but were nevertheless somehow assimilated after arriving in Brazil. One answer to this puzzle is that Africans expanded their pharmacopoeia by learning from Brazil's native Tupinambá as well as from their Portuguese captors. Indeed, in spite of the cultural divisions and language barriers that separated Indians and Africans in the early years of colonization, their collective needs and mutual misery surely encouraged some degree of sharing and exchange of medicinal knowledge. During the first century at least of the slave trade, both Indians and Africans lived and worked as captive laborers, were housed in the same squalid shacks, and toiled side by side in the forests and cane fields, sometimes forming family units and *zambos* (African-Amerindian children) (Schwartz 1992, 111). Opportunities for the exchange and hybridization of plant knowledge, at least in the early years of captivity, certainly existed.

However, a more common conduit for the early colonial assimilation and hybridization of the various Amerindian, African, and European medicinal plant traditions was the strategic actions of the newly founded Society of Jesus. The Jesuits were present in Brazil and numerous colonial outposts in "the Orient" from the very earliest days of exploration and settlement. More than any other political or ministerial entity, they were in intimate contact with the indigenous populations as they sought to bring them into the Christian fold and to block their enslavement on the sugar plantations. In addition to their ecclesiastical mission, the Jesuits were scholars, representing the de facto botanists, linguists, geographers, ethnographers, and even astronomers of the era. Most importantly, they engaged in an active and near global campaign to absorb and circulate the medicinal knowhow of native shamans and healers. In a world with few formally trained physicians, the Jesuits were often the front line of colonial medicine in Brazil as well as the other far flung Spanish and Portuguese colonies and trading partners—the Philippines, Peru, Angola, Mozambique, China, India, and many others (Anagnostou 2007; Maia 2012; Walker 2013).

The Jesuits were responsible for the discovery of many of the tropical world's most widely used medicinal plants (see chapter 4). And they were enthusiastic participants in the floristic homogenization that took place throughout the early modern era. Some of their acquired knowledge was kept secret, such as the Brazilian theriac (*triaga Brasílica*), the hugely profitable herbal panacea for poison and snakebite that the Jesuits peddled in their international chain of medicine shops (F. Santos 2009). The theriac included a host of Mediterranean plants, many of which had been naturalized in the Americas, as well as Brazilian, Asian, and African species that had been dispersed around the tropical latitudes (Pereira et al. 1996). But most of the Jesuit's medicinal knowledge entered into mainstream community healing practices, as evidenced by the widespread knowledge of Tupinambá medicinal plants among Brazilians of European and African descent. Because Africans had a lengthy tradition of employing plants to treat what ailed them in their homeland, and because they lacked access to Portuguese-trained physicians in Brazil, they naturally had a deeply vested interest in acquiring whatever knowledge and skills existed, regardless of origin, on the healing properties of nature.

A final and crucial strategy for "learning" the healing properties of Brazil's native flora was profoundly cultural in origin—the one-to-one correspondence between Candomblé plant species and Yoruba deities in the pantheon. As noted earlier, the legend of Ossâim revealed that his sacred leaves, after being blown about by the winds of Iansá, came to rest in the habitat possessed by each of the *orixás*. Consequently, as earthly embodiments of their Yoruba spirits, these plants nearly always exhibit one or more features that link them with their owner-deity. Four examples among the many are illustrative: Oxum, Ogun, Oxóssi, and Exú. Oxum is the feminine *orixá* of running water. She is a voluptuous fertility figure who is anatomically associated with the female organs and the stomach. Vain and narcissistic, Oxum adores gold, jewelry, and perfume. Reflective of her love of perfume, nearly all of her healing leaves and flowers are sweetly fragrant. These aromatic plants are added to the healing baths of her devotees for their soothing properties. The mint family (Lamiaceae), noted for its essential oils, is particularly well represented in her pharmacopoeia.[8] Oxum's remaining flora reveals her obsession with wealth, especially gold, symbolized by the bright yellow flowers of numerous species.[9] These archetypal plants contrast markedly with those possessed by the male spirits Oxóssi and Ogun, brother gods according to legend. Oxóssi holds sway over the hunt and the forest, and his preferred colors are green or red. Ogun is the god of iron and war. He is the consummate warrior figure, keeping evil at bay, opening up passages, and winning battles. His colors are green or blue. The sacred leaves of these brother deities, many of which they share, are characterized by long blades and pointed tips representing the spears and arrows of these warrior spirits. These species are typified by *peregun* (*Dracaena fragrans*), and the two bowstring hemps, *espada-de-Oxóssi* and *espada-de-Ogun* [Oxóssi's sword and Ogun's sword] (color varieties of *Sansevieria aethiopica*) (Fig. 7.6). The common names of Ogun and Oxóssi's leaves evoke their aggressive archetypes and their ability to solve the problems of their followers. These species include *abre caminho* (*Baccharis* sp.)—"open the way," *vence tudo* (*Rolandra fruticosa*)—"conquers everything," *comigo ninguém pode* (*Dieffenbachia seguine*)—"no one overpowers me,"

7.6 *Espada-de-Oxóssi* (*Sansevieria aethiopica*). Photo credit: Robert Voeks.

tira teima (*Zanthoxylum* sp.)—"take away stubbornness," and *vence demanda* (*Lepidaploa cotoneaster*)—"achieve objectives." Some of these enter into power baths, whereas others are cultivated in front of the house as prophylaxis against the negative energies of visitors and passersby.

And finally there is Exú, the deity of passageways and crossroads. Capricious and at times malicious, he serves as messenger to the *orixás*, the conveyor of spiritual energy from mortals to gods. Although Exú is a notorious troublemaker, he is also the catalyst that makes things happen. Properly appeased, Exú clears obstructions to human wants and desires; ignored, he brings calamity with a vengeance. Symbolic of his temperament, Exú's colors are crimson and black, and his associated plant species are as threatening as his personality. Many impart a burning sensation when touched. And most are employed for malevolent ends, such as dissolving relationships, bringing bad lu' ‾ and fomenting general chaos. Covered with spines and prickly pubesc:nce, several species are painful to the touch. The razor-sharp leaf margii. ; of *tiririka* (*Scleria* sp.) slice into the skin of those who brush against it, while the prickly burrs of *carrapicho* (*Bidens pilosa*) cling to the legs of passersby, finding by this means transport to trails and roads, the preferred haunts of Exú. And the awful poisoned hairs of *cansanção* (*Cnidoscolus urens*) are not soon forgotten for those unfortunate enough to touch it. Exú's leaves are as menacing as his temperament.

The flexibility of this binary system of plant-deity correspondence is based on the physical and sensory (organoleptic) features of the plant and the archetype of the deity, including his or her colors, habitat preference, or behavior. Lacking a better explanation, it is reasonable to assume that priests and priestesses of the religion employed the Doctrine of Signatures, perhaps while in trance, to select aromatic and golden-colored flowers to belong to Oxum, sword-shaped species to belong to Ogun and Oxóssi, thorny plants to belong to Exú, and so on. Thus, whether exotic or native, Brazil's wild and domesticated flora provided a range of suitable candidates for enslaved and free Africans to incorporate into their healing and magical recipes. In the case of Candomblé, time and geography presented only minor impediments to continuity and adaptation of a diaspora's plant pharmacopeia.

This last point is central to understanding the nature of immigrant ethnobotany, as well as the question of stasis in the jungle medicine narrative. Because immigrants are exposed to new cultures and healing traditions in their adopted homes, there is the distinct possibility that their traditional healing flora and associated traditions will expand in size and conceptual space over time, rather than contract. Immigrants interact with local people, and they exchange ideas about health and healing. It's an ongoing and very human process. In the case of African forced immigrants in Brazil, they developed over time a rich and highly creoled plant pharmacopoeia, some via chance, others via intentional assimilation. In so many ways, the Candomblé pharmacopoeia is a mirror image of most tropical pharmacopoeias. It is successional, domesticated, exotic, weedy, and often herbaceous, suggestive of the highly humanized spaces into which the black diaspora found themselves. And as one of the world's principal "victim diasporas," similar to the Jewish experience, Africans and their identity in the Americas were tied profoundly to the degree to which they could preserve the traditions of their various homelands (Cohen 1996). This of course included healing plants and healing traditions.

The Brazilian experience, in which Africans and Europeans lived side by side in highly humanized habitats, is not the only model for African diaspora in the Americas. The situation with enslaved Africans in Dutch Suriname was vastly different, and this is reflected in their ethnobotanical acquisition process and outcome.

Maroon Magic and Medicine

One of the prominent features of African slavery in the Americas was the flight of the enslaved from their European captors and the consequent formation of escaped slave (maroon) communities. Maroon enclaves, known variously as *quilombos, palenques, mocambos*, and *cumbes*, were common cultural landscape features of Mesoamerica, South America, and the Caribbean's Greater Antilles. Most were small and transitory, but some emerged as large and powerful political units that challenged European colonial hegemony. The maroon community at San Basilio, Colombia, for instance, survived for sixty years and eventually contained three thousand inhabitants (Landers 2002). Brazil's Palmares *quilombo* existed nearly a century and reached a population of perhaps 20,000 (Price 1996).

The largest and most enduring maroon settlements were established by escaped slaves in Suriname, formerly Dutch Guiana. After a protracted war with the Dutch military, Surinamese maroons ultimately signed peace treaties in the 1760s with the colonial authorities, becoming essentially independent African states deep in the South American rainforest. Eventually absorbed by independent Suriname during the twentieth century, these black "tribes" continue to exist today, including the Aukan, Saramacca, Aluku, Paramaka, Kwinti, and Matawai. These were not "lost African tribes in the Americas," as they are sometimes portrayed, since their occupants hailed from several African ethnic groups and they maintained varying degrees of contact with Europeans over the years (Price 1996 and 2007; Thoden van Velzen and van Wetering 2004). Moreover, although it is true that African priests and their associated traditions and beliefs arrived in Suriname, the institutional structures necessary to rigidly perpetuate their cultural roles in an alien cultural landscape did not. Nevertheless, although maroon settlements were not mirror images of African societies in the Americas, their isolation from European coastal influence meant that considerable cultural elements of African origin survived and adapted.

By fleeing to isolated and inaccessible locations, Surinamese maroons had much more contact with local Indians and their traditions than with Europeans. Indeed, with the exception of their African-derived religion, most of the material features of Surinamese maroon technology— forest farming, fishing and hunting traditions, and pottery making— were clearly assimilated from local indigenous groups during the early years of contact. Africans and their descendants had contentious and often violent confrontations with Amerindians over the centuries, and bad blood between these groups was the rule rather than exception until recent times. But there was still ample opportunity for exchange of plant knowledge and skills, as Indians and Africans on many occasions intermarried or lived side-by-side in the same community (N. Z. Davis 2016, 13; Price 2010). The ultimate consequence of this geographical isolation, intimate contact with neighboring indigenous peoples, and association with old-growth rather than humanized forests, was the acquisition of ethnobotanical knowledge and traditions that are different in many ways from those that developed among Brazil's non-maroon African diaspora.

Afro-Surinamers practice a West African-derived religion known as Winti, or "wind." Similar to Brazilian Candomblé, followers recognize a pantheon of supernatural entities that are honored by a combination of music, dance, costumes, prayers, and magical plants. Cultivating a harmonious relationship with one's ancestors and guardian spirits can bear the precious fruit of health and prosperity. But these same spirits, if not properly appeased with appropriate offerings, can deliver disease and ill fortune. During the Dutch colonial period, African dances and rituals were generally forbidden, and Winti was specifically prohibited by Surinamese law. After independence from the Netherlands in 1975, however, prohibitions against African traditions were lifted, and Winti was more openly practiced by maroons in Suriname and by their diaspora in the Netherlands. In both cases, medicinal and magical plants came to play a pivotal role in religious practice.

Over 400 species of plants enter into Winti healing and magical ceremonies. As in Candomblé, most plants are employed in healing baths (*wasi*). *Wasi* include up to fifteen different species that are mixed in water and poured over the client's head with a calabash gourd (*Crescentia cujete*). These baths serve to appease the myriad malevolent spirits that plague maroon community members, such as *papa*, a snake spirit associated with boa constrictors, *bakru*, a short and dark evil demon, and *akantasi*, a fierce spirit that inhabits termite mounds. Many other species are used as magical charms, such as to bring good luck, to avoid black magic, and to attract a partner.

The similarities and differences in the flora that Surinamese maroons and adherents to Brazilian Candomblé employ are highly instructive of the processes of immigrant ethnobotanical acquisition. In the case of Winti, selection of medicinal and liturgical species was guided by two processes. First, a spiritual entity informs a mortal practitioner which plant to use while he or she is in trance. I heard this also from several Candomblé priests, who reported sprinting into the woods at the insistence of a spirit, where the correct species were revealed by the appropriate deity. Perhaps more important, however, was the use of the "signature" of the species. Over half (210) of the documented Winti species appear to have been identified by means of the perceived healing or supernatural features associated with the plant's organoleptic qualities, that is, growth form, color, shape, or aroma. And many of these can only be understood in the context of enslaved men and women in their quest for freedom. For example, numerous species are known as one or another manifestation of *kromanti*, a general term for spirits of the sky, thunder, war, and medicinal knowledge. The name refers to a Dutch fort in distant Ghana from which multitudes of enslaved Africans were shipped to Suriname. People from Ghana were known as fierce warriors and powerful healers, and were believed capable of preparing magic that made them immune to bullets, fire, and machetes. In colonial Suriname, these warrior-magicians formed the backbone of armed resistance against the planters. Today, *kromanti* are known as flying spirits that accompanied the enslaved to Suriname. Their various names and properties reflect their bellicose past and flight to freedom. Many *kromanti* species are air plants (epiphytes) or parasitic plants, which are seèn as powerful botanical entities, descending from the sky and able to envelop and kill trees with their roots. Their life-forms are highly symbolic of the legendary power of their Ghanian ancestors. The growth form of other species led maroons to use them to manage *yorkas*, angry spirits of the dead. Ghosts of the deceased often haunt family members, particularly those that offended them in some way. *Yorkas* "blow spines on your path" to make your life more difficult. These metaphoric spines can include fevers, sexually transmitted diseases, and other medical maladies. To address *yorka* afflictions, healers prepare baths from plants with unpleasant barbs or irritating hairs to "blow the spines away" and help pacify the offended ghost (Van Andel et al. 2013, 259–260).

One of the most unique elements of the Surinamese maroon ethnoflora is the array of species that served slaves in their early escape to freedom. For instance, during their wars with the Dutch, Africans cultivated a cosmopolitan herb for its spherical, rock-hard seeds known as *weglopershagel*, or "runaway hail." These tiny seeds were coated in poisonous substances and blasted like shotgun pellets at enemy Dutch forces, aiding the runaways in their getaway. Today the seeds are used in herbal baths as magical protection against bullets and general violence. This and other "battle" species have attained almost sacred status among Surinamese maroons. Another example is *kibriwiwiri*, or hiding herb (*Psychotria ulviformis*), the most expensive magical plant sold in Surinamese markets. This creeping herb with purple-brown leaves blends in so perfectly with the forest floor litter that it is almost undetectable (Fig. 7.7). This concealing property was mobilized early by rebel slaves, who deployed it as a magical, leafy cloak to hide and flee from their owners. Today *kibriwiwiri* is still a popular plant for making the user invisible, such as cheating on your woman, slipping by your enemy, or smuggling drugs (pers. comm. Hoffman 2017; Van Andel et al. 2013).

Surinamese forced immigrants, like their Brazilian counterparts, incorporated Old World crop plants and weeds into their healing and magical ethnoflora. Among the foods were yams (*Dioscorea* sp.), pigeon peas (*Cajanus cajan*), sesame seeds, Bambara groundnut (the "peanut of the ancestors"), and African rice (*Oryza glaberrima*), all of which were and are used in ancestor worship ceremonies. The discovery of African rice in the interior of Suriname is especially significant, as it represents an early introduction to South America from West Africa, likely the present-day Ivory Coast, and is the last known site in the Americas where it is cultivated (Van Andel 2010; Van Andel et al. 2016). Many of these plants are associated with *mama aisa*, the Winti goddess of homegardens. But as important to Surinamese maroons as these crops and weeds of African provenance are, the vast majority of Winti magical and medical plants are in fact of New World origin. Most are native to Suriname, and nearly half inhabit relatively undisturbed rainforest. And unlike in Brazil, where only a tiny fraction of Candomblé plants are trees, in Suriname over 30% of their ethnoflora is arborescent. This is at considerable variance with the Candomblé pharmacopoeia, which is dominated by weedy and otherwise disturbance plant species. Why the difference? The answer may be the degree of geographical and cultural isolation experienced by Surinamese maroons after their flight from servitude.

Confronted with several thousand mostly unknown trees, shrubs, and herbs, Surinamese maroons did what all immigrants do; they exercised

7.7   The hiding plant, *kibriwiwiri* (*Psychotria ulviformis*) in central Suriname. Photo credit: Bruce Hoffman.

the principle of attachment, seeking out what was familiar in the landscape and codifing it in terms of what they knew from their homeland. As a result, many plants currently known and used by Surinamese maroons retain West African names, or reference to West Africa, especially Gabon and Angola. But they are in fact South American in origin. Thus, New World plants that were sufficiently similar to species left behind in their distant homeland, often at the rank of genus or family, were assigned African names. Unlike in Brazil, however, this substitution process included not only weedy and otherwise disturbance species, but also a large number of botanically related trees from South America's old-growth tropical forests. These plant families inhabit both sides of the Atlantic, including legumes (Fabaceae), figs (Moraceae), sapodillas (Sapotaceae), and custard apples (Annonaceae), and in a sense were preadapted to being recognized and cognitively assimilated by newly arrived Africans. Maroons recognized similar-appearing species, and applied the names that they knew to them (Van Andel et al. 2014). This process of botanical substitution of tree species was particularly pronounced among Surinamese maroons due to their considerable isolation from the European-dominated plantation zone. Unlike in the anthropogenic cane fields, kitchen gardens, and human settlements of the coastal zone, which were choked with Old World crops and weeds, Suriname's maroons lived and worked in intimate association with native primary rainforests. And they employed much the same mode of subsistence as neighboring Amerindian communities. Unlike Brazil's enslaved population, maroons were at liberty to wander at will in the forest, to study and assimilate its various useful properties, including food, fiber, fuel, craft, and medicine. Time, space, and exchange of knowledge with Amerindians rather than Europeans, provided Suriname's maroons a distinctly unique environment for botanical knowledge acquision. These features, in the end, produced a deep understanding of the magical and medicinal properties of the local flora, one that was oriented towards old-growth rather than disturbed vegetation, towards native rather than exotic species, and one that owed more to interactions with local indigenous people than to the Dutch. African forced immigrants in Brazil and Suriname both developed profound understanding of the healing properties of nature, but the legacy of maroon ꜜ in Suriname differed markedly from the story of Africans in coastal Braz'l, and their plant pharmacopoeias and cognitive understanding of nature are highly reflective of these differences.

Over the course of roughly three centuries, Surinamese maroons amassed an immense assemblage of magical and medicinal plant species. Some were exotic, many were native, and most were occupants of old-growth rainforest. But how does present-day maroon ethnomedical knowledge compare with that of local Amerindian communities, who have in principle called these forests their home for millennia? This question is at the crux of the "time contingency" question, that is, is the quality and quantity of a group's ethnobotanical knowledge a function of duration of occupancy? Like other diaspora communities, African immigrants had the motive and the means to build an arsenal of medicinal and magical plants. But how and in what ways does their ethnobotanical knowledge differ from that of longer-term residents? Fortunately, this question has been explored using a quantitative examination of the differences in plant knowledge between a Surinamese maroon community (Saramacca) and the local Amerindian group, the Trio Indians (Hoffman 2013). The working hypothesis of the study was that the longer-term residence of the Trio Indians would translate to a greater knowledge of the local biodiversity. The results are illuminating.

Plot sampling of old-growth and second-growth tropical forest was established in and around a Saramacca and a Trio village. The useful species in each plot were surveyed by three Trio and four Saramaccan males. Females could not participate due to cultural restrictions. An important feature of the study was that only trees and lianas with a diameter-at-breast-height of ten cm or greater were included; this unfortunately excluded small shrubs, epiphytes, and herbaceous plants, the sorts of plants associated with disturbance regimes. The data were later separated into use categories. The medicine category included all plants that were used for healing and ritual purposes, as well as for poison and magic. The plot data revealed that the Trio knew overall a greater percentage of plant species than the Saramaccans. The Trio named 13.4% more species, and identified the uses of 20.8% more species. But whereas the Trio knew and used more plants than Saramaccans in the old-growth forest plots, this difference disappeared in the second-growth (fallow) plots. There the quantitative knowledge of African maroons was equal to or greater than that of the Trio. The take-home point is that the Trio, who had occupied these forests for the longest period, had a greater understanding of relatively undisturbed forests than the more recently arrived African maroons. These Surinamese maroons, in turn, maintained a greater quantitative knowledge of relatively undisturbed forest than their Brazilian Candomblé counterparts, who had lived for a similar time in South America, but had less opportunity to interact with old-growth tropical forest. Ethnobotanical knowledge, at least in these cases, is both time and space contingent.

Examined in greater detail, the medicine category is seen to represent the greatest difference in ethnobotanical knowledge between the Trio and the maroons. The Trio use-value[10] for medicine was twice as high as that of the Saramaccan maroons, meaning that each indigenous participant knew on average twice as many medicinal uses for the plants encountered in the forest plots. Overall, the indigenous Trio "know more" about the local flora than the shorter-resident, African diaspora (Hoffman 2013, 361). Medicinal plant knowledge in this well documented study is indeed time contingent. And the details of this cognitive disparity are important. Maroon medicinal plant species often covered a range of generalized health problems, and there was a focus on baths and other ritualized healing methods. Among the Trio, however, most medicinal plants were employed to treat specific medical ailments, and many more were taken internally. This result speaks directly to the issue of the biochemical safety of experimenting with new medicinal plant species. Plants that are well known and consumed, such as medicinal foods and spices, or those that are applied externally, such as baths, have

a much lower likelihood of causing poisoning or other deleterious outcomes compared to species employed for their biochemical properties.
And they are easiest and most likely to add to the pharmacopoeia. Plant
medicines derived from poorly known forest plants that are taken internally (concoctions and infusions), however, likely require more time to
learn and assimilate. Perhaps they are observed being used by mammals
or birds, as many early chroniclers suggested. Or perhaps their experimentation is encouraged by the plant's similarity to an illness or ailment—
white latex from ficus suggesting a cure for poor lactation, or red resin
exudate from dragon's blood suggesting a cure for blood disease. But by
whatever means, medicinal plant acquisition is a time-dependent process.

Enslaved Africans began arriving in the Americas almost five centuries ago. They readily adopted the medicinal and ritual foods that they
recognized, either African native species, or American species that had
been introduced and adopted in Africa so long ago that they were perceived as native. They likewise recognized medicinal weeds that had
preceded them to the New World, and quickly added them to their pharmacopoeia. Over time they employed the Doctrine of Signatures to add
previously unknown New World taxa to their healing arsenal, often for
healing rituals or magical purposes. These were often applied by baths
or other topical means. But like most immigrants with a legacy of herbal
healing traditions, they were also keen to learn the pharmacological
healing properties of the local flora. And the most direct route to this
body of knowledge was through interaction with locals. Africans in
Brazil assimilated considerable indigenous medicinal plant knowledge
through the Jesuits, who in turn amassed this from the native Indians
early in the colonial enterprise. Africans also likely acquired some medicinal plant skills directly from their early experience of shared enslavement with native peoples. In Suriname, the Saramaccan maroons had a
similar learning curve, but theirs was informed less by association with
Europeans and anthropogenic space, and more with long-term interaction with Amerindians. The pharmacopoeia that the Saramaccans compiled over time was similar in many respects to that acquired by Brazilians, but it had a much more primary rainforest and arboreal life-form
focus, reflecting their differing circumstances. Both diasporas developed
a thorough knowledge of the healing properties of their tropical homes,
but it was quantitatively and qualitatively different from each other, as
well as that of indigenous residents.

The examples of the African diaspora of Suriname and Brazil are highly
suggestive of the relationship between relatively long-term residence

and ethnobotanical knowledge acquisition. In both instances, forced migrants were transported from tropical and subtropical Africa to tropical and subtropical South America. At least in terms of climate and vegetation, the opportunities for plant transference and substitution were legion. Moreover, because African forced immigrants represented victim diaspora, there were powerful impulses to recreate their botanical traditions in an alien landscape. Naturalized plants from their homeland came to represent compelling cultural markers for Brazilians, Surinamese, and many other dispossessed African people. By embracing their ethnobotanical traditions, Africans declared their separateness from their European oppressors, and in so doing reclaimed a significant measure of their cultural identity. No wonder so many of these edible, magical, and medicinal plants, which in their African homeland were simply part of their day-to-day experience, attained nearly sacred status in the Americas. The plant knowledge harbored by African forced immigrants differs considerably from that of Amerindians, but it is nevertheless sufficiently rich to merit the interests of ethnobotanists and bioprospectors, just as it did during the colonial period.[11] The exclusion of the descendants of enslaved Africans from the jungle medicine narrative speaks more to our preconceptions of Mother Africa as an intellectual backwater than to a realistic appraisal of the botanical knowledge of her displaced daughters and sons.

The cognitive contributions of indigenous and diaspora people to decoding the healing properties of tropical nature were and continue to be considerable. But what matters most to tropical communities is their ability to treat what ails their families and their neighbors, not the possible contributions of their medicinal plant knowledge to the development of Western pharmaceuticals. Indeed, whatever wonder drugs that were in the past, or might be in the future, developed as a result of ethnobotanical leads, will seldom be available to the people who shared their knowledge. In the twenty-first century, this obvious and cruel contradiction is eclipsed in the jungle medicine narrative only by the apparently declining knowledge and use of plants by tropical folk. In communities across the equatorial latitudes, the ethnomedical wisdom upon which discoveries have been made in the past is eroding, in many cases precipitously. The main sources of this downward spiral are purported to be tropical deforestation, unsustainable harvest of medicinal plants, and, most importantly, cultural decay (devolution).The latter process in particular, according to the jungle medicine narrative, has accelerated dramatically due to the forces of globalization and land use change. What are the effects of increasing formal education, religious conversion, and access to pharmaceutical drugs on traditional relations with tropical nature? Does commodification of medicinal plants undermine culturally defined plant traditions? And does deforestation and overexploitation of botanical resources in the less developed world signal the end of ethnobotany? These topics are explored in the next chapter.

EIGHT

Forgetting the Forest

Many years ago, when I was living in Brunei Darussalam (northern Borneo), I took my geography class on a short fieldtrip to the coast. My students were a mix of young Malays, the numerical majority in the country, a few Chinese, the children of immigrants from several generations past, and one young Dusun man named Narak.[1] The Malays were mostly the descendants of fishing folk, but were by this time largely employed by the state bureaucracy. The Chinese students were mostly the children of businessmen and shopkeepers. Neither group historically had an intimate connection with the biologically rich forests that cloak the interior of the island. Narak, on the other hand, was the son of rainforest hill-rice farmers, an ancient indigenous Bornean society with an enduring subsistence and spiritual dependence on the native forests and fields, and an impressive ethnobotanical repertoire. I had carried out medicinal plant research with Dusun elders in the past (Voeks and Nyawa 2001), and as we wandered along the dunes and wetlands facing the South China Sea, I was curious to discover how knowledgeable a more acculturated member of the group was regarding the coastal flora. So, as we walked along, I stopped and asked him if he knew the name or uses of a small climbing pitcher plant (*Nepenthes* spp.). "No, sir" he responded. Like my other university students, Narak was very polite. I later asked him about a beach morning glory (*Ipomoea pes-caprae*), a very common and easily identified coastal runner, and again he replied, "No, sir." Finally, there was a stand of casuarina (*Casuarina equiseti-*

folia), an oddly shaggy coniferous tree species that inhabits local sandy soils (spodosols). Receiving the same negative response, I told him I was a bit surprised that he didn't know any of these common plants. "Sir," he responded, "I don't want to know any of these plants."

This wasn't the first time I had encountered young people who seemed wholly disinterested in the ethnobotanical heritage of their elders. In Brazil, both in cities and in rural areas, I'd heard from my older field collaborators that, from their perspective at least, knowledge of useful plants was quickly dying out. The younger generation simply wasn't interested in the old ways. But this was different. Narak wasn't simply uninterested in the Dusun's plant traditions; he seemed to be making a conscious effort to forget the botanical legacy of his parents and grandparents. In his effort to escape his "primitive" ancestry, Narak wanted desperately to forget the forest.

The phenomenon of ethnobotanical erosion (devolution)[2] is a crucial element in the jungle medicine narrative. However many generations it may have required forest folk to discover, label, and culturally assimilate healing plants, there is the real possibility that much of this knowledge domain could disappear in one or two generations. This is on the one hand certainly not a novel phenomenon. Ethnobotanical decline has been commented upon by scientists for many years. At the first annual meeting of the Society for Economic Botany in 1960, Richard Evans Schultes, who probably spent more time living and working with tropical forest people than any other researcher, lamented "the progressive divorcement of primitive peoples. . . . from dependence upon their immediate environment." He challenged his audience to get to work and "salvage some of the medico-botanical lore before it shall be forever entombed with the culture that gave it birth" (Schultes 1960, 257). Schultes himself had been inspired to do the same twe⸱ 'y years earlier by his own mentor, Harvard professor Oakes Ames, who encᵒuraged him to investigate these "treasured traditions" with haste, befoɾᵉ they were entirely forgotten (Balick 2012, 6–7). In subsequent years, more and more scientists took up the call, reporting from Africa, Asia, Latin America, and Oceania that traditions were being "forgotten or ignored," "treasures of medicinal knowledge" were being lost, and the stewards of medical knowledge were dying out "often without passing on their knowledge to the young, mainly because the latter are not interested" (Muzik 1952, 251; Stopp 1963, 16; Weiss 1979, 35). But the pace of globalization is orders of magnitude greater today than in Schultes' time. Human knowledge of nature, according to many, is in freefall. And because the early discovery of pharmaceutical drug plants was clearly dependent upon the botanical knowledge of forest people (see chapter 4), the possible loss of this cognitive library in many respects surpasses in significance all the other elements of the jungle medicine narrative.

What Is Traditional Plant Knowledge?

Declining ethnobotanical knowledge has been observed and commented upon for many years. But the process was seldom documented, and for good reason. For what seems like a manageable question—is the plant knowledge sustained by indigenous and other tropical forest societies disappearing?—is in fact difficult to measure. Consider some of the challenges. First, there is the issue of what is traditional indigenous knowledge. In general, it constitutes practical understanding of the environment that is rooted in place, is transmitted orally or by imitation, is often empirical rather than theoretical, and is subject to day-to-day reinforcement. It is "time tested and wise," using "head and heart together." Most importantly, it is malleable, "constantly changing, being produced as well as reproduced, discovered as well as lost" (Berkes 2012, 3–4; Ellen and Harris 2000, 4–5). Traditions by their nature are retained as long as they contribute to the individual's and the group's adaptability and resilience. When they do not, they are abandoned. This includes knowledge and use of plants. What is considered traditional and useful today may well be arcane and redundant tomorrow. Traditions change.

It is, moreover, often presumed that there is a common baseline of knowledge that is shared in a culture, that is, a community consensus of what does and does not constitute a medicinal plant. The sum total of these shared ideas constitutes the traditional plant pharmacopoeia. But there are complications. First, there is considerable variation in plant knowledge in each community (intracultural), and it can be surprisingly idiosyncratic. Many people will know some common medicinal plants, often medicinal foods (nutraceuticals) and other garden cultivars, and women and men may well have contrasting knowledge profiles (Quinlan et al. 2016; Voeks 2007a). But there are also often a few specialists in the community, perhaps shamans or herbalists, or just wise and experienced older women, whose understanding of the healing properties of the flora far surpasses that of other community members. These specialists often possess sharply diverging and sometimes secretive sources of plant knowledge that is not shared with novices in the community. Thus, the more species that one or a few specialists knows and uses compared to other community members, the less their knowledge can be considered a shared commodity (Vandebroek 2010). Indeed, widely ranging botanical knowledge may in itself be traditional (Quinlan and Quinlan 2007). For example, among the indigenous Fulni-ô people of Northeast Brazil, there is an immense disparity of medicinal plant knowledge among participants. Some know only one or two species, but one respondent knows the names and medicinal uses of 75 species (Albuquerque et al. 2011). Similarly in highland Peru, the mean number of known medicinal species at the household level is about 47, but the range of values per household varies from 12 to 99. And in Bolivia, the average number of medicinal plants known at the household level is 28, but individual households vary from 15 to 50 (Mathez-Stiefel and Vandebroek 2012). Clearly, pinpointing precisely what is and is not shared ethnobotanical knowledge in a community is no easy task.

The issue of intracultural variation is complicated further by the time-constrained perspective of most ethnobotanical inquiry. When a researcher begins a study in a particular area, he or she identifies with the quantity and quality of knowledge maintained by the community at that point in time. This baseline of information becomes the ideal against which change over time—losses or gains—is measured into the future. The complication is that the ethnobotanist usually knows very little about the social, economic, or environmental changes that have occurred in the past, and how these played out in terms of community knowledge of nature. Thus, a plant species that was relatively common in the time of the elders of the community may have been overexploited over time, to the point that it is now rare and nearly impossible to find. Older people still consider this species as part of their arsenal of healing plants, but for the younger generation, it is just folklore with little or no relevance to their lives. For the researcher, however, this age-related disparity in knowledge can easily be interpreted as ethnobotanical erosion, when in fact it reflects community adaptability and resilience (Hanazaki et al. 2013; McCarter and Gavin 2015). For example, the heart of palm from súrtuba (*Geonoma edulis*) is widely appreciated for its unique flavor in Costa Rica, especially during Holy Week, the week immediately before Easter Sunday. According to many local residents, "It is not Holy Week without palm heart from súrtuba" (Sylvester and Avalos 2009, 184). But súrtuba has become a victim of its own popularity and growing commercial value. It has been severely depleted over much of its range, with most of the remaining populations now occurring in national parks.

Because of its high commercial demand during Holy Week, most of the supply is now filled by professional harvesters working illicitly in parks. Over time, this shift in focus from subsistence to the economic value of the species has led to diminished interest in other uses of the palm, especially its medicinal qualities. Today the species' healing properties are recollected by only a few elderly community members. Is this cultural loss, or simply sensible adaptation to a shifting social and economic reality? Similarly, an elderly woman named Dona Belinha in the northeast of Brazil some years ago told me about a plant that was used extensively in the 1930s during an epidemic (possibly yellow fever) that ravaged the region, and had taken the life of her sister. She seemed to be the only remaining person in the community that still remembered this use for the species, and with her passing a few years later, knowledge of the plant's healing properties disappeared. This certainly constitutes ethnobotanical erosion, but since knowledge of the plant was employed to treat a disease that is now easily prevented with a vaccine, its "loss" in fact represents an adaptive community response to a problem that no longer exists (Voeks 2010).

The question of shifting baselines and what does or does not constitute traditional plant knowledge and use becomes particularly vexing when a deeper time frame, such as decades or centuries, is invoked. Over time, social and environmental shifts are bound to occur, and these in many cases act as powerful drivers of change in people's relations with nature. Where written records exist, the process of change over time is fairly straightforward to monitor. An excellent example is provided by the European medicinal plant remedies documented first by the ancient Greeks and Romans, going back more than two millennia. The humoral system of health and healing as it was recorded and refined by Galen and Hippocrates, and the associated 500-plus healing plant species assembled by Dioscorides in his *De materia medica*, were hugely influential throughout the Mediterranean region and eventually all of Europe (see chapter 4). Wherever Europeans journeyed, for many centuries to come, they carried both their healing concepts and medicine chests full of these venerable botanical remedies. Well into the European period of exploration and colonization, settlers and the clergy deployed many of the same plants for injury and disease that had been used for close to two thousand years, having weathered the effects of wars, pestilence, and other major watersheds in history. The great breach from these traditions finally occurred in the mid-nineteenth century, with the development of germ theory, the creation of synthetic drugs, and the rise of pharmaceutical companies. From that time forward, almost none of the ancient plant remedies were included in national pharmacopoeias (De Vos 2010), although many continued to be employed in folk medicine. But even these adapted and changed over time.

A good example is provided by rue (*Ruta* spp.), a rank-smelling woody herb that was used extensively in the time of Hippocrates to cure an assortment of medical maladies, including pulmonary and throat problems, and especially gynecological issues, such as uterine and menstrual disturbance. It was also widely employed as an abortifacient. In the Middle Ages, it "sharpened eyesight and dissipates flatulence" as well as "augments the sperm and dampens the desire for coitus" (Arano 1976) (Fig. 8.1). Today, rue is still used throughout Europe in home healing, but many of its applications have changed over the ages. It continues to be recommended as an abortifacient, but most of the other gynecological and reproductive uses spoken of in the past have disappeared. And the means of preparing rue as a medicine has changed markedly over the centuries. Originally it was most often prepared with wine, but today it is almost always concocted with hot water extracts (Leonti et al. 2009; Pollio et al. 2008). In this case, the species has persisted through over two thousand years of medicinal application in Europe, but it would be a considerable exaggeration to interpret this as continuity of a traditional ethnomedical practice.

In the rural tropical realm, where most knowledge of nature is transmitted orally or by demonstration rather than by written text, change over time is less easy to chronicle. Most of our understanding of this process comes from comparing archival descriptions by early colonial observers, often clergy, with current indigenous use patterns. For example, of the documented medicinal plants that were used in colonial Peru and Ecuador, roughly half continue to be employed today. Depending on one's perspective, this suggests either a remarkable level of species use persistence, or on the contrary, that roughly half of the colonial pharmacopoeia has been lost over time. Interestingly, in the case of Peru, considerable local experimentation and growth in its pharmacopoeia has occurred over the past 500 years, whereas in neighboring Ecuador, medicinal plant knowledge has remained fairly stagnant. Although this hints that social conditions in Ecuador somehow fostered medicinal plant continuity, in fact indigenous herbalism was outlawed in Ecuador until recently. This suggests, paradoxically, that social or political processes that discourage ethnobotanical innovation over time may actually encourage retention of traditional plant use, but at the expense of long-term healthcare resilience (Bussmann and Sharon 2009).

Ethnobotanical traditions are also responsive to environmental change, and none was more sweeping over the previous five centuries than the Columbian Exchange (see chapters 5 and 7). The European Age of Exploration initiated a global-scale reshuffling of the world's plants, animals, and microbes and, in so doing, facilitated monumental innovations in the composition of indigenous healing floras. For just as novel infectious diseases were decimating the native populations of the Americas and Oceania, newly arrived medicinal foods, herbs, ornamentals, and weeds offered infinite opportunities for experimentation and incorporation into traditional pharmacopoeias. Some of these exotic plants may have filled vacant biochemical niches, providing unique classes of compounds to combat local illnesses (Alencar et al. 2010). Others were tested by local people and added over time to their healing pharmacopoeias. In northern Amazonia, as noted earlier, indigenous and mestizo people have incorporated over two hundred non-native species into their pharmacopoeias. Most were introduced by outsiders principally for their food or ornamental value, but over time came to be appreciated for their healing properties as well. Others, such as aloe (*Aloe vera*), belladonna (*Atropa belladonna*), and lemon grass (*Cymbopogon citratus*) were introduced specifically for their medicinal qualities (Bennett and Prance 2000). Likewise in the Hawaiian Islands, which endured profound social, economic, and environmental upheaval following the arrival of Europeans and Americans, native people assimilated many newly arrived plants into their traditional pharmacopoeia. Today roughly half of their medicinal species are exotics (Palmer 2004). In far west Nepal, introduced species are occupying an increasing share of local plant pharmacopoeias. Many are foods, ornamentals, and weeds that thrive in disturbed habitats. And these are gradually replacing the indigenous healing flora, which is becoming rare due to deforestation and climate change (Kunwar et al. 2015).

Whether newly arrived plants are or are not adopted by local people depends on their cultural traditions and needs. Some species retain the perceived medicinal virtues of their homelands whereas others, such as the moon-flower (*Ipomoea alba*), do not. Native to the New World, this attractive ornamental was employed in the colonial Americas by Native Americans as a purgative, for inflammation, to treat asthma, and for a host of other medical ends. By the 1600s, the species had already become widely dispersed throughout the tropical world, where it took on local and completely unique medicinal uses. In India, for instance, it became a snakebite remedy, a use that was never recorded in the Americas. Similarly in the Samoan Islands, moon-flower is used to treat boils,

A

8.1 A. Rue (*Ruta graveolens*). In: Luisa Cogliati Arano. 1976. *The Medieval Health Handbook: Tacuinum Sanitatis.* Plate 35. 15th century by unknown author. RB 622160. The Huntington Library, San Marino, California. Digital Image by Robert Voeks. B. Brazilian woman wearing rue in her hair as prophylaxis against evil eye. Photo credit: Robert Voeks.

filariasis, and stomach ache, again ailments unrecorded in its place of origin (Austin 2013). Traditional pharmacopoeias, it seems, should be conceived of as multidimensional moving targets, cognitive domains that adapt to social and environmental changes over time, growing, stagnating, or declining, depending on the unique set of drivers and cultural responses.

Many ethnobotanists will claim, and with considerable justification, that the difficulty of pinning down exactly what constitutes a traditional pharmacopoeia does not detract from the fact that ma   re at present

B

8.1   *continued*

in precipitous decline, quantitatively and qualitatively. It is one thing to consider gradual adjustments in plant knowledge and use in response to periodic social and environmental changes. But it is quite another to be witness to the total abandonment of a sphere of knowledge that has served rural tropical societies for many centuries and that was crucial to the development of many life-saving pharmaceutical drugs. The principal drivers of this loss appear to be radical changes in the social and economic environment, especially globalization and the increasing commercialization of medicinal plants (Balick 2007; Stanley et al. 2012). Globalization is a particularly insidious force in terms of traditional relations with nature as it links the lives and livelihoods of rural people and communities to places and processes occurring hundreds or even thousands of kilometers away (Latorre and Latorre 2012). And while it is true that even the remotest tropical communities have never really been completely isolated, the pace of penetration of market-based economies and values, formal education, Western medicine, and technology over the past few generations is unprecedented. Rural global peripheries are increasingly drawn into global systems, often without the knowledge and consent of local people, and frequently with tragic cultural and environmental consequences. As early noted by Schultes and others, the consequences of globalization in terms of people-plant relations are profoundly negative.

Plants have myriad material and cultural uses and values for tropical people, including timber, fiber, fuel, crafts, magic, and many others, and all are subject to transformation in the face of globalizing influences. But knowledge and use of medicinal plants, in particular, may be "uniquely vulnerable to acculturation." In a pioneering use of quantitative methods to explore indigenous ethnobotanical knowledge, the relationship between medicinal plant knowledge and age of participants was explored in lowland Peru. Forest children were found to learn the identities and uses of foods, fibers, and other plant use categories early in life, quickly developing knowledge profiles that were similar to that of older members of the community. But community knowledge of medicinal plants had the steepest learning curve, meaning that if dramatic socioeconomic changes occurred, such as sudden access to pharmaceutical drugs or immersion in formal education, medicinal plants would be the first domain of traditional plant knowledge to disappear (Phillips and Gentry 1993, 41). Subsequent research over the past two decades, however, has shown that the process of medicinal plant erosion is more nuanced than first envisioned.

Ethnobotanical Change

Several years ago, Angela Leony and I explored the question of ethnobotanical decline in the rural Brazilian community of Lençóis, Bahia (Voeks and Leony 2004). The town was an ideal location for the study. There was considerable reliance on medicinal plants by local residents, at least until recently, and the entire region was in the grip of major social and economic disruption. It was not the first time that globalizing influences had left their mark in the region. During the great Brazilian diamond rush of the early nineteenth century, thousands of prospectors had streamed into the mountainous region seeking their fortunes in alluvial diamonds. When mining finally declined due to overexploitation and competition from newly discovered African deposits, people continued to eke out a living cultivating cassava and other subsistence crops, and extracting wild plant and animal resources. A few determined miners continued to search for diamonds. Far from the bustling coastal cities, Lençóis lacked the basics of modern life well into the 1970s for most residents, including reliable electricity, indoor plumbing, schools, and access to medical clinics. Then things changed. A national park was established adjacent to the town in 1985, the Diamond Plateau National Park (*Parque Nacional da Chapada Diamantina*), and ecotourists discovered the rugged beauty of the surrounding landscape. Beginning with a few hundred backpacker tourists in the 1970s, visitation to the region grew exponentially to over 60,000 per year by the year 2000. By the time I arrived, nearly everyone was directly or indirectly employed by the tourist industry, whether in the numerous locally owned bed and breakfasts, or as nature guides leading tourists on journeys into the park. Nearly all the children attended school, and there was a small pharmacy. For young men and boys, most of whom by this time were self-described nature guides, the hardscrabble existence of the previous generation, cultivating crops and grubbing for minerals, held little appeal.

We set out a medicinal plant trail on the edge of town. It began on a public trail beside some homegardens, passed through some weedy areas and second-growth forest, and continued along the banks of a tannin-rich black water stream to a local clothes-washing and bathing section of the stream (Fig. 8.2). The trail was used by most members of the community, and on any sunny day, the streamside rocky outcrops appeared as garlands of drying clothes and sheets. Our aim was to identify the medicinal plants on the trail, using local specialists, and then determine how many of these species were known by local residents, especially

A

8.2   Study area. A. Trail through weeds and second-growth forest. B. Trail along the blackwater
Lençóis River. Photo credits: Robert Voeks.

younger people. We employed two local people, a man and a woman
known for their medicinal plant expertise, to first census the medicinal
plants on the trail. We marked and identified a total of 45 medicinal species, ranging from common fruit trees and weeds to a few less common
plants. We then led people one at a time along the trail, pointing out

229

B

8.2 *continued*

each of the species and asking "do you know the name of this plant?" and "do you know of a use for this species?" Before we started, we asked each participant a series of questions to get a sense of their social and economic status, and their overall worldliness. These questions included "how many years did you go to school?," "do you have indoor plumbing?," "were you born within 25 kilometers of Lençóis?," "what is the furthest you have traveled in your life from home?," and many others. A total of 67 men and women, boys and girls, ranging in age from 10 to 82 years, participated in the study.

Data gathering revealed some of the challenges inherent in trying to measure ethnobotanical change. For instance, all of the participants could identify the most common fruit tree species, such as papaya, cashew, and pitanga (*Eugenia uniflora*), although not all could identify a medicinal use for them. Most could identify the showy climber estradeira-vermelha (*Periandra coccinea*), but only three elderly people knew of its use by women experiencing conception problems. Most women could identify jaborandi (*Pilocarpus pennatifolius*), which is used to remove lice from children's hair, but men were mostly ignorant of this use (Fig. 8.3) There were some unforeseen ambiguous answers that required subjec-

A

8.3 Medicinal species on the Lençóis River study trail. A. *Estradeira-vermelha* (*Periandra coccinea*). B. *Jaborandi* (*Pilocarpus pennatifolius*). Photo credits: Robert Voeks.

tive decisions on our part. For example, young people sometimes knew the name of a plant and how it was used, but in fact had never actually employed it themselves. They noted that they had seen it prepared by their mothers or grandmothers. In retrospect, it would have been useful to have included a question regarding when was the last time the participant had actually prepared and used the plant. In addition, several participants supplied common names for species that had not originally been furnished by our specialists. Most were provided by young boys, and several were quite vulgar. This suggested that perhaps new labels were in the process of being created or, just as likely, that this was generational knowledge, retained at one point in life, but abandoned with a bit of maturity. Finally, a few of our informants were formally educated naturalists who had migrated into the region to work in the ecotourist trade. Some were Brazilians, but others hailed from other parts of South America, Europe, and North America. These participants sometimes were not familiar with the local common name of one or another species, but they did know its scientific name, at least to the rank of genus. Several

B

8.3 *continued*

were aware of their medicinal uses, which they often shared with tourists on guided walks, but most had never actually used the plants to treat an illness themselves, nor were they likely to do so.

The survey results confirmed some of our hypotheses, but contradicted others. First, we failed to find a significant association between knowledge of the names and/or medicinal uses of plants on the trail and a person's relative economic prosperity. None of the individual economic variables, such as ownership of a television, access to indoor plumbing, or number of rooms in their home, correlated with level of medicinal plant knowledge.[3] This was unexpected, given that the more prosperous participants in the study were further along the road towards modernization and were more likely to have access to Western doctors. The relative worldliness of participants also failed to factor into medicinal plant knowledge. People that had flown in an airplane, at that time quite rare for the average Brazilian, were no more or less knowledgeable about the healing flora than those who had not. Nor did distance traveled away from the region, or even distance to birthplace, correlate with plant knowledge. Thus, several of the most likely drivers of acculturation seemed irrelevant to this community in terms of knowledge of nature.

Level of formal education, on the other hand, varied considerably among participants in the study, from those who had never attended school, to a few with university degrees. And this variable exhibited a strong association with medicinal plant knowledge. Increasing formal education, measured in terms of number of years attending school, was negatively correlated with aptitude in the area of medicinal plant identification. Likewise, participants who were literate knew much less about the medicinal flora than those who could neither read nor write. Finally, as anticipated, participant age was the strongest indicator of medicinal plant knowledge. There was a clear increase in knowledge from the very young to the very old, with the largest upsurge occurring in those over 50 years of age.

We concluded that formal education in this region was having an adverse effect on local knowledge of medicinal plants. In the past, medicinal plant skills represented a crucially important survival skill. Healthcare was a constant concern, and local people dealt with day-to-day maladies, and even life-threatening emergencies, with an arsenal of plant-based remedies they had learned over the years from family and friends. Today, young people in the community are educated by public school teachers following a nationally mandated curriculum of math, history, science, and a host of subjects meant to prepare them for the twenty-first-century market-based economy. The career aspirations of young people today have little in common with the agrarian and extractive lifestyles of their parents and grandparents. Consequently, for the younger generation, the healing properties of nearby forests and fields seem old fashioned and culturally rather pointless.

Numerous other researchers in the subsequent decade or so have used quantitative methods to explore the question of ethnobotanical change in rural communities. Many of their findings are similar to ours, but some are not. From all points in the geographical compass—South and Middle America, sub-Saharan Africa, South and Southeast Asia, and Oceania—nearly all research reports a striking association between a participant's age and medicinal plant knowledge. And many view this as evidence of ethnobotanical erosion (cf. Albuquerque et al. 2011; Begossi et al. 2002; Estrada-Castillón et al. 2014; McCarter and Gavin 2014; Müller et al. 2014; Teklehaymanot 2009; Zent 2001). Older people are knowledgeable about the healing properties of tropical nature; young people much less so. Anecdotal observations of children's lack of interest in medicinal plants are often taken as evidence of ethnobotanical erosion. In Boca de Toro, Panama, for instance, the observation that schoolchildren cite Coca-Cola as the best remedy for stomach problems suggests that ethnobotanical knowledge is on the road to extinction (Ramirez 2007, 247).

But age-based results need to be considered with caution as indicators of change because people accumulate knowledge and experience over time. It should come as no surprise that older people know more about certain topics than younger people, and vice versa. Additionally, the task of healthcare in many traditional cultures falls on older members of the community, often women, who have the time and necessary lifetime experience to focus their efforts on remembering and applying medicinal plant formulas. Any child can quickly learn that a succulent purple fruit is good to eat, but not many will be sufficiently motivated to learn the healing properties, mode of preparation, and dosage of one or another bitter-tasting herb (Quinlan et al. 2016). Documenting loss in plant knowledge over time requires more than simply noting that older people know more than younger people (Godoy et al. 2009; Hanazaki et al. 2013).

Modernization, access to new technologies, and changing career aspirations are important erosive agents in traditional plant knowledge. And one way to measure this relationship is by identifying the distance and/or access of rural villages to more modern city centers as the independent globalization variable (Reyes-García et al. 2013a). In Burkina Faso, for example, people living in the remotest households are more reliant on traditional medicine than those living closer to cities with ready access to Western medicine (Pouliot 2011). Similarly in Mayan communities in Mexico, "knowledge devolution" is occurring along a rural to urban transition. Children in the rural areas know more plants than their urban counterparts, and they gain this knowledge at an earlier age (Shenton et al. 2011). In Papua New Guinea, there is a negative correlation between distance from various villages to the nearest town and medicinal plant knowledge. People living in town also know significantly fewer medicinal species than their village counterparts, as the region's healing flora is in the process of being replaced by imported pharmaceutical drugs (Case et al. 2005). The subtle nature of these changes is illustrated by the situation in southern Benin in West Africa. People cultivate species of amaranth (*Amaranthus* spp.) and collect other species of the genus from the wild for food and medicine. The plants have highly nutritious seeds, and their leaves are used as a vegetable and in medicinal preparations. But along the modernization transition from rural to urban Benin, the use of wild, mostly medicinal amaranth declines pre-cipitously along the demographic gradient, replaced by cultivated and edible amaranth. Wild amaranth species that are considered as food and medicine in the countryside are treated in the city as just useless weeds (Sogbohossou et al. 2015). In these cases, distance decay is strongly as-sociated with ethnobotanical devolution.

Not every rural to urban transition follows these patterns, however. In the move from the Dominican Republic to New York City, immigrants over time are abandoning the use of many non-food species, but at the same time increasing their reliance on medicinal food plants, such as lemon, lime, cinnamon, garlic, and coconut (Vandebroek and Balick 2012; 2014). Similarly in Brussels, Belgium, West African immigrants have not completely abandoned their reliance on homeland healing. How-ever, they are of necessity employing more food-medicines, which are easier to source, at the expense of herbal remedies, which are increas-ingly difficult and expensive to import (Van Andel and Fundiko 2016). In the Solomon Islands, there is no relationship at all between distance to the nearest town and level of ethnobotanical knowledge. In this case, the communities have been able to absorb aspects of the modern world and continue to employ their traditional plant knowledge (Furusawa 2009). Similarly in southern Brazil, Caiçara people continue to rely on traditional plant medicine, regardless of proximity to urban areas, in this case because industrialized drugs are just too expensive (Begossi et al. 2002).

In some cases, increasing access to formal education is a critical fac-tor in the loss of plant knowledge. Venezuela's indigenous Piaroa, for example, have experienced massive social and economic changes since the 1960s. They transitioned from an isolated, foraging and farming subsistence in Amazonian uplands, to sedentary riverside dwellers, avid consumers of foreign material goods, and integration into the regional and national economy. Access to education among the young, combined with the ability to speak both Piaroa and Spanish, are strongly asso-ciated with declining ethnobotanical knowledge, including of medici-nal plants (Zent 2001). But access to formal education in other cultures is not associated with loss of medicinal plant knowledge. In the Do-minican Republic, people living in more modernized households and with more commercial-oriented vocations, know relatively more bush medicine, not less as predicted. In this case, when better educated par-ticipants become parents, their knowledge of medicinal plants actually increases. Modernization, it seems, does not erode medicinal plant use in a straightforward way. Rather, the individual personality traits of the person matter most; their general "resourcefulness, exactness," and "interest in the subject matter" are better predicators of plant knowledge. In addition, modernized and commercially oriented individuals are savvier in regards to plant medicine, reflecting the fact that they are worldlier. They have come into contact with more and different sorts of people from different regions and as a result have had the opportunity to learn about new healing traditions and new plant therapies (Quinlan and Quinlan 2007). Similarly in rural Bolivia, level of medicinal plant knowledge is a result of "one's own personal active quest." In this case, differences in knowledge are not related to level of education, age, migration, or any of the other usual suspects, but rather to the inherent diversity of knowledge within the group. Plant knowledge depends on individual motivation, experience, and personality (Mathez-Stiefel and Vandebroek 2012).

Historically, one of the first and most ruthless globalizing influences to severely undermine indigenous relations with nature was the brutal imposition of monotheistic religions of salvation. Throughout the European era of exploration and colonization, rituals with pagan overtones were banned, sacred texts were burned, and healer-shamans were subjugated (Hemming 1978a). How many medicinal species and their uses were lost in these early encounters is unknown. A few were purposely rescued, however, as members of the Jesuit order were keenly interested in acquiring new and more effective remedies (see chapters 4 and 7). But the overall effect of colonial proselytizing activities must be viewed as a biocultural catastrophe. These days, missionaries mostly concentrate their faith-based efforts on community health and poverty alleviation. But some continue to target those few isolated groups that managed to evade early religious conversions. In some cases, conversion of native peoples is accelerated by distributing Western drugs, which often work more quickly than plant cures, leading to loss of confidence in indigenous medicine. In the process, traditional healing ceremonies, which often have direct connections to indigenous cosmology and spirit worship, have been hybridized with or replaced by Christian morality, prayer, and often pharmaceutical drugs (Godoy et al. 2009; Luzar and Fragoso 2013). There are rare cases, such as the Maasai of East Africa, where Christian conversion seems not to have affected knowledge and use of plant medicines (Kiringe 2005). But these cases are the exception. On Malekula Island (Vanuatu island archipelago), for instance, local people perceive that there is much less medicinal plant knowledge now than in the past, and that this is especially due to the "psychological warfare" carried out by missionaries against traditional knowledge (Mc-Carter and Gavin 2014, 293). Similarly among the Tsimane' of Bolivia, Catholic and Protestant missions in 1940s effectively banned the role of forest shamans. As a result, this cultural dimension of healthcare and medicinal plant knowledge has now disappeared (Godoy et al. 2009). My own experience among the Penan hunter-gatherers of Borneo supports the largely negative impacts of "invasive belief systems." In Brunei Darussalam, the only Penan community in the country was under considerable pressure in the 1990s to convert to Islam. As one after another family accepted the faith, they were forbidden to engage in certain traditional rituals, especially those involving alcohol or pork, as well as healing rituals that invoked forest spirits. To the west, across the Baram River in neighboring Malaysia (Sarawak), however, the struggle for the souls of the Penan was being won not by Muslims, but by Christians of the Borneo Evangelical Mission. On one particularly memorable visit to a remote Penan longhouse, linguist Peter Sercombe and I attempted to engage local residents in a conversation about their knowledge of the spirit realm and healing traditions. It was a brief conversation, as they informed us that they no longer maintained any of these "primitive" beliefs. In our discussion, we learned that the local evangelical missionary had schooled them against using or even discussing their traditional plant pharmacopoeia, since these plants and their uses were all manifestations of Satan. When we inquired about how they treated illness these days, they showed us a big tin of aspirin that the missionary had left (Voeks and Sercombe 2000).

This leads us to what many believe is the primary driver of medicinal plant devolution in rural tropical regions—the seductive influence of Western pharmaceutical drugs. When these become available and affordable, the allure of injections and "little green packages" is terrifically inviting (Shanley and Rosa 2004, 152). In the remotest rural settings, access to Western drugs is rare. And when they are available, modern drugs are usually outside of the financial wherewithal of local people (Begossi et al. 2002). Even when pharmaceutical drugs are donated by foreign aid agencies, they often arrive intermittently and, due to the tropical heat and moisture, expire quickly. Some of these are "dumped" in rural tropical communities only because they've expired and can't be sold in Europe or North America. In addition, these donated drugs are often self-administered, leading to improper dosages, and are used to treat the wrong illness (Carlson 2001, 495).

Access to pharmaceutical drugs in rural areas is usually a function of proximity to towns and cities, where most of these services are located. In the Bolivian Amazon, the closer geographically the village is to doctors and healthcare centers, the less likely people are to use medicinal plants (Vandebroek et al. 2004b). Similarly in rural China, only the poorest people in remote areas with little to no access to modern medicine still rely on medicinal plants (Huber et al. 2010). But there are exceptions. Among the Mazatec people of southern Mexico, access to pharmaceuticals not only does not lead to ethnobotanical erosion, but those who know more about pharmaceutical drugs paradoxically also know more about medicinal plants. In this case, people make informed decisions concerning the various treatment options that are available. Medicinal plants are considered the best treatment for common and easily treated illnesses, therefore fulfilling the role of over-the-counter drug plants, whereas pharmaceuticals are employed when the patient fails to respond to herbal treatments (Giovannini et al. 2011). This is the case as well in southern Africa, where HIV/AIDS patients often begin treatment with a traditional healer (Audet et al. 2012; Fasinu et al. 2016). If this is not successful, they move on to a Western doctor and antiretroviral therapies.[4]

The dynamic and adaptive nature of traditional pharmacopoeias is illustrated by a study carried out in Panama. In this case, village people generally view health and healing from the perspective of the situation in the past and in the present (*antes* and *hoy*). Some medicinal plants are no longer being used (*antes*), but others are being learned and added to the pharmacopoeia (*hoy*). This includes *noni* (*Morinda citrifolia*), a Pacific Island tree that has wide usage for food and especially medicine in its native range, and was introduced to Panama by outside development agencies. Because younger people operate in a different social and economic environment than their parents and grandparents, they tend to be more open to new ideas, including adding new plants to their repertoire. But although the village is increasingly "modern," its extreme level of material poverty still leads people to depend on nature for many of their basic needs. Pharmaceutical drugs are available, but are mostly unaffordable. Thus, knowledge of the healing power of nature is not so much eroding in this community as it is adjusting and transitioning into a new social and economic reality. Plants with currently perceived value are retained, but others that no longer have cultural meaning or value are discarded (Müller-Schwartz 2006).

The complex nature of ethnobotanical change over time is especially well documented among the Tsimane' people of Bolivia. This culture group, like so many others in the past half-century, has undergone major social and economic transformations. In addition to conversion to Christianity beginning in the 1950s, the Tsimane's relationship with nature became more and more commodified as forest and agricultural products were increasingly destined for market. But 'e many Tsimane' have taken jobs in logging camps and commercial agricultural operations, they are still in many respects self-sufficient. Anu unlike so many other marginalized people, their schools are taught in the Tsimane' language (Godoy et al. 2009). Thus, at least some of the cultural features that foster the survival of traditional ecological knowledge remain. Regarding the question of ethnobotanical erosion, however, results to date are mixed. In one case, there were no significant differences in overall knowledge of useful species between people born in the 1920s and those born in the 1980s. Rather than globalization as a driver of cultural decline, the increasing mobility of Tsimane' due to improved roads is exposing them to new and adaptively useful realms of botanical information. Indeed, in some areas, their overall knowledge of plants may well be growing. Because ethnobotanical knowledge is dynamic, "one must be open to the possibility that it may increase, and not that it inevitably wanes" (Godoy et al. 2009, 64). A more recent study, however, turned up quite different results. Knowledge of useful plants among the Tsimane' has declined 20% from 2000 to 2009, translating to a 2.2% loss per year. This drop in usage is occurring in all age groups, but is most pronounced among the very young. Thus, this forest community is rapidly abandoning its traditional relationship with nature, which from their perspective, "[does] not equip them well to deal with the new socioeconomic and cultural conditions they face" (Reyes-García et al. 2013a, 256; Reyes-García et al. 2013b). Finally, separating out the various plant use categories—timber, fuel, craft, and others—the Tsimane' appear to be losing their knowledge of medicinal plants, but not of species used for other purposes, such as construction. Because building materials represent an increasingly important economic activity, knowledge of these species appears to be surviving and even expanding (Gómez-Baggethun and Reyes-García 2013). Declining interest in medicinal plants among the Tsimane' is associated with increasing access to Western physicians, vaccinations, pharmacies, and medicines. It is also caused by the increasing social stigma associated with adherence to traditional healing practices.

To conclude, with some notable exceptions, globalizing influences such as formal education, alien religion, Western medicine, modern technology, and market economies are undermining indigenous knowledge of medicinal plants. Isolation is no longer a barrier to these devolutionary forces. As predicted by the jungle medicine narrative, the collective and individual knowledge and use of medicinal plants in most cases decline as the tentacles of the modern world penetrate even the remotest rural communities. And like other elements in the story—the virgin tropical forests, the noble and innocent savages, the wise and wily shamans, and the prospect of solving some of society's most troubling diseases— the concept of ethnobotanical erosion is rooted in evidence-based science and social science. Dig deeply enough into any of these subjects, and you can uncover the quantitative data or the ethnographic observations necessary to either support or refute one or another element in the storyline. But like most environmental narratives, this one gains strength and credibility more by how many times it has been repeated in college texts, popular magazines, Hollywood productions, and increasingly social media. In the final chapter, I explore the nature of environmental narratives, and why they are so much more compelling than real science.

NINE

Environmental Narratives

The first clue that the pristine tropical forests of Africa, Asia, and the Americas were being razed entered the Western consciousness around the mid-1970s. Of the lengthy list of environmental calamities that seemed to threaten the very existence of humankind—explosive population growth, pesticides, nuclear power, air and water pollution, and many others—tropical deforestation was portrayed as somehow the most ignominious and senseless. Global indignation reached an early apex in 1975 when NASA scientists detected a single monstrous fire in Amazonia, some 25,000 hectares in extent (61,776 acres), being used to clear primary rainforest for pasture development. At the time, the press declared it to be the largest intentional conflagration in world history. Adding insult to injury, much of the lands were owned and operated by a foreign corporation—Volkswagen. The German company, which was then the largest car manufacturer in Brazil, had acquired vast Amazonian lands for pasture development, a seemingly bizarre venture for a car manufacturer. But as ecologist Norman Myers reported at the time, "Volkswagen believes that, although people may come to purchase fewer cars [as the price of fuel increases] they will hardly be inclined to eat less beef" (Myers 1980). The environmental community was quick to denounce these actions, claiming that the German car manufacturer had illegally destroyed a patch of paradise roughly "the size of Lebanon" (Acker 2014, 23). Brazilian environmentalist Jose Lutzenberger termed this and other ongoing multinational exploits a "Holocaust in Amazonia" (Lutzenberger 1982, 249). But while the actual involvement of Volkswagen in the great fire was considerably exaggerated in the press, the seeds of an internationally directed Armageddon in the Amazon were firmly sown into the West's environmental consciousness.

The Volkswagen debacle was followed shortly by the emergence of a new and especially compelling environmental narrative—the hamburger connection. Geographer James Parsons was one of the first to inform the scientific community that massive swaths of old-growth rainforest in Central America were being cut, burned, and transformed into permanent pasture for livestock grazing. Cattle culture from Mediterranean Europe and tenacious grass species from the savannas of Africa were quickly and silently supplanting the majestic evergreen rainforests that mantled much of Central America. This forest to pasture process, which he termed "an almost mindless mania," was occurring at the same time that the per capita consumption of beef for most poor Central Americans was in a steady decline. Whatever benefits were accruing from the "grassification" process were not being realized by the average Central American (Parsons 1976, 122). Much of this pasture growth was attributed to the rapid expansion of the American fast food industry in the 1970s which, in the midst of a deep national recession, seemed to be the only economic bright spot. North Americans were turning to fast-food in droves, fueled by the success of McDonalds, Burger King, and Kentucky Fried Chicken (KFC). At the spiritual center of this metastasizing obsession was the American hamburger. Long maligned by the unsanitary images penned by Upton Sinclair's The Jungle, the humble hamburger was resurrected in the gleaming, stainless steel assembly lines of fast-food kitchens. But being cheap was one of the central hallmarks of the fast-food movement, and the industry's continued success depended on the availability of inexpensive locker-grade beef. And that was a problem. The cost of cattle in the United States was at the time climbing steadily, with no end in sight. In response, the American fast-food industry turned its gaze to its southern neighbors, in particular Costa Rica, Nicaragua, Honduras, and Guatemala. In some ways, this region seemed to be a sensible supplier. It had a rich history of cattle culture, and it was free of the devastating effects of foot-and-mouth disease (aftosa) that plagued South America's bovines. Most importantly, from the perception of international lenders, capitalist-style development in the form of pasture creation and cattle exports would help to fend off peasant interest in the communist model that at the time was thriving in nearby Cuba (Myers 1981). On its surface, replacing useless forest with productive pasture seemed like a win-win.

But the ecological consequences of converting Central America's forests to artificial grasslands were a different matter. For an environmentally conscious public just beginning to care about the fate of the world's tropical forests, the hamburger connection narrative resonated deeply. It contained a host of interwoven storylines—environmental destruction, grating Third World poverty, fast-food addicted North Americans, and greedy multinational corporations. By the mid-1980s, several environmental organizations began to agitate for change, including the nascent Rainforest Action Network (RAN). Working originally (and ironically) out of an old Volkswagen van, organizers staged protests in front of various Burger King franchises, which at the time were believed to import more Central American beef than any of its competitors. McDonald's asserted that they used only American beef, but this claim was widely questioned. In any case, Americans quickly learned about the hamburger connection through their news outlets, as protests were widely publicized. At a Burger King in San Francisco, protesters organized a rally and guerrilla theater demonstration "in which two activists dressed in cow suits ate rainforest tree leaves and defecated the likeness of Whoppers, either cardboard containers or Styrofoam hamburgers." And RAN organized "Whopper-Stopper Month"[1] to publicize the plight of the rainforest, and at each venue sang the "Whopper Song."

Well they come into the forest with machines with giant teeth
Knock a nickel off every whopper they make with cheap imported beef
So won't you lay down your whopper baby,
Lay down your whopper and your fries
Save the rainforest baby
Before the rainforest dies.
—BILL OLIVER 1986

The hamburger connection narrative served the purpose of its proponents. Many millions of North Americans were introduced to a pressing environmental problem as well as a practical strategy for how to fix it—boycott the worst of the fast-food offenders. And their efforts seemed to pay off. Sales at Burger Kings declined significantly, although whether this was a direct result of the boycott is contested. Regardless, the company reluctantly agreed to stop importing Central American beef, and beef exports to the United States declined. Meanwhile, Costa Rica gradually turned to nature tourism as its central development strategy, and in time became a global flagship for balancing environmental protection and economic development (Honey 2008). And shortly, the alleged negative environmental impacts of Central American beef exports were challenged. Biologist Dan Janzen, who had spearheaded tropical forest conservation efforts in Costa Rica for many years, introduced an equally persuasive counter-narrative: "buy Costa Rican beef." His reasoning was sound, and based on the realities of land and life in Central America. The outcome of boycotting beef, he argued, would be to drive local people to clear the forest for some other purpose, such as bananas, coffee, or cassava. Costa Rica was a tiny country struggling to justify allocating its limited resources for nature conservation, and boycotting beef imports directly undermined these efforts. Indeed, the pressure to cut the remaining bits of tropical forest was logically reduced by continuing to graze cattle on lands that had already been cleared. "If you really want to help Costa Rica save its remaining forests," Janzen argued, "buy every Costa Rican product you can find—its bananas, its beef, its coffee and its airplane tickets" (Janzen 1988, 258).

There are three points to take away from the short history of the hamburger connection. First, although the scientific facts that underpin an environmental issue may be understood by a small group of researchers, they gather the power of public support only when the disparate threads are woven together into a good story. Otherwise, the problem remains unknown and unresolved in the arcane universe of science and scientists. Second, fledgling environmental narratives are necessarily constructed with equal portions of fragmentary scientific evidence, personal prejudices, and nostalgic sentimentalism. Whatever the good intentions of the storytellers, they seldom let competing facts or complex data get in the way of a compelling tale. Finally, once an environmental story enters the realm of received wisdom, it is very difficult to dislodge, even as better data and more cogent concepts appear. One need look no further than the enduring myths of environmental determinism, the pristine rainforest, and the noble savage to appreciate how great stories can become calcified over time. All this leads to an obvious question— what is it about environmental problems in particular that lend themselves to storytelling?

A Forest of Fables

People are natural storytellers. They love to tell stories, and they love to hear them. And for good reason. Storytelling is a cross-cultural characteristic found in all human societies, from hunting and gathering communities to the information society. It is not just a form of entertainment:

it is very likely an evolved feature that confers Darwinian fitness on the speaker and on the audience. Stories and storytelling are as much a part of our evolutionary history as speech and monogamy. People have been telling stories since the dawn of language, perhaps 100,000 years ago or earlier, often huddled around the flickering flames of the evening fire. For our distant ancestors, night-talk around the hearth differed from day-talk, which had a more immediate economic and pragmatic focus such as tracking game or foraging for plants. As individual images of observers softened in the warm glow of the embers, storytellers employed their narrative skills to relate and embellish events that members of the audience did not need to personally experience. Listeners could vicariously sense the terror of being stalked by a lion or charged by a bear without experiencing the actual life and death event. Practical conversation was replaced by social conversation, and the narrative could trespass freely into abstract and conceptual areas of thought. Empirical observations and experiences, recent or in the past, coul ' be articulated, massaged, exaggerated, and strung together into narrativ's that were culturally meaningful, arresting, and entertaining. Stories ai ut distant ancestors and spiritual experiences took on lives of their own. Compelling stories would be remembered and retold to future generations, some even entering into the realm of myth. Most importantly, the storyteller could modify the significance of events in the narrative in order to manipulate the meaning or moral that he or she wished to convey. There was a "bigger picture" to the story, and it was often interest-laden with the perspective of the storyteller (Dunbar 2014; Sugiyama 2001; Wiessner 2014). Because storytelling penetrates so deeply into our DNA, it should come as no surprise that environmental narratives, which seek to translate complex science and social science into comprehensible stories, have found such a wide and accepting audience.

Environmental problems lend themselves to narratives and storytelling, more so than social and economic crises, because their pace of change is so gradual. Economic downturns, social upheaval, religious conflicts, and other problems that negatively affect the lives and livelihoods of people are salient, often immediate, and therefore reasonably accessible to cause-and-effect analysis. But not environmental changes. Most are incremental and barely perceptible in a human lifetime. Soil erosion, ocean pollution, climate change, and many others are not easily observable and seldom actively and immediately harm people and property. They either unravel too slowly or are too tangential to day-do-day experience to merit attention. And when they do lead to catastrophic events, they don't connect well in the minds of most people with long-term environmental drivers, such as automobile use or forest fires. Seen in this light, environmental problems are a form of "slow violence." They are "incremental and accretive" and "spectacle deficient"; only over time do they lead to calamitous repercussions (Nixon 2011, 2, 47). How to awaken a busy and distracted public to the slow violence of environmental threats that are complex, difficult to pin down, and full of scientific uncertainties? The answer is to craft a slowly unfolding narrative, one that vividly describes the root of the problem, paints a persuasive picture of the consequences of inaction, and provides a practical path to a successful resolution.

But environmental narratives, particularly for the purpose of informing policy and management of resources, have a problematic legacy. Africa in particular has borne the brunt of a series of mostly "degradation narratives," devised by colonial authorities and serving ultimately to deprive indigenous people of their lands and resources. In North Africa, for instance, French colonial authorities surmised from ancient Greek and Roman texts that Morocco and adjacent territories had once constituted some of the richest forests and most productive granaries in their respective empires. Over the centuries, it was believed, this latter day Eden had been deforested and turned to desert by the unsustainable livestock grazing practices of its nomadic Arabs. This desertification narrative informed and misdirected the early forest and range science carried out in this semi-arid landscape, so much so that by about 1940 the "literary story" crafted from the libraries of the ancients had been replaced by "a scientific story complete with ecological statistics and maps that helped to justify policy making." Much of this narrative, however, appears to have been a myth. As in so many cases, when long-term longitudinal evidence of environmental change is assessed, casual observations turn out to be erroneous. In this case, fossil pollen evidence (palynology) revealed that forest extent and tree species composition in the region had in fact changed little in the past two millennia (D. K. Davis 2005, 226). Nevertheless, good stories die hard, and the received wisdom that Africans and other marginalized peoples are somehow incapable of sustainably managing their own resources persists and continues to inform the opinions of policymakers.

Degradation narratives are employed widely to frame problematic nature-society relations, particularly in the less developed world. They have been used to connect population growth and deforestation in the Himalaya with catastrophic flooding along the Ganges and Brahmaputra Rivers. And as in many other environmental narratives, it is "The ignorant and fecund subsistence farmer . . ." [who becomes] "a convenient scapegoat" (Blaikie and Muldavin 2004; Ives 1987, 193). In Laos, environmental narratives are represented as a "chain of degradation" to connect upland poverty and intensified shifting cultivation with downstream siltation and flooding (Lestrelin et al. 2012). In Cote d'Ivoire, Ethiopia, and elsewhere in Africa, environmental narratives have been deployed to link shifting cultivation, brush fires, and pastoralism with continent-wide desertification (Bassett and Zuéli 2000; McCann 1997). And throughout the developing world, the standard wildlife conservation narrative was predicated on the assumption that animal protection was incompatible with the subsistence needs of indigenous people. "Local people do not value wildlife, at least not in the same ways that outside and often Northern people do." The usual solution was to expel people from their ancestral lands in order to provide protected space for wildlife as well as tourist dollars. Local people that continued to hunt in the area were labeled as poachers (L. M. Campbell 2002, 30; Maddox 2003, 253). The results of these negative portrayals of nature-society relations, supported by good science or contrived, led to catastrophic consequences for local populations. The famine and total social collapse experienced by the Ik people of Uganda following their expulsion to create the Kidepo Valley National Park in the 1960s was a highly publicized result of such actions (Turnbull 1972). And while the park minus people approach is seldom invoked today, the impacts of past actions are still being felt. In South Africa's Kruger National Park, for instance, one of the showcases of African wildlife conservation, a reported 500 Mozambican poachers were shot and killed by park rangers just between 2000 and 2014 (Reuters 2015).

Degradation narratives follow a common storyline. They begin in an idealized golden age, often in an imagined pristine landscape that existed in the distant past, or perhaps still exists today. Through long-term cultural adjustment and adaptation, native people were able to carve out a harmonious and sustainable relationship with nature. At this point, a disruptive force causes a breach in the system, and the perceived nature-culture balance cascades into chaos. The environment, which is the principal concern of outside observers, is as a consequence in a state of collapse. The cause of the crisis is usually traced to one of two possible sources. Either the problem is Malthusian, such that the introduction of Western biomedicine has led to a population explosion. Or the problem stems from the seductive influence of outside markets and resource commercialization, which are driving local stakeholders to exploit resources beyond their carrying capacity. In either case, the cost of complacency is catastrophic, both for local people, whose livelihoods depend on the soils, forests, and wildlife, and for outsiders, who ascribe considerable existence value to the natural world. But in reality, this belief in the pristine ideal, as well as the source of the threat, may be based on little more than cursory observations, hearsay, and the fertile imaginations of visitors and government administrators. Regardless, like all good stories, these gain the patina of legitimacy over time through statement and restatement by the storytellers, at first scientists and resource managers, later through films and the popular press. Eventually, the narrative becomes orthodoxy, an unfalsifiable "just-so" story that is difficult to challenge, even when pesky facts threaten to undermine the moral of the story.

Degradation narratives have expanded in recent years into the realm of ethnobotany. Although these stories are often based on limited knowledge of the intensity and frequency of plant harvest, the shift from subsistence to commercial exploitation of non-timber forest products (NTFPs) is frequently assumed to be unsustainable until proven otherwise. Forest-dwelling collectors are depicted as powerless pawns in the greater web of commercial desires by outsiders. The great paradox of this dialectic, however, is that throughout the colonial period and well into the twentieth century, unsustainable harvest of useful plants in the tropical realm was the direct result of species "management" by outsiders (chapter 4). In particular, given the very recent history of extractive enterprise in tropical Africa and America, it is hypocritical to instinctively blame indigenous populations for the ills of unsustainable rainforest management. Some of the darkest pages in recent human history are associated with the management of tropical nature by outsiders. In Amazonia, the quest for latex from the Brazilian rubber tree (*Hevea brasiliensis*) led to the genocidal exploitation and near extermination of indigenous people well into the twentieth century. Because the global West needed a reliable source of wild rubber, commercial enterprise procured the precious white resource by whatever means were deemed necessary. In the infamous case of the Putumayo people of northwestern Amazonia, agents of the Amazon Rubber Company regularly tortured native men and women to death, "their bodies were used for food for their dogs. . . . the Indians were mutilated in the stocks, cut to pieces with machetes, crucified head downwards . . . and burned alive, both men and women." These and more horrors were meted out to any "lazy" natives and their children who failed to fulfill their latex quota (Hardenburg and Enock 1912, 28–29, 203–210). And this was hardly an isolated case. A few years earlier in the great Congo watershed of Africa, the carnage carried out under the auspices of King Leopold II of Belgium in the procurement of wild

Congo rubber (*Landolphia* spp.) rivaled that of Nazi Germany. Through torture, kidnapping, murder, and starvation, an estimated ten million Africans sacrificed their lives in the European pursuit of latex-rich rainforest vines. The few observers who felt compelled to report these atrocities were jailed or publicly vilified by high-ranking politicians and even the Catholic Church (Hochschild 1999). Again, the justification for such barbarity was the phlegmatic stereotype that had been effectively deployed against indigenous people by outsiders since the earliest years of colonization (chapter 3). It is against this ignoble backdrop that current degradation narratives in ethnobotany are constructed, often with indigenous people as the perceived perpetrators.

Today, a bounty of previously subsistence forest products have entered into the global commercial marketplace, including foods, fibers, latex, cosmetics, and especially medicinal plants. The tyrannical NTFP managers of the colonial period and the previous century are gone, replaced by resource agencies, aid agencies, and non-governmental agencies attempting to reconcile rural economic development with habitat conservation. Distant First World markets, drawn often to the romantic allure of rainforest exotica, are increasingly the drivers of useful plant harvest and management. But while NTFPs are often touted as a successful strategy for rural development and resource conservation success, there is also the lingering expectation on the part of resource managers that the commercialization of extractive plant products will lead to overharvest and eventual species extirpation. Indeed, some have argued that external demand for NTFPs necessarily leads to unsustainable "boom and bust" economic cycles, further marginalization of rural communities, and overexploitation of botanical resources to the point of extinction (Crook and Clapp 1998). To date, however, and with some notable exceptions, the data do not support this assumption. Whether at the individual, population, community, or ecosystem level, most NTFP extraction is both ecologically and economically sustainable (Stanley et al. 2012). Nevertheless, perhaps because of the ghost of earlier degradation narratives, land use managers often conclude that NTFP harvest should be limited or curtailed entirely, even when ecological studies demonstrate clearly that they are being harvested sustainably. Many researchers, it seems, have been "indoctrinated to view all harvesting as detrimental," and have unconsciously bought into the prevailing narrative that indigenous people are powerless to regulate resource consumption in the face of international demand (Shackleton et al. 2015, 6). It is within this equivocal legacy of environmental storytelling that the jungle medicine narrative appeared.

Jungle Medicine Revisited

The jungle medicine narrative was born with the primary objective of saving the world's tropical rainforests from destruction. It was a strategy deployed by well-intentioned scientists and environmentalists to address a pressing environmental problem. And it appeared at a propitious point in history, just as the Western world was beginning to take a more globalized view of environmental issues. The realization that DDT, chlorofluorocarbons, and nuclear radiation were having effects many thousands of kilometers from their sources forced us to view the world as a single integrated entity. At the same time that magazines and documentaries were introducing us to the rainforest's wondrous biological bestiary, the hamburger connection narrative confronted us with the realization that these precious "global resources" were being sacrificed for the sake of America's obsession with fast food. Finally, the narrative arrived during a period of heightened global health concerns. The world was in the early stages of the AIDS epidemic, and horrific descriptions of the effects of Ebola coming out of West Africa signaled a renewed interest in the ancient quest for miracle cures from God's mythical medicine chest.

This latter day "green rush" for nature's healing plants was constructed along similar lines as other environmental chains of degradation. In the abbreviated version, tropical forests were being destroyed by the usual cast of characters, especially population growth, rural poverty, and neoliberal economic policies. Resultant intermediate impacts included loss of indigenous healers and extinction of medicinal plants. The final downstream effects included pharmaceuticals that would never be developed, diseases that would never be cured, and hefty profits that would never be realized. The twentieth century's search for the healing power of nature was, in several regards, similar to the great colonial quest for medicinal plants. It too was motivated by waves of new diseases, the possibility of prodigious profits, and ancient theories about the relationship between people and the curative properties of tropical nature. But there were differences. In terms of the environment, the degradation narrative was deployed during earlier centuries only as a rationalization for genetic theft, as in the case of Andean cinchona (chapter 4). Most importantly, although the modern jungle medicine narrative was supported by the legitimacy of science and its practitioners, each of the elements of the story was underpinned by a persuasive assemblage of myth, sentimentality, and nostalgia for a long-lost equatorial Eden.

The tropics have always constituted more of a dreamlike metaphor than a geographical reality for Westerners. Early on, the mysterious space between the tropics of Cancer and Capricorn was imagined to be the location of the biblical Garden of Eden. Later myths placed long lost civilizations and unfathomable riches in the bosom of the jungle. In the coming centuries, as earlier visions proved illusory, the warm and wet tropics were sexualized and sentimentalized by legions of explorers, exploiters, naturalists, and artists. The forests were seen simultaneously as sublime manifestations of God's creative energy and as satanic expressions of disease and deadly creatures. One particularly stubborn myth involved the notion that the tropics represented primordial expressions of wild nature, pristine forests and fields scarcely touched by the hand of humankind. Decades of research by anthropologists, biologists, and geographers, however, have not been kind to the pristine myth. At least in the Americas, this vision has been shown to be largely an illusion. Domesticated and mostly cultural in origin and composition, tropical forests are now seen as mosaics of more or less humanized habitats, grading from marginally managed in especially wet interfluves and highlands, to almost complete products of human modification near lakes and rivers. Science has largely abandoned the myth of the pristine, but it continues to thrive in the minds of many, reinforced by film, fiction, and tourist promotion.

The West's view of tropical forest-dwellers has likewise wavered wildly over the ages. Much as they have perceived tropical landscapes, outsiders have maintained concurrent and often conflicting stereotypes of its indigenous occupants—good, bad, or both simultaneously. In the time of the ancient Greeks, the blistering soils and boiling seas of the equatorial zone were imagined to be inhabited by Dog-Men, Amazons, and other brutish beasts. If humans survived, they reasoned, their geographic separation from the stimulating climate of the Mediterranean zone predestined them to a life of cannibalistic barbarism, fit only for enslavement. During the European Age of Exploration, these armchair prophecies had to be reconciled with the observations of Columbus, Vespucci, Cabral, and others. But although these early mariners put to rest the theory that the Torrid Zone was uninhabitable by humans, they nevertheless saddled tropical forest-dwellers with some of the same myths and stereotypes dreamed up by Strabo, Pliny, and Aristotle. Tropical primitives were man-eating barbarians, they reported, but they also cast them as licentious, godless, indolent, and surprisingly long-lived. The nineteenth-century arrival of European naturalists and artists, however, introduced a romanticized turn from which tropical peoples have never completely recovered. No longer depicted as primitive half-wits good only for forced labor, the forest-dwellers of tropical America and Oceania metamorphosed into noble savages, living languidly and harmoniously in the forest's Edenic plenty. Over time, this ennobled image became hopelessly intertwined with the romanticized depiction of tropical nature that prevailed into the mid-nineteenth century—exalted nature and natural man—the latter living in transcendent harmony due to the luxuriant fecundity provided by the former. In the early twentieth century, proponents of climatic determinist theories sought in vain to recast tropical people as lazy and unintelligent. But the age of environmentalism and ecological awareness, fortified by the mass media, has successfully reinstated the concept of tropical forest folk as the noble guardians of the forest primeval.

Fascination with the medicinal properties of tropical nature has a considerable legacy in the Western world. Some was fueled by the astronomically expensive medicinal spices that had flowed into Europe from mystical antipodal sources since at least the Middle Ages. But it was the European encounter with the Americas, as well as contact and colonization in sub-Saharan Africa and Asia, that forever cemented the association of efficacious healing pharmacopoeias with the warm and wet equatorial climes. Confronted by an armada of life-threatening worms and germs, many of which were spread by European travelers, colonial bioprospectors drew on ancient ethnobiological axioms to mine the healing floras of newly encountered lands and peoples. Among these was the theory that technologically primitive people were closest to nature, and therefore privy to an instinctive understanding of its medicinal virtues. The biblically inspired theory that diseases and their cures had similar birthplaces further stimulated the scouring of tropical latitudes for healing plants. Humoral medical theory from the time of Galen, as well as the ubiquitous Doctrine of Signatures, served as rough guideposts as to which new species served for which ailment. Driven by their sense of cultural and racial superiority, Europeans felt entitled to plunder the knowledge of healing plants from native peoples by whatever means were necessary, often coercion and trickery, sometimes torture and death. The notion that indigenous people had rights to their genetic and intellectual property was still centuries away. These early attitudes served to encourage and even promote the brazen biopiracy plots that were hatched in the eighteenth and nineteenth centuries. The plunder of the fabulously profitable nutmeg trees from the Banda Islands was followed shortly by a coordinated British conspiracy to steal quinine-yielding cinchona from Bolivia and Peru (chapter 4). It was within this tangled legacy of archaic theory, human exploitation, and bottomless greed that a new generation of bioprospectors began to plumb the pharmacological potential of tropical forests.

The 1960s discovery of a cure for childhood leukemia from the extracts of an obscure tropical shrub jump-started the jungle medicine narrative. If one plant could produce a blockbuster drug worth hundreds of millions of dollars, how many other unknown species awaited identification in the mysterious tropics? The unthinkable prospect of sacrificing future miracle cures for cancer, AIDS, and other diseases in order to graze a few scrawny cattle or to harvest a few hardwood trees was a call to action. Lacking evidence to the contrary, it was presumed that these unknown plants were to be found in old-growth rainforest, the object of growing environmentalist passion. But this turned out to be a premature assumption. Subsequent research coming out of Asia, Africa, Oceania, and the Americas revealed that most medicinal plants inhabit anthropogenic space—homegardens, trails, swiddens, and second-growth forest—not the primary rainforests so often alluded to. Indeed, in addition to native ruderals and weeds and garden cultivars, much of the tropical world's current plant pharmacopoeias are the accidental outcome of the global relocation and range expansion of crops and weeds that took place as part of the Columbian Exchange. Thus, although deforestation surely compromises many of the forest's biological values and ecosystem services, its impact on the ecological status of medicinal plants is relatively minor.

The inclusion of the secretive and knowledgeable shaman served to put a human face on the jungle medicine narrative. But it was always a man's face. In the early years of ethnobotanical inquiry, time constraints and cultural norms translated to mostly male ethnobotanists working with male field collaborators. This gender-imbalanced research made it difficult to discern the profound division of labor and resultant gendered knowledge of nature that characterizes so many traditional communities. With notable exceptions, men are engaged in subsistence and increasingly commercial activities that take them to less disturbed habitats, especially forests, whereas women are more likely to master the resources associated with anthropogenic space, exactly the habitats within which most medicinal plants grow wild or are tended. Empirical research in recent years suggests that women's quantitative medicinal plant knowledge is overall about the same as that of their male counterparts—sometimes greater, sometimes less, often just different. Contrary to the images transmitted by films and fictional accounts, tropical plant pharmacopeias are as much a province of informal female herbalists as they are of specialized male shamans. And in the years to come, as forest-dwelling men increasingly direct their day-to-day activities away from subsistence and towards commercial exploitation of nature, and move to cities for wage-earning jobs, the ethnobotanical knowledge and skills of mothers and grandmothers may come to play an even greater role in tropical communities.

Just as women were largely excluded from the jungle medicine narrative, so too were non-indigenous forest dwellers. Botanical knowledge was long conceived as the trial-and-error outcome of permanent residence and gradual familiarity with the local flora. The assumption of space and time contingency, however, left little room for the healing plant wisdom of immigrants and their offspring. This untested belief has been explored recently in Brazil and Suriname, both of which witnessed large arrivals of forced African immigrants during the slave trade. In both regions, enslaved women and men and their descendants went to great lengths to craft healing ethnofloras in their alien adopted landscapes. This was accomplished by means of species continuity, species substitution, and species acquisition. In Brazil, the African diaspora was mostly associated with anthropogenic space, especially towns and plantations. Their resultant healing floras were as a result largely derived from cultivated plants and successional species, many of Old World origin. In Suriname, however, where many Africans escaped to freedom to form maroon communities in the less-disturbed interior, their resultant plant pharmacopoeias included considerable successional species, but also a healthy sampling of native old-growth forest trees. For unlike Brazil's enslaved population, Surinamese maroons were able to wander freely in the forest and study its useful properties. And new species acquisition was guided more by association with Amerindians with whom they shared the forest, than with coastal Europeans. As victim diasporas, Africans in Brazil and Suriname were powerfully motivated to recreate their botanical traditions in the Americas. In both cases, the exclusion of the African diaspora from the jungle medicine narrative speaks more to our racial preconceptions than to a balanced appraisal of their ethnbotanical knowledge and skill.

Finally, the purported loss of shamans and their cognitive libraries of botanical knowledge is best understood in terms of the general devolution of traditional societies. The question of culture change and ethnobotanical erosion has been explored throughout the Old and New World tropics, and the results are nearly always the same. The principal globalizing influences, especially access to formal education, monotheistic religion, Western medicine, and market economies, are undermining indigenous knowledge and use of medicinal plants. There are some notable exceptions, and the nuances of change vary widely from one geographical location to another. But the wider message is that traditional knowledge of the healing properties of tropical nature, however that may be conceived, is indeed declining at a precipitous rate. On this issue the jungle medicine narrative was spot on; whatever potential miracle cures may be awaiting discovery from the traditional pharmacopoeias of tropical forest-dwellers are dwindling rapidly as modernization reaches the most isolated recesses of the equatorial zone.

Epilogue

Metaphors and narratives are effective platforms for translating scientific ideas to a largely science-phobic public. Through persuasive syntax and finely crafted stories, they make arcane and complex concepts both digestible and engaging. But storytelling in science is not without its hazards. And in the case of degradation narratives, the more often that projections of Armageddon fail to materialize, or appear so slowly as to be imperceptible to the senses, the less credibility science and scientists will have with the public and with policymakers (Kueffer and Larson 2014). If the nuances and uncertainty of environmental claims are not firmly embedded in the narrative, it runs the hazard of causing more harm than good. The jungle medicine narrative may well fall into this category. Scientists and environmentalists had their own reasons for promoting rainforest protection, but the deployment of the narrative, with its cast of environmental, economic, and social characters, steeped in ancient metaphor and nineteenth-century sentimentalism, proved to be an inflated and ultimately transitory conservation strategy.

The jungle medicine narrative made for a particularly compelling story because it effectively integrated such a diverse array of values and victims—pristine tropical forests, innocent indigenous societies, miracle cures, and obscene profits. It occurred at an opportune time in environmental history, as people were awakening to the idea that many environmental problems have global dimensions. And for Westerners, it was an easy target; they could emotionally invest in the concept of preserving distant tropical lands and peoples without any real personal sacrifice. But over two decades later, it is evident that all did not go as hoped. In the beginning, the primary drivers of the narrative were saving tropical forests and discovering novel drug plants. Towards this end, INBio (Instituto Nacional de Biodiversidad) in Costa Rica signed an agreement with Merck Pharmaceuticals to search for and share any royalties that came from development of new drugs. The first installment included one million US dollars, which was earmarked for research and forest protection. At the time, the project received considerable positive publicity. Similar efforts were launched by Shaman Pharmaceuticals, a private startup out of San Francisco, and the Periwinkle Project under the American NGO Rainforest Alliance. Both employed the ethnobotanical method to search for new drug plants, and both were dedicated to benefit sharing with indigenous collaborators. Optimistic articles in scientific journals with enticing titles like "Western Medicine Men Return to the Field" sang the praises of these newly formed collaborative arrangements (Joyce 1992). Unfortunately, all these efforts failed within two decades. Shaman Pharmaceuticals, the "tree huggers dream," filed for bankruptcy in 2001 (Anonymous 1999; Clapp and Crook 2002). INBio and the Periwinkle Project met similar fates. Although the reasons for the demise of each were complex, the bottom line was that not a single blockbuster drug was developed (Desmarchelier 2010; J. S. Miller 2011). Within a few years, most of the big pharmaceutical companies, including Bristol Myers, Squibb, Merck, Johnson and Johnson, Pfizer, and GlaxoSmithKline, had pared back or completely terminated their natural products programs (David et al. 2015).

There were other unexpected collateral impacts, the most troubling among these being the emergence of the biopiracy counter-narrative. In this new story, the objective of bioprospecting efforts in the tropics was not to incentivize nature protection and solve public health issues. Rather, it was part of a century's long conspiracy meant to deprive indigenous people and the developing world of its intellectual and genetic property. Big Pharma and its ethnobotanical errand boys were not the saviors of the forest and its noble denizens; they were the enemy. And much like the early history of medicinal plant exploitation, this newest wave of bioprospectors was oblivious to the concept of fair and equitable compensation. Some of these arguments were based on questionable claims, such as the idea that the inhabitants of Madagascar should be paid royalties for the development of vincristine (Frisvold and Day-Rubenstein 2008; see chapter 5). Such compensation claims, in turn, were underpinned by the vastly inflated figures produced early in the jungle medicine narrative, suggesting that pharmaceutical drugs were (or would be in the future) yielding billions of dollars in revenue.

Although none of these profits actually materialized, this factoid was repeated so many times that it became part of received wisdom. In the process, the biopiracy narrative assumed the moral high ground from the jungle medicine narrative.

The enduring legacy of exploitation of native peoples and the deeply conspiratorial tone of the biopiracy narrative assured its success. People are drawn to conspiracy theories as much as they are good stories, particularly where these involve big drug companies. In a recent survey, for example, 37% of respondents believed that big Pharma has pressured the FDA into preventing the public from getting natural cures for cancer and other diseases (Oliver and Wood 2014). Just since the year 2000, several hundred books and scholarly articles have appeared with the word "biopiracy" in the title. It has become a hugely popular topic, particularly among human rights advocates and lawyers. In the meantime, believing that highly profitable drug plants were being spirited out of their forests, developing world countries dramatically tightened the requirements for any sort of ethnobotanical inquiry. Even local scientists were precluded from field research. But this widely reported biodiversity "pot of gold at the end of the rainbow" failed to materialize (Antonelli and Rodriguez 2009; Clement 2007). Not a single blockbuster drug was developed. And even when equitable benefit-sharing agreements were negotiated with local indigenous communities, there was no guarantee that these would weather the sorts of misinformation being broadcast by some NGOs (Berlin and Berlin 2004). Unrealistic financial expectations filtered down to indigenous stakeholders in the field as well. Many had heard rumors about the fortunes being made, and everyone wanted a piece of the pie. In Cameroon, for instance, a promising research project with an anti-AIDS plant stumbled in part because the Cameroonians "were all wanting to buy themselves new Mercedes" (cited in Conniff 2012, 2). But there was no pie. No blockbuster drugs were developed, no diseases were cured, no money was made, and no tropical forests were protected.

But it was not all bad news. Of all the justifications put forth for protecting tropical forests, the jungle medicine narrative resonated most with the general public. And in spite of the rise of biopiracy narratives, most educated individuals are at least familiar with the idea that there are many healing plants in the tropical realm. Introductory college textbooks still refer to the cure for childhood leukemia by the Madagascar periwinkle as a powerful social and financial justification for forest protection. Developing world governments have eased their restrictions on ethnobotanical inquiry, and bioprospecting for novel compounds has returned to most tropical countries, now often directed by local scientists. It may be that there are some miracle cures awaiting discovery using the ethnobotanical method, but I'm inclined to believe that most of the low-hanging fruit was harvested many years ago. But there are still tens of thousands of compounds that were never identified by indigenous healers awaiting discovery, from plants, insects, amphibians, reptiles, and increasingly microbes. Who can say what researchers may turn up in the future? Most importantly, now that the rush to find a miracle cure has subsided, ethnobotanists are increasingly focusing their efforts on identifying the healing properties of tropical nature as a means of improving the lives of indigenous and immigrant people. There can be no better outcome than continued collaboration between ethnobotanists and tropical forest elders, documenting their medicinal plant knowledge, and supporting their efforts to collect and cultivate their healing flora for the benefit of their communities.

Notes

1. Plants also produce a large number of physical mechanisms to deter herbivory and pathogens, including fibrous leaves, spines, stinging hairs, latex and resin ducts, and many others. See: Coley and Barone 1996; Kaplan et al. 2008; Wink 2008. Claims of increased herbivory at low latitudes, and that plants have responded by developing higher levels of chemical defenses, have been challenged by Moles et al. 2011 and by Marquis et al. 2011. Lim et al. 2015 found that it applies to evergreen but not deciduous trees. The subject is not yet settled.

2. This frequently cited figure is just an estimate, as the company has not divulged actual profit figures for the drug.

3. Personal communication with Walter Lewis, February 2011.

4. The random testing of yew bark (*Taxus brevifolia*) in Oregon in the 1960s as part of NCI's plant screening program, and its eventual development into the powerful anti-cancer treatment taxol in the 1990s, represents a notable exception. However, as Walter Lewis notes (pers. comm. 2011), the expensive random approach would not have been necessary if researchers had paid attention to earlier published works showing the ethnobotanical uses of *Taxus baccata* in Asia. See also Walsh and Goodman 1999.

5. When I collected *Pilocarpus pennatifolius* in 2002, it was used by Afro-Brazilians in the state of Bahia in a bath as a treatment for head lice, and as a leaf infusion taken internally to treat constipation (Voeks and Leony 2004).

6. When I was diagnosed with open-angle glaucoma over thirty years ago, pilocarpine was the drug of choice used to lower my eye pressure.

CHAPTER TWO

1. According to Humboldt and Bonpland, this is a direct translation of Linnaeus's original statement, made in *Systema Naturae* in 1735, "vescitur palmis," roughly "palm eaters."
2. The ubiquity of humanized Amazonian landscapes is disputed by various authors. Bush et al. (2015), for example, argue that Amazonia should be viewed as a heterogeneous mosaic of landscapes, from significantly affected by human actions, especially close to rivers, to barely affected habitats deep in the interfluves.
3. According to P. A. Metcalf (1976), my field collaborator was mistaken, as the Kenyah did not practice "nulang," or "secondary burial." Rather it was another local linguistic group, the Berawan or perhaps the Tring.

CHAPTER THREE

1. The authenticity of André Thevet's observations is questioned in: Forsyth 1983, 147–178.
2. This phenomenon found its parallel in North America during the birth of the environmental movement in the1960s and 1970s, as the Noble Savage stereotype transitioned to the Ecological Indian (Krech 1999).
3. The human-induced collapse of Easter Island's forest ecosystems and human population is hotly contested. Alternative explanations for the demise include a mix of climate change, agricultural expansion, and invasive species introduction (cf. Rull et al. 2016).
4. According to Friedman (1981, 1), although some of these beings simply differed in physical appearance and social practice, writers in the Middle Ages referred to them as "monstrous."
5. These views reappear periodically, most recently with the publication of Jared Diamond's *Guns, Germs and Steel*, but now lacking the racist implications. See also Judkins et al. 2008.

CHAPTER FOUR

1. See also Warren Dean 1995, 179, who notes that in spite of the efforts of Melo Palheta, Brazil was still a net importer of coffee a century later.
2. Newson 2006 argues that physicians in the New World, away from the watchful eye of the Holy Inquisition, may not have felt so constrained by these ancient and dated ideas.
3. See also Drayton 2000, 9; Wear 2004, 330; Boomgaard 1996, 48; and Trapham 1679, 93–94: "the overflowing bounty of the great healer of us all, who hath given a balm for every Sore, and that not to be far sought and dear bought, but neer at hand, were our eyes opened like Hagars to see the thirsted for Remedy."

4. The affliction was later called chlorosis, and was a form of anemia.
5. Richard Evans Schultes shared this view. From his own considerable personal experience, he argued that if researchers treated indigenous healers and shamans with appropriate respect, his or her natural curiosity will lead to "an easier exchange of conversation and ideas concerning plant uses" (Schultes 1960, 261). The considerable reticence noted by so many travelers and physicians may well have been due to their attitudes of superiority.
6. This transaction is reported by various authors, but there is suspiciously no mention of it in Rolander's diary of his time in Suriname (June 1755–January 1756), although he discusses Quassi at length (Rolander 2008 [1755]; Van Andel et al. 2012, and pers. comm. Tinde van Andel 2013). Davis (2016, 15) states that the cure was "extracted" from Quassi by a Swedish settler and taken to Linnaeus in Sweden. See also: Moraes 2012, 28.
7. According to Henry Bruman (1948), there were likely other similar discoveries of how to pollinate vanilla by hand.
8. See also, Garcia de Orta's 1563 *Colloquies on the simples and drugs of India*, which similarly broke with European tradition by privileging indigenous healing knowledge over European (Gunn 2003, 160–161).
9. The most notorious act of biopiracy did not involve a medicinal species, but rather the Amazonian rubber tree (*Hevea brasiliensis*). Wild rubber was once the region's primary source of wealth, but thousands of seedlings were smuggled out of Brazil to Kew Gardens in the United Kingdom by Henry Wickham. They were later transported to Southeast Asia, where they went on to dominate the globe's supply of the precious resource, and thereby destroy Brazil's famous rubber boom. See Jackson 2008.
10. Variations of these accounts are provided by numerous authors, including: Honigsbaum 2002, 22–23; Lloyd 1911, 29; Sherman 2011, 23–25.
11. Brockway 1979, 103–139, argues that suggesting that Indians were not aware of the healing properties of their own flora simply legitimized the British authority to exploit and to transfer the plant to other regions and further inflated their own sense of superiority.

CHAPTER FIVE

1. These are stored in the herbarium of the Centro de Pesquisa do Cacao, in Itabuna, Bahia.
2. On the island of Vanuatu, healers use plants that are easily accessible and available, but more of these are trees than herbs (Bradacs et al. 2011). There are numerous surveys of useful species in tropical landscapes not included here because the researchers only censused trees and large lianas.
3. I include only a sampling of this literature here: Albuquerque et al. 2011; Begossi et al. 2002; Chazdon and Coe 1999; Furusawa 2016; Gavin 2009; Giday et al. 2009; Heinrich and Barerra 1993; Junsongduang et al. 2013; Kohn 1993; Kunwar et al. 2016; Lyon and Hardesty 2012; Marshall and

Hawthorne 2012; Mutheeswaran et al. 2011; Poderoso et al. 2012; Rodenburg et al. 2012; Stagegaard et al. 2002; Stepp and Moerman 2001; Thomas
et al. 2011; Toledo et al. 1992; Towns et al. 2014.

4. Most of these studies have been carried out in arid or semi-arid environ-
    ments, however, rather than moist tropical forests.

5. It should be noted that in general, the presence of allelochemicals (which
    often don't taste good) decreases along the gradient from wild to domes-
    ticated species. However, whether this principle applies to the part of the
    plant that people use medicinally—fruit, grain, seed, or leaf—has not been
    investigated.

6. Cardenas (2013) points out, however, that the Greeks did not go so far as to
    conflate food and medicine. They were quite different features of the heal-
    ing process. Thus, the much quoted Hippocratic phrase "let food be thy
    medicine and medicine be thy food" appears to be a popular fabrication.

7. The noteworthy exception to this are spices, many of which have medici-
    nal benefits. Although most originated in the tropical realm, they have
    been widely incorporated into temperate zone cuisines. See Sherman and
    Billing 1999.

CHAPTER SIX

1. I only include studies that employed some measure of quantitative meth-
    ods to explore the question. The numerous anecdotal passages about one
    or another gender's superiority in medicinal plant knowledge are omitted.

2. I exclude studies that by their nature explored medicinal plant knowledge
    from one or another gender perspective, such as plants used to treat exclu-
    sively male or female health complaints.

CHAPTER SEVEN

1. The term diaspora referred originally to scattered Jewish groups that lived
    among minorities, but in recent years it has been generalized to include
    all voluntary and forced migrants, regardless of race, religion, profession,
    or economic standing. The concept involves a sense of being at home in a
    new place, but still retaining significant connections with a distant locale,
    often considered a homeland (Dufoix 2008).

2. I am not discounting the cultural and biological impacts of early Indian
    Ocean plants and people movements, but they are not part of this story.
    See for example: Boivin et al. 2013.

3. For example: *pau pombo* (*Tapirira guianensis*), Brazilian pepper-tree (*aroeira*—
    Schinus terebinthifolia), and *canela-de-velho* (*Miconia* sp.).

4. Terborgh et al. 2016 challenge the abiotic hypotheses for the origin of bio-
    diversity differences between Africa and South America.

5. These included *Aristolochia, Bauhinia, Casearia, Ficus, Mimosa, Piper, Sida, Vernonia, Vitex*, all of which enter into the oral myths of Yoruba *babalaô* (Yoruba father of secrets) and, in time, were incorporated into the Candomblé spiritual flora via species substitution. See Mori et al. 1983; Verger 1967.

6. The origin of the species is still disputed, with evidence supporting both South Asia and West Africa (Boivin et al. 2014)

7. The "apostles" were mostly botany and zoology students who traveled to the far corners of the globe under the tutelage of Linnaeus. Many lived and collected in tropical Asia, Africa, and the Americas, and some died for their efforts. Swedish-born Linnaeus himself never traveled further south than Paris.

8. These include *alecrim* (*Hyptis fruticosa*), *poejo* (*Mentha pulegium*), and *patchulí* (*Pogostemon cf. cablin*). Other aromatics associated with Oxum include *sabugueiro* (*Sambucus australis*), *macaça* (*Brunfelsia uniflora*), *jasmin do brejo* (*Hedychium coronarium*), and *beti cheiroso* (*Piper* sp.).

9. These include *arrozinha* (*Zornia* cf. *gemella*), *mal-me-quer* (*Wedelia paludosa*), *maravilha* (*Caesalpinia pulcherrima*), and several others.

10. There are various calculations of use value (UV) in ethnobotany (chapter 8). In this case, UV is calculated as the total number of use citations (how many times a medicinal use is cited), divided by the total number of people participating in the census. Thus, if 4 people identified a total of 50 medicinal uses for species in a plot, the UV would be 50/4, or 12.5.

11. In most instances, enslaved and free blacks were given free rein to tend to their own ailments in the Americas, and in many instances, attained considerable fame as herbalists. This was especially true in the area of magic and aphrodisiacs, in which they were often considered great masters (Voeks 2009).

CHAPTER EIGHT

1. Not his true name.

2. Wolff and Medin (2001) use the term devolution to describe diminishing knowledge of nature as a result of modernization. For various proximate reasons, this causes people to have less contact with nature, and hence less opportunity and reason to sustain and create knowledge of nature.

3. Because very few participants had wage employment, we used these features as culturally relevant proxies for economic standing in the community.

4. In the specific case of HIV/AIDS, the use of some medicinal plants (*Hypoxis* sp.) may produce an unfavorable herb-drug interaction with antiretroviral therapy. In addition, time spent consulting traditional healers rather than Western doctors may increase the risk of HIV patients developing AIDS.

CHAPTER NINE

1. The "whopper" is Burger King's signature hamburger.

References

Achaya, K. T. 1998. "Bounty from the New World." In *European Intruders and Changes in Behaviour and Customs in Africa, America and Asia before 1800. An Expanding World*, vol. 30, edited by M. J. MacLeod and E. S. Rawski, 259–282. Brookfield, VT: Ashgate.

Acker, A. 2014. " 'O Maior Incêndio do Planeta': Como a Volkswagen e o Regime Militar Brasileiro Acidentalmente Ajudaram a Transformar a Amazônia em Uma Arena Política Global." *Revista Brasileira de História* 34 (68): 13–33.

Acosta, J. 1970 [1604]. *The Natural & Moral History of the Indies*. Translated by E. Grimston. Vol. 61 (2). Hakluyt Society. New York: Burt Franklin.

Adanson, M. 1759. *A Voyage to Senegal, the Isle of Goree, and the River Gambia*. London: J. Nourse.

Aguilar-Støen, M., Moe, S. R., and Camargo-Ricalde, S. L. 2009. "Home Gardens Sustain Crop Diversity and Improve Farm Resilience in Candelaria Loxicha, Oaxaca, Mexico." *Human Ecology* 37: 55–77.

Ahmad, S., Pillai, S., and Yusof, N. 2011. "Rehabilitating Eden: Archetypal Images of Malaya in European Travel Writing." *Journeys* 12: 22–45.

Albuquerque, U., Andrade, L., and Caballero, J. 2005. "Structure and Floristics of Homegardens in Northeastern Brazil." *Journal of Arid Environments* 62: 491–506.

Albuquerque, U., Soldati, G. T., Sieber, S. S., Ramos, M. A., de Sá, J. C., and de Souza, L. C. 2011. "The Use of Plants in the Medical System of the Fulni-ô people (NE Brazil): A Perspective on Age and Gender." *Journal of Ethnopharmacology* 133: 866–873.

Alencar, N. L., de Sousa Araújo, T. A., de Amorim, E. L. C., and de Albuquerque, U. P. 2010. "The Inclusion and Selection of Medicinal Plants in Traditional Pharmacopoeias—Evidence in Support of the Diversification Hypothesis." *Economic Botany* 64: 68–79.

Almeida, L. F. de. 1975. "Aclimatação de Plantas do Oriente no Brasil Durante os Séculos XVII e XVIII." *Revista Portuguesa da História* 15: 339–481.

Alpern, S. B. 2008. "Exotic Plants of Western Africa: Where They Came from and When." *History in Africa* 35: 63–102.

Alvard, M. 1993. "Testing the 'Ecologically Noble Savage' Hypothesis: Interspecific Prey Choice by Piro Hunters of Amazonian Peru." *Human Ecology* 21: 355–387.

Anagnostou, S. 2007. "The International Transfer of Medicinal Drugs by the Society of Jesus (Sixteenth to Eighteenth Centuries) and Connections with the Work of Carolus Clusius." In *Carolus Clusius*, edited by F. Egmond, P. Hoftijzer, and R. Visser, 293–313. Amsterdam: Royal Netherlands Academy of Arts and Sciences.

Andrews, J. 1995. *Peppers: The Domesticated Capsicums.* Austin: University of Texas Press.

Anonymous. 1665. *A True Relation of the Unjust, Cruell, and Barbarous Proceedings against the English at Amboyna in the East-Indies.* 3rd ed. London: William Hope.

———. 1993. "New in the Botanical Book of Records: Highest Tree Diversity in the World." *Field Notes from the New York Botanical Garden* 2: 1–2.

———. 1999. "Shaman Loses Its Magic." *Economist*, February 18.

Antonelli, A., and Rodriguez, V. 2009. "Brazil Should Facilitate Research Permits." *Conservation Biology* 23 (5): 1068–1074.

Applequist, W., Ratsimbason, M., Kuhlman, A., Rakotonandrasana, S., and Kingston, D. 2017. "Antimalarial Use of Malagasy Plants Is Poorly Correlated with Performance in Antimalarial Bioassays." *Economic Botany* 71 (1): 75–82.

Arano, L. C. 1976. *The Medieval Health Handbook: Tacuinum Sanitatis.* New York: George Braziller.

Araújo, F., and Lopes, M. 2012. "Diversity of Use and Local Knowledge of Palms (Arecaceae) in Eastern Amazonia." *Biodiversity and Conservation* 21: 487–501.

Arnold, A., and Lutzoni, F. 2007. "Diversity and Host Range of Foliar Fungal Endophytes: Are Tropical Leaves Biodiversity Hotspots?" *Ecology* 88: 541–549.

Arnold, D. 1996. *The Problem of Nature: Environment, Culture and European Expansion.* Oxford, UK: Blackwell.

———. 2000. "Illusory Riches: Representations of the Tropical World, 1840–1950." *Singapore Journal of Tropical Geography* 21: 6–18.

Asiimwe, S., Kamatenesi-Mugisha, M., Namutebi, A., Borg-Karlsson, A. K., and Musiimenta, P. 2013. "Ethnobotanical Study of Nutri-medicinal Plants Used for the Management of HIV/AIDS Opportunistic Ailments among the Local Communities of Western Uganda." *Journal of Ethnopharmacology* 150 (2): 639–648.

Atkins, J. 1737. *A Voyage to Guinea, Brazil, and the West Indies in His Majesty's Ships, etc.* London: C. Ward and R. Chandler.

Atkinson, J. M. 1992. "Shamanism Today." *Annual Review of Anthropology* 21: 307–330.

Audet, C. M., Blevins, M., Moon, T. D., Sidat, M., Shepherd, B. E., Pires, P., Vergara, A., and Vermund, S. H. 2012. "HIV/AIDS-Related Attitudes and Practices among Traditional Healers in Zambézia Province, Mozambique." *Journal of Alternative and Complementary Medicine* 18 (12): 1133–1141.

Austin, D. F. 2013. "Moon-Flower (*Ipomoea alba*, Convolvulaceae)—Medicine, Rubber Enabler, and Ornamental: A Review." *Economic Botany* 67 (3): 244–262.

Avohou, H. T., Vodouhe, R. S., Dansi, A., Kpeki, B., and Bellon, M. 2012. "Ethnobotanical Factors Influencing the Use and Management of Wild Edible Plants of Agricultural Environments in Benin." *Ethnobotany Research & Applications* 10: 571–592.

Baker, H. G. 1974. "The Evolution of Weeds." *Annual Review of Ecology and Systematics* 5: 1–24.

Balée, W. 2013. *Cultural Forests of the Amazon: A Historical Ecology of People and their Landscapes*. Tuscaloosa: University of Alabama Press.

Balée, W., and Gély, A. 1989. "Managed Forest Successions in the Amazon: The Ka'apor Case." *Advances in Economic Botany* 7: 149–173.

Balick, M. 1990. "Ethnobotany and the Identification of Therapeutic Agents from the Rainforest." In *Bioactive Compounds from Plants*, edited by D. J. Chadwick and J. Marsh, 22–39. Chichester: Wiley.

———. 2007. "Traditional Knowledge: Lessons from the Past, Lessons for the Future." In *Biodiversity & the Law: Intellectual Property, Biotechnology & Traditional Knowledge*, edited by C. R. McManis, 280–296. London: Earthscan.

———. 2012. "Reflections on Richard Evans Schultes, the Society for Economic Botany, and the Trajectory of Ethnobotanical Research." In *Medicinal Plants and the Legacy of Richard Evans Schultes*, edited by B. Ponman and R. Bussmann, 3–21. St. Louis: Missouri Botanical Garden.

Balick, M., and Cox, P. 1996. *Plants, People, and Culture: The Science of Ethnobotany*. New York: Scientific.

Ban, N., and Coomes, O. 2004. "Home Gardens in Amazonian Peru: Diversity and Exchange of Planting Material." *Geographical Review* 94: 348–367.

Barbhuiya, A. R., Sharma, G. D., Arunachalam, A., and Deb, S. 2009. "Diversity and Conservation of Medicinal Plants in Barak Valley, Northeast India." *Indian Journal of Traditional Knowledge* 8 (2): 169–175.

Barker, G., Barton, H., Beavitt, P., Bird, M., Daly, P., Doherty, C., Gilbertson, D., Hunt, C., Krigbaum, J., Lewis, H., and Manser, J. 2002. "Prehistoric Foragers and Farmers in South-east Asia: Renewed Investigations at Niah Cave, Sarawak." In *Proceedings of the Prehistoric Society*, 68: 147–164. Cambridge, UK: Cambridge University Press.

Barnes, L. L. 2005. *Needles, Herbs, Gods and Ghosts: China, Healing and the West to 1848*. Cambridge, MA: Harvard University Press.

Barros, J. F. P. de. 1983. Ewé o Osányin: Sistema de classificação de vegetais nas casas de santo Jêje-Nagô de Salvador, Bahia. Unpubl. PhD diss., Universidade de São Paulo, Brazil.

Bassett, T. J., and Zuéli, K. B. 2000. "Environmental Discourses and the Ivorian Savanna." *Annals of the Association of American Geographers* 90 (1): 67–95.

Bates, H. W. 1864. *The Naturalist on the River Amazons.* Vol. 1. London: John Murray.

Bates, M. 1952. *Where Winter Never Comes: A Study of Man and Nature in the Tropics.* New York: Charles Scribner's Sons.

Bauer, I. L. 2013. "Candiru—A Little Fish with Bad Habits: Need Travel Health Professionals Worry? A Review." *Journal of Travel Medicine* 20: 119–124.

Bedoya, L. M., Sanchez-Palomino, S., Abad, M. J., Bermejo, P., and Alcami, J. 2001. "Anti-HIV Activity of Medicinal Plant Extracts." *Journal of Ethnopharmacology* 77: 113–116.

Beekman, E. M. 1981. *The Poison Tree: Selected Writings of Rumphius on the Natural History of the Indies.* Amherst: University of Massachusetts Press.

Begossi, A., Hanazaki, N., and Tamashiro, J. Y. 2002. "Medicinal Plants in the Atlantic Forest (Brazil): Knowledge, Use, and Conservation." *Human Ecology* 30: 281–299.

Bennett, B. C. 2007. "Doctrine of Signatures: An Explanation of Medicinal Plant Discovery or Dissemination of Knowledge?" *Economic Botany* 61: 246–255.

Bennett, B. C., and Prance, G. T. 2000. "Introduced Plants in the Indigenous Pharmacopoeia of Northern South America." *Economic Botany* 54: 90–102.

Ben-Noun, L. L. 2003. "Figs—The Earliest Known Ancient Drug for Cutaneous Anthrax." *Annals of Pharmacotherapy* 37: 297–300.

Bergreen, L. 2003. *Over the Edge of the World: Magellan's Terrifying Circumnavigation of the Globe.* New York: Perennial.

Berkes, F. 2012. *Sacred Ecology.* 3rd ed. New York: Routledge.

Berlin, B. 1992. *Ethnobiological Classification: Principles of Categorization of Plants and Animals in Traditional Societies.* Princeton: Princeton University Press.

Berlin, B., and Berlin, E. A. 2004. "Community Autonomy and the Maya ICBG Project in Chiapas, Mexico: How a Bioprospecting Project that Should Have Succeeded Failed." *Human Organization* 63 (4): 472–486.

Bernal, R., Torres, C., García, N., Isaza, C., Navarro, J., Vallejo, M. I., and Balslev, H. 2011. "Palm Management in South America." *Botanical Review* 77: 607–646.

Biwolé, A. B., Morin-Rivat, J., Fayolle, A., Bitondo, D., Dedry, L., Dainou, K., Hardy, O., and Doucet, J. L. 2015. "New Data on the Recent History of the Littoral Forests of Southern Cameroon: An Insight into the Role of Historical Human Disturbances on the Current Forest Composition." *Plant Ecology and Evolution* 148 (1): 19–28.

Blaikie, P.M., and Muldavin, J. S. 2004. "Upstream, Downstream, China, India: The Politics of Environment in the Himalayan Region." *Annals of the Association of American Geographers* 94 (3): 520–548.

Blégny, N. 1682. *The English Remedy, Or Talbor's Wonderful Secret for Cureing of Agues and Feavers.* London: F. Wallis.

Blench, R. 2009. "Bananas and Plantains in Africa: Re-interpreting the Linguistic Evidence." *Ethnobotany Research and Applications* 7: 363–379.

Bode, A. M., and Dong, Z. 2011. "The Two Faces of Capsaicin." *Cancer Research* 71: 2809–2814.

Boivin, N., Crowther, A., Helm, R., and Fuller, D. Q. 2013. "East Africa and Madagascar in the Indian Ocean World." *Journal of World Prehistory* 26: 213–281.

Boivin, N., Crowther, A., Prendergast, M., and Fuller, D. Q. 2014. "Indian Ocean Food Globalisation and Africa." *African Archaeological Review* 31 (4): 547–581.

Bontius, J. 1769. *An Account of the Diseases, Natural History, and Medicines of the East Indies.* Translated from the Latin of James Bontius. London: T. Noteman.

Boomgaard, P. 1996. "Dutch Medicine in Asia: 1600–1900." In *Warm Climates and Western Medicine: The Emergence of Tropical Medicine 1500–1900,* edited by D. Arnold, 42–64. Amsterdam: P. V. Rodopi.

Bory, S., Grisoni, M., Duval, M., and Besse, P. 2008. "Biodiversity and Preservation of Vanilla: Present State of Knowledge." *Genetic Resources and Crop Evolution* 55: 551–571.

Bosman, W. 1721. *A New and Accurate Description of the Coast of Guinea, Divided into the Gold, the Slave, and the Ivory Coasts.* London: James Knapton.

Bourgeois, N. L. 1788. *Voyages Intéressans dans Différentes Colonies Françaises, Espagnoles, Englaises, etc.* London: Jean-François Bastien.

Bouton, J. 1640. *Relation de l'etablissement des François depuis l'an 1635.* Paris: NP.

Boxer, C. R. 1963. *Race Relations in the Portuguese Colonial Empire 1414–1825.* Oxford, UK: Clarendon.

Boyle, R. 1696. *Medicinal Experiments: Or A Collection of Choice and Safe Remedies, The Most Part Simple and Easily Prepared.* 3rd ed. London: Samuel Smith and B. Walford.

Bradacs, G., Heilmann, J., and Weckerle, C. S. 2011. "Medicinal Plant Use in Vanuatu: A Comparative Ethnobotanical Study of Three Islands." *Journal of Ethnopharmacology* 137 (1): 434–448.

Brazeal, B. 2007. *Blood, Money and Fame: Nago Magic in the Bahian Backlands.* PhD diss., University of Chicago.

Breton, R. 1978 [1647]. *Relations de l'Île de la Guadeloupe.* Basse-Terre: Société d'histoire de la Guadeloupe.

Brncic, T. M., Willis, K. J., Harris, D. J., and Washington, R. 2007. "Culture or Climate? The Relative Influences of Past Processes on the Composition of the Lowland Congo Rainforest." *Philosophical Transactions of the Royal Society B* 362: 229–242.

Brockway, L. H. 1979. *Science and Colonial Expansion: The Role of the British Royal Botanic Gardens.* New York: Academic.

Brookfield, H. 2007. "Working with and for Plants: Indigenous Fallow Management in Perspective." In *Voices from the Forest: Integrating Indigenous Knowledge in Sustainable Upland Farming,* edited by M. Cairns, 8–15. Washington, DC: Resources for the Future.

Brosius, J. P. 1991. "Foraging in Tropical Rainforests: The Case of the Penan of Sarawak, East Malaysia." *Human Ecology* 19: 123–150.

Browner, C. 1991. "Gender Politics in the Distribution of Therapeutic Herbal Knowledge." *Medical Anthropology Quarterly* 5: 99–132.

Bruman, H. 1948. "The Culture History of Mexican Vanilla." *Hispanic American Historical Review* 28: 360–376.

Bry, T. de. 1592. *Americae Tertia Pars.* Part III. 1st ed., 2nd issue. Latin. NP.

Bunbury, E. H. 1959. *A History of Ancient Geography: Among the Greeks and Romans from the Earliest Ages till the Fall of the Roman Empire.* New York: Dover.

Bush, M. B., McMichael, C. H., Piperno, D. R., Silman, M. R., Barlow, J., Peres, C. A., Power, M., and Palace, M. W. 2015. "Anthropogenic Influence on Amazonian Forests in Pre-history: An Ecological Perspective." *Journal of Biogeography* 42 (12): 2277–2288.

Bussmann, R. 2012. "I Know Every Tree in the Forest: Reflections on the Life and Legacy of Richard Evans Schultes." In *Medicinal Plants and the Legacy of Richard E. Schultes,* edited by B. Ponman and R. Bussmann, 13–21. St. Louis: Missouri Botanical Garden.

Bussmann, R., and Sharon, D. 2009. "Shadows of the Colonial Past—Diverging Plant Use in Northern Peru and Southern Ecuador." *Journal of Ethnobiology and Ethnomedicine* 5: 4. doi:10.1186/1746-4269-5-4.

Cabrera, L. 1971. *El Monte.* Miami, FL: NP.

Callaway, E., and Cyranoski, D. 2015. "China Celebrates First Nobel." *Nature* 526 (7572): 174–175.

Camargo, M. 2005. "Estudo Etnobotanico da Mandioca (*Maniot esculenta* Crantz Euphorbiaceae) na Diaspora Africana." In *Anais do Seminário Gastronomia de Gilberto Freyre,* 22–30. Recife, Pernambuco: Fundacão Gilberto Freyre.

Caminha, P. Vaz de. 1963 [1500]. *Carta a El Rei D. Manuel.* São Paulo, Brazil: Dominus Editora.

Camou-Guerrero, A., Reyes-García, V., Martínez-Ramos, M,. and Casas, A. 2008. "Knowledge and Use Value of Plant Species in a Rarámuri Community: A Gender Perspective for Conservation." *Human Ecology* 36: 259–272.

Campbell, L. M. 2002. "Conservation Narratives in Costa Rica: Conflict and Coexistence." *Development and Change* 33 (1): 29–56.

Campbell, M. O. N. 2005. "Sacred Groves for Forest Conservation in Ghana's Coastal Savannas: Assessing Ecological and Social Dimensions." *Singapore Journal of Tropical Geography* 26 (2): 151–169.

Caniago, I., and Siebert, S. F. 1998. "Medicinal Plant Ecology, Knowledge and Conservation in Kalimantan, Indonesia." *Economic Botany* 52: 229–250.

Cañizares-Esguerra, J. 2006. *Nature, Empire, and Nation: Explorations of the History of Science in the Iberian World.* Stanford, CA: Stanford University Press.

Cardenas, D. 2013. "Let Not Thy Food Be Confused with Thy Medicine: The Hippocratic Misquotation." *e-SPEN Journal* 8: 260–262.

Cardim, F. 1939 [1584]. *Tratados da Terra e Gente do Brasil.* São Paulo: NP.

Carey, C. 2003. "The Political Economy of Poison: The Kingdom of Makassar and the Early Royal Society." *Renaissance Studies* 17: 517–543.

Carlson, T. J. 2001. "Language, Ethnobotanical Knowledge, and Tropical Public Health." In *On Biocultural Diversity. Linking Language, Knowledge and the Environment*, edited by L. Maffi, 489–502. Washington, DC: Sm·⁺hsonian Institution Press.

Carneiro, E. 1948. *Candomblés da Bahia*. NP: Editora Tecnoprint.

Carney, J. 2013. "Seeds of Memory: Botanical Legacies of the African Diaspora." In *African Ethnobotany in the Americas*, edited by R. Voeks and J. Rashford, 13–33. New York: Springer.

Carney, J., and Rosomoff, R. 2011. *In the Shadow of Slavery: Africa's Botanical Legacy in the Atlantic World*. Berkeley: University of California Press.

Carney, J., and Voeks, R. 2003. "Landscape Legacies of the African Diaspora in Brazil." *Progress in Human Geography* 27 (2): 139–152.

Cartas do Brasil do Padre Manuel Nóbrega. 1886 [1549–1560]. *Materaes e Achêgas para a História e Geografia do Brasil No. 1*. Rio de Janeiro: Imprensa Nacional.

Cartas Jesuíticas. 1933. *Cartas, Informacões, Fragmentos Históricas e Sermões do Padre Joseph de Anchieta, S. J. (1554–1594)*. Rio de Janeiro: Civilização Brasileira.

Casas, A., Otero-Arnaiz, A., Pérez-Negrón, E., and Valiente-Banuet, A. 2007. "In Situ Management and Domestication of Plants in Mesoamerica." *Annals of Botany* 100: 1101–1115.

Case, R. J., G. F. Pauli, and Soejarto, D. D. 2005. "Factors in Maintaining Indigenous Knowledge among Ethnic Communities of Manus Island." *Economic Botany* 59: 356–365.

Castaneda, C. 1968. *The Teachings of Don Juan: A Yaqui Way of Knowledge*. New York: Simon and Schuster.

Chabrán, R. 2000. "The Classical Tradition in Renaissance Spain and New Trends in Philology, Medicine, and Material Medica." In *Searching for the Secrets of Nature: The Life and Works of Dr. Francisco Hernández*, edited by S. Varey, R. Chabrán, and D. Weiner, 21–32. Stanford, CA: Stanford University Press.

Chambers, N. 2000. *The Letters of Sir Joseph Banks. A Selection 1768–1820*. London: Imperial College Press.

Chappell, B. S. 1975. "Dr. Caesar and His Cure for Poisons and Rattle-snake Bites." *Journal of the South Carolina Medical Association* 71: 183–187.

Chazdon, R., and Coe, F. G. 1999. "Ethnobotany of Woody Species in Second-Growth, Old-Growth, and Selectively Logged Forests of Northeastern Costa Rica." *Conservation Biology* 13: 1312–1322.

Christensen, H. 2003. "Fallows and Secondary Forests—A Primary Resource for Food." In *Local Land Use Strategies in a Globalizing World: Shaping Sustainable Social and Natural Environments: Proceedings of the International Conference, August 21–23, 2003*. Institute of Geography, University of Copenhagen, Denmark. Vol. 1, edited by O. Mertz, R. Wadley, and A. E. Christensen, 235–263. Denmark: DUCED SLUSE.

Cieza de León, P. 1918 [1518–1554]. *Civil Wars of Peru. Part 4, Book 2. The War of Chupas*. Translated by Sir Clements Markham. London: Hakluyt Society.

Clapp, R. A., and Crook, C. 2002. "Drowning in the Magic Well: Shaman Pharmaceuticals and the Elusive Value of Traditional Knowledge." *Journal of Environment & Development* 11 (1): 79–102.

Clark, C. 2017. "Brazilian Peppertree Packs Power to Knock out Antibiotic-Resistant Bacteria." eScience Commons. February 10. https://esciencecom mons.blogspot.com/2017/02/brazilian-peppertree-packs-power-to.html

Clark, V., Raxworthy, C., Rakotomalala, V., Sierwald, P., and Fisher, B. 2005. "Convergent Evolution of Chemical Defense in Poison Frogs and Arthropod Prey between Madagascar and the Neotropics." *Proceedings of the National Academy of Sciences* 102: 11607–11622.

Clarke, R., and Merlin, M. 2013. *Cannabis: Evolution and Ethnobotany.* Berkeley: University of California Press.

Clement, C. R. 1999. "1492 and the Loss of Amazonian Crop Genetic Resources. I. The Relation between Domestication and Human Population Decline." *Economic Botany* 53 (2): 188–202.

———. 2007. "Um Pote de Ouro no Fim do Arco-íris? O Valor da Biodiversidade e do Conhecimento Tradicional Associado, e as Mazelas da Lei de Acesso: Um Visão e Proposta a Partir da Amazônia." *Amazônia: Ciencia e Desenvolvimento* 3 (5): 177–198.

Clement, C. R., Denevan, W. M., Heckenberger, M. J., Junqueira, A. B., Neves, E. G., Teixeira, W. G., and Woods, W. I. 2015. "The Domestication of Amazonia before European Conquest." *Proceedings of the Royal Society B* 282 (1812). doi:10.1098/rspb.2015.0813

Coe, F. G., and Anderson, G. J. 1996. "Ethnobotany of the Garífuna of Eastern Nicaragua." *Economic Botany* 50: 71–107.

Cohen, R. 1996. "Diasporas and the Nation-State: From Victims to Challengers." *International Affairs* 72 (3): 507–520.

Coles, W. 1657. *Adam in Eden or, Natures Paradise. The History of Plants, Fruits, Herbs and Flowers.* London: J. Streater.

Coley, P. D., and Aide, T. M. 1991. "Comparison of Herbivory and Plant Defenses in Temperate and Tropical Broad-leaved Forests." In *Plant-Animal Interactions: Evolutionary Ecology in Tropical and Temperate Regions,* edited by P. W. Price, T. M. Lewinsohn, G.W. Fernandes, and W. W. Benson, 25–49. New York: Wiley.

Coley, P. D., and Barone, J. A. 1996. "Herbivory and Plant Defenses in Tropical Forests." *Annual Review of Ecology and Systematics* 27: 305–335.

Coley, P. D., Heller, M., Aizprua, R., Araúz, B., Flores, N., Correa, M., Gupta, M., Solis, P., Ortega-Barría, E., Romero, L., Gómez, B., Ramos, M., Cubilla-Rios, L., Capson, T., and Kursar, T. 2003. "Using Ecological Criteria to Design Plant Collection Strategies for Drug Discovery." *Frontiers in Ecology and the Environment* 1: 421–428.

Colfer, C., and Minarchek, R. D. 2013. "Introducing 'The Gender Box': A Framework for Analysing Gender Roles in Forest Management." *International Forestry Review* 15 (4): 411–426.

Colfer, C., Sheil, D., and Kishi, M. 2006. *Forests and Human Health: Assessing the Evidence*. Vol. 45. Jakarta, Indonesia: Center for International Forestry Research.

Columbus, C. 1932 [1498]. "Letter to Queen Isabella Describing the 4th Voyage." In *Selected Documents Illustrating the Four Voyages of Columbus. Vol II. The Third and Fourth Voyages*, edited by C. Jane. London: Hakluyt Society.

Condamine, C. M. de. 1738. "Sur l'arbre Quinquina." In *Historie de l'académie Royale des Sciencies*, 226–243.

Conniff, R. 2012. "A Bitter Pill." *Conservation Magazine* 13 (1): 1–4.

Conrad, J. 2012 [1899]. *Heart of Darkness*. London: Sovereign.

Cook, N. 1998. *Born to Die: Disease and New World Conquest, 1492–1650*. Cambridge, UK: Cambridge University Press.

Coomes, O. 2010. "Of Stakes, Stems, and Cuttings: The Importance of Local Seed Systems in Traditional Amazonian Societies." *Professional Geographer* 62 (3): 323–334.

Coomes, O., and Ban, N. 2004. "Cultivated Plant Species Diversity in Home Gardens of an Amazonian Peasant Village in Northeastern Peru." *Economic Botany* 58 (3): 420–434.

Cosgrove, D. 2001. *Apollo's Eye: A Cartographic Genealogy of the Earth in the Western Imagination*. Baltimore: Johns Hopkins University Press.

Costa Lima, V. 1963. "Notas Sobre uma Farmacopeia Africana." *Diaria de Noticias (Salvador)*, November 3.

Costa-Neto, E. M. 1999. "Barata é um Santo Remédio." In *Introdução à Zooterapia Popular no Estado da Bahia*. Universidade Estadual de Feira de Santana, Brazil.

Court, W. E. 1985. "The Doctrine of Signatures or Similitudes." *Trends in Pharmacological Sciences* 6: 225–227.

Cox, P. A. 1999. *Nafanua: Saving the Samoan Rain Forest*. New York: W. H. Freeman.

———. 2000. "Will Tribal Knowledge Survive the Millennium?" *Science* 287: 44–45.

Cox, P. A., and Balick, M. 1994. "The Ethnobotanical Approach to Drug Discovery." *Scientific American* 270 (6): 60–65.

Cox, P. A., Sperry, L. R., Tuominen, M., and Bohlin, L. 1989. "Pharmacological Activity of the Samoan Ethnopharmacopoeia." *Economic Botany* 43: 487–497.

Crawford, M. J. 2007. "'Para Desterrar las Dudas e Adulterasiones': Scientific Expertise and the Attempts to Make a Better Bark for the Royal Monopoly of Quina." *Journal of Spanish Cultural Studies* 8: 193–212.

Crook, C., and Clapp, R. 1998. "Is Market-Orientated Forest Conservation a Contradiction in Terms?" *Environmental Conservation* 25: 131–145.

Crosby, A. W. 1993. *Ecological Imperialism: The Biological Expansion of Europe, 900–1900*. Cambridge, UK: Cambridge University Press.

Cruz-Garcia, G. S., and Price, L. L. 2011. "Ethnobotanical Investigation of 'Wild Food' Plants Used by Rice Farmers in Kalasin, Northeast Thailand." *Journal of Ethnobiology and Ethnomedicine* 7 (1): 1.

Cruz-Garcia, G. S., and Struik, P. C. 2015. "Spatial and Seasonal Diversity of Wild Food Plants in Home Gardens of Northeast Thailand." *Economic Botany* 69 (2): 99–113.

Dalby, A. 2000. *Dangerous Tastes: The Story of Spices*. Berkeley: University of California Press.

Dalziel, J. M. 1948. *Useful Plants of West Tropical Africa*. London: Crown Agents for the Colonies.

Dampier, W. 1703. *A New Voyage around the World*. Vol. 1. London: James Knapton.

Darwin, C. 1832. "Letter from Charles Darwin to Henslow, 12 May." In *The Autobiography of Charles Darwin and Selected Letters*, edited by Francis Darwin. New York: Dover.

———. 1871. *The Descent of Man and Selection in Relation to Sex*. Vol. 1. New York: D. Appleton.

Daunay, M. C., and Janick, J. 2007. "History and Iconography of Eggplant." *Chronica Horticulturae* 47 (3): 16–22.

David, B., Wolfender, J. L., and Dias, D. A. 2015. "The Pharmaceutical Industry and Natural Products: Historical Status and New Trends." *Phytochemistry Reviews* 14 (2): 299–315.

Davis, D. K. 2005. "Potential Forests: Degradation Narratives, Science, and Environmental Policy in Protectorate Morocco, 1912–1956." *Environmental History* 10 (2): 211–238.

Davis, M., and 17 other authors. 2011. "Don't Judge Species on their Origins." *Nature* 474 (7350): 153–154.

Davis, N. Z. 2016. "Physicians, Healers, and Their Remedies in Colonial Suriname." *Canadian Bulletin of Medical History* 33 (1): 3–34.

Dawson, W. R. 1929. "Studies in Medical History: (a) The Origin of the Herbal. (b) Castor-oil in Antiquity." *Aegyptus* 10 (1): 47–72.

Dean, W. 1995. *With Broadax and Firebrand: The Destruction of the Brazilian Atlantic Forest*. Berkeley: University of California Press.

DeClerck, F. A., Chazdon, R., Holl, K. D., Milder, J. C., Finegan, B., Martinez-Salinas, A., and Ramos, Z. 2010. "Biodiversity Conservation in Human-Modified Landscapes of Mesoamerica: Past, Present and Future." *Biological Conservation* 143 (10): 2301–2313.

De Léry, J. 1625. "Extracts Out of the Historie of John Lerivs, a Frenchman who Lived in Brazil (1557 and 1558)." In *Purchas His Pilgrimes, Contayning a History of the World, in Sea Voyages & Lande-Travells, by Englishmen & Others*, edited by S. Purchas. London: William Stansby.

———. 1990 [1557]. *History of a Voyage to the Land of Brazil, Otherwise Called America*. Translated by Janet Whatley. Berkeley: University of California Press.

Delumeau, J. 1995. *History of Paradise: The Garden of Eden in Myth and Tradition*. Translated by M. O'Connell. New York: Continuum.

De Marees, P. 1987 [1602]. *Description and Historical Account of the Gold Kingdom of Guinea (1602)*. Translated by A. van Dantzig and Adam Jones. London: Oxford.

Denevan, W. M. 1992. "The Pristine Myth: The Landscape of the Americas in 1492." *Annals of the Association of American Geographers* 82: 369–385.

———. 2011. "The 'Pristine Myth' Revisited." *Geographical Review* 101: 576–591.

————. 2016. "After 1492: Nature Rebounds." *Geographical Review* 106: 381–398.

Denevan, W. M., Treacy, J. M., Alcorn, J. B., Padoch, C., Denslow, J., and Paitan, S. F. 1985. "Indigenous Agroforestry in the Peruvian Amazon: Bora Indian Management of Swidden Fallows." In *Change in the Amazon Basin*, vol. 1, *Man's Impact on Forests and Rivers*, edited by J. Hemming, 137–155. Manchester, UK: Manchester University Press.

D'errico, F., Backwell, L., Villa, P., Degano, I., Lucejko, J. J., Bamford, M. K., Higham, T. F. G., Colombini, M. P., and Beaumont, P. B. 2012. "Early Evidence of San Material Culture Represented by Organic Artifacts from Border Cave, South Africa." *Proceedings of the National Academy of Sciences* 109: 13214–13219.

Desmarchelier, C. 2010. "Neotropics and Natural Ingredients for Pharmaceuticals: Why Isn't South American Biodiversity on the Crest of the Wave?" *Phytotherapy Research* 24: 791–799.

Desowitz, R. S. 1997. *Who Gave Pinta to the Santa Maria? Torrid Diseases in a Temperate World*. New York: W. W. Norton.

Detienne, M. 1981. "Between Beasts and Gods." In *Myth, Religion and Society*, edited by R. L. Gordon, 215–228. Cambridge, UK: Cambridge University Press.

De Vos, P. 2010. "European Materia Medica in Historical Texts: Longevity of a Tradition and Implications for Future Use." *Journal of Ethnopharmacology* 132 (1): 28–47.

Diamond, J. 2005. *Collapse: How Societies Choose to Fail or Succeed*. New York: Penguin.

Dicke, M. 2010. "Behavioural and Community Ecology of Plants that Cry for Help." *Plant, Cell and Environment* 32: 654–665.

Disney, A. 2007. "Portuguese Expansion, 1400–1800: Encounters, Negotiations, and Interactions." In *Portuguese Oceanic Expansion, 1400–1800*, edited by R. Bethencourt and D. Curto, 283–313. Cambridge, UK: Cambridge University Press.

Dounias, E. 2001. "The Management of Wild Yam Tubers by the Baka Pygmies in Southern Cameroon." *African Study Monographs* (Suppl.) 26: 135–156.

Dovie, D. B., Witkowski, E. T. F., and Shackleton, C. M. 2008. "Knowledge of Plant Resource Use Based on Location, Gender and Generation." *Applied Geography* 28 (4): 311–322.

Drayton, R. 2000. *Nature's Government: Science, Imperial Britain, and the "Improvement" of the World*. New Haven, CT: Yale University Press.

Driver, F. 2004. "Imagining the Tropics: Views and Visions of the Tropical World." *Singapore Journal of Tropical Geography* 25 (1): 1–17.

Du Bartas, G. 1613. *His Divine Weekes and Workes*. Translated by Joshua Sylvester. Madrid: Humfrey Lownes.

————. 1979 [1604]. *The Divine Weeks and Works of Guillaume de Saluste, Sieur du Bartas*. Translated by Joshua Sylvester, edited by S. Snyder. Oxford: Clarendon.

Du Puis, F. M. 1972 [1652]. Relations de l'etablissement d'une Colonie Françoise dans la Guadeloupe Isle de l'Amérique, et des Moeurs des Sauvages. Basse-Terre: Société d'histoire de la Guadeloupe.

Dudney, K., Warren, S., Sills, E., and Jacka, J. 2015. "How Study Design Influences the Ranking of Medicinal Plant Importance: A Case Study from Ghana, West Africa." *Economic Botany* 69 (4): 306–317.

Duffin, J. 2000. "Poisoning the Spindle: Serendipity and Discovery of the Anti-tumor Properties of the Vinca Alkaloids." *Canadian Bulletin of Medical History* 17: 155–192.

Dufoix, S. 2008. *Diasporas*. Berkeley: University of California Press.

Duke, J. A. 2010. *Duke's Handbook of Medicinal Plants of the Bible*. New York: CRC.

Dumbacher, J., Wako, A., Derrickson, S., Sameulson, A., Spande, T., and Daly, J. 2004. "Melyrid Beetles (*Choresine*): A Putative Source for the Batrachotixin Alkaloids Found in Poison-Dart Frogs and Toxic Passerine Birds." *Proceedings of the National Academy of Sciences* 101: 15857–15860.

Dunbar, R. I. 2014. "How Conversations around Campfires Came to Be." *Proceedings of the National Academy of Sciences* 111 (39): 14013–14014.

Duncan, R. P., Boyer, A. G., and Blackburn, T. M. 2013. "Magnitude and Variation of Prehistoric Bird Extinctions in the Pacific." *Proceedings of the National Academy of Sciences* 110 (16): 6436–6441.

Duquet, M. 2003. "The Timeless African and the Versatile Indian in Seventeenth Century Travelogues." *Journal of the Canadian Historical Society* 14: 23–44.

Ecott, T. 2004. *Vanilla: Travels in Search of the Ice Cream Orchid*. New York: Grove.

Edwards, B. 1793. *The History, Civil and Commercial, of the British Colonies in the West Indies*. Vol. 1. London: Mundell & Son.

Edwards, S. E., and Heinrich, M. 2006. "Redressing Cultural Erosion and Ecological Decline in a Far North Queensland Aboriginal Community (Australia): The Aurukun Ethnobiology Database Project." *Environment, Development and Sustainability* 8 (4): 569–583.

Eekelen, A. 2000. "The Debate about Acclimatization in the Dutch East Indies (1840–1860)." *Medical History Supplement* 20: 70–85.

Egerton, F. 2012. "History of Ecological Sciences, Part 41: Victorian Naturalists in Amazonia—Wallace, Bates, Spruce." *Bulletin of the Ecological Society of America* 93: 35–59.

Ellen, R. 1998. "Indigenous Knowledge of the Rainforest: Perception, Extraction, and Conservation." In *Human Activities in the Tropical Rainforest: Past, Present, and Possible Future*, edited by B. Maloney, 87–99. Dordrecht: Kluwer.

———. 2007. "Local and Scientific Understandings of Forest Diversity on Seram, Eastern Indonesia." In *Local Science versus Global Science: Approaches to Indigenous Knowledge in International Development*, edited by P. Sillitoe, 41–74. Oxford, UK: Berghahn.

Ellen, R., and Harris, H. 2000. "Introduction." In *Indigenous Knowledge and its Transformation: Critical Anthropological Perspectives*, edited by A. Bicker, R. Ellen, and P. Parkes, 1–29. Amsterdam: Harwood.

Eltis, D. 2001. "The Volume and Structure of the Transatlantic Slave Trade: A Reassessment." *William and Mary Quarterly* 58 (1): 17–44.

Eltis, D., and Richardson, D. 1997. "West Africa and the Transatlantic Slave Trade: New Evidence of Long Term Runs." In *Routes to Slavery: Direction, Ethnicity and Mortality in the Transatlantic Slave Trade*, edited by D. Eltis and D. Richardson, 16–35. London: Routledge.

———. 2010. "The Transatlantic Slave Trade Database." Atlanta, GA: Emory University.

Emboden, W. 1979. *Narcotic Plants*. New York: Macmillan.

Endara, M. J., and Coley, P. D. 2011. "The Resource Availability Hypothesis Revisited: A Meta-analysis." *Functional Ecology* 25 (2): 389–398.

Enright, K. 2009. *The Maximum of Wilderness: Naturalists & the Image of the Jungle in American Culture*. PhD diss., Rutgers University.

Estes, J. W. 2000. "The Reception of American Drugs in Europe, 1500–1650." In *Searching for the Secrets of Nature: The Life and Works of Dr. Francisco Hernández*, edited by S. Varey, R. Chabrán, and D. Weiner, 111–121. Stanford, CA: Stanford University Press.

Estrada-Castillón, E., Garza-López, M., Villarreal-Quintani' J. Á., Salinas-Rodríguez, M. M., Soto-Mata, B. E., González-Rodríguez, H., Gonzàlez-Uribe, D., Cantú-Silva, I., Carrillo-Parra, A., and Cantú-Ayala, C. 2014. "Ethnobotany in Rayones, Nuevo León, México." *Journal of Ethnobiology and Ethnomedicine* 10 (1): 62.

Etkin, N. 2009. *Foods of Association: Biocultural Perspectives on Foods and Beverages that Mediate Sociability*. Tucson: University of Arizona Press.

Fabricant, D., and Farnsworth, N. 2001. "The Value of Plants Used in Traditional Medicine for Drug Discovery." *Environmental Health Perspectives* 109: 69–75.

Fadiman, M. 2005. "Cultivated Food Plants: Culture and Gendered Spaces of Colonists and the Chachi in Ecuador." *Journal of Latin American Geography* 4 (1): 43–57.

Farnsworth, N. 1988. "Screening Plants for New Medicines." In *Biodiversity*, edited by E. O. Wilson, 83–97. Washington, DC: National Academy Press.

Fasinu, P. S., Gurley, B. J., and Walker, L. A. 2016. "Clinically Relevant Pharmacokinetic Herb-Drug Interactions in Antiretroviral Therapy." *Current Drug Metabolism* 17 (1): 52–64.

Fett, S. 2002. *Working Cures: Healing, Health and Power on Southern Slave Plantations*. Chapel Hill: University of North Carolina.

Fiaschetti, G., Grotzer, M., Shalaby, T., Castelletti, D., and Arcaro, A. 2011. "Quassinoids: From Traditional Drugs to New Cancer Therapeutics." *Current Medicinal Chemistry* 18: 316–328.

Filgueiras, T., and Peixoto, A. 2002. "Flora e Vegetação do Brazil na Carta de Caminha." *Acta Botanica Brasileira* 16: 263–272.

Finucci, V. 2008. "There's the Rub: Searching for Sexual Remedies in the New World." *Journal of Medieval and Early Modern Studies* 38: 523–557.

Flint, V. 1992. *The Imaginative Landscape of Christopher Columbus*. Princeton: Princeton University Press.

Forsyth, D. W. 1983. "The Beginnings of Brazilian Anthropology: Jesuits and Tupinambá Cannibalism." *Journal of Anthropological Research* 39: 147–178.

Forzza, R., and 25 additional authors. 2012. "New Floristic Brazilian List Highlights Conservation Challenges." *BioScience* 62: 39–45.

Foster, S. 2010. "From Herbs to Medicines: The Madagascar Periwinkle's Impact on Childhood Leukemia: A Serendipitous Discovery for Treatment." *Alternative and Complementary Therapies* 16 (6): 347–350.

Frampton, J. 1580. *Joyfull Newes out of the New Found World, Wherein is Declared the Rare and Singular Virtues of Diverse and Sundrie Hearbes, Trees, Oyles, Plantes, and Stone.* Translated by N. Monardes. London: William Norton.

Freedman, B. 2012. "Shamanic Plants and Gender in the Healing Forest." In *Plants, Health and Healing: On the Interface of Ethnobotany and Medical Anthropology*, edited by E. Hsu and S. Harris, 135–178. London: Berghahn.

Freedman, P. 2008. *Out of the East: Spices and the Medieval Imagination.* New Haven, CT: Yale University Press.

Friedman, J. B. 1981. *The Monstrous Races in Medieval Art and Thought.* Cambridge, MA: Harvard University Press.

Frisvold, G., and Day-Rubenstein, K. 2008. "Bioprospecting and Biodiversity Conservation: What Happens When Discoveries Are Made?" *Arizona Law Review* 50: 545–576.

Furusawa, T. 2009. "Changing Ethnobotanical Knowledge of the Roviana People, Solomon Islands: Quantitative Approaches to its Correlation with Modernization." *Human Ecology* 37 (2): 147–159.

———. 2016. *Living with Biodiversity in an Island Ecosystem.* Singapore: Springer.

Fusée-Aublet, J. C. B. 1752. *Histoire des Plants de la Guiane Françoise.* London: Chez Pierre-François.

Gadgil, M., and Vartak, V. D. 1976. "The Sacred Groves of Western Ghats in India." *Economic Botany* 30 (2): 152–160.

Gallagher, D. 2016. "American Plants in Sub-Saharan Africa: A Review of the Archaeological Evidence." *Azania: Archaeological Research in Africa* 51: 24–61.

Gareis, I. 2002. "Cannibals, Bons Sauvages, and Tasty White Men: Models of Alterity in the Encounter of South American Tupi and Europeans." *Medieval History Journal* 5: 247–266.

Gascoigne, J. 2009. "The Royal Society, Natural History and the Peoples of the 'New World(s)', 1660–1800." *British Journal for the History of Science* 42: 539–562.

Gavin, M. C. 2009. "Conservation Implications of Rainforest Use Patterns: Mature Forests Provide More Resources but Secondary Forests Supply More Medicine." *Journal of Applied Ecology* 46 (6): 1275–1282.

Gentry, A. H. 1993. "Tropical Forest Biodiversity and the Potential for New Medicinal Plants." In *Human Medicinal Agents from Plants*, edited by A. D. Kinghorn and M. F. Balandrin, 13–24. Washington, DC: American Chemical Society.

Gheorghiade, G., van Veldhuisen, D., and Colucci, W. 2006. "Contemporary Use of Digoxin in the Management of Cardiovascular Disorders." *Circulation* 113: 2556–2564.

Giday, M., Asfaw, Z., Woldu, Z., and Teklehaymanot, T. 2009. "Medicinal Plant Knowledge of the Bench Ethnic Group of Ethiopia: An Ethnobotanical Investigation." *Journal of Ethnobiology and Ethnomedicine* 5 (1): 34.

Gils, C., and Cox, P. A. 1994. "Ethnobotany of Nutmeg in the Spice Islands." *Journal of Ethnopharmacology* 42 (2): 117–124.

Giovannini, P., Reyes-García, V., Waldstein, A., and Heinrich, M. 2011. "Do Pharmaceuticals Displace Local Knowledge and Use of Medicinal Plants? Estimates from a Cross-Sectional Study in a Rural Indigenous Community, Mexico." *Social Science & Medicine* 72 (6): 928–936.

Glacken, C. 1967. *Traces on the Rhodian Shore: Nature and Culture in Western Thought from Ancient Times to the End of the Eighteenth Century.* Berkeley: University of California Press.

Godoy, R., Reyes-García, V., Broesch, J., Fitzpatrick, I. C., Giovannini, P., Rodríguez, M. R. M., Huanca, T., Leonard, W., McDade, T., Tanner, S., and TAPS Bolivia Study Team. 2009. "Long-Term (Secular) Change of Ethnobotanical Knowledge of Useful Plants: Separating Cohort and Age Effects." *Journal of Anthropological Research* 65: 51–67.

Golden, C. D., Rasolofoniaina, B. R., Anjaranirina, E. G., Nicolas, L., Ravaoliny, L., and Kremen, C. 2012. "Rainforest Pharmacopeia in Madagascar Provides High Value for Current Local and Prospective Global Uses." *PLOS ONE* 7 (7): e41221.

Gómez-Baggethun, E., and Reyes-García, V. 2013. "Reinterpreting Change in Traditional Ecological Knowledge." *Human Ecology* 41 (4): 643–647.

Gómez-Pompa, A., Flores, J. S., and Sosa, V. 1987. "The 'pet kot': A Man Made Tropical Forest of the Maya." *Interciencia* 12 (1): 10–15.

Goor, A. 1965. "The History of the Fig in the Holy Land from Ancient Times to the Present Day." *Economic Botany* 19: 124–135.

Gottlieb, O., and Borin, M. 2009. "Insights into Evolutionary Systems via Chemobiological Data." *Ethnopharmacology* 1: 1–25. http://www.eolss.net/Sample-Chapters/C03/E6-79-02.pdf

Gradé, J., Tabuti, J., and van Damme, P. 2009. "Four Footed Pharmacists: Indications of Self-Medicating Livestock in Karamoja, Uganda." *Economic Botany* 63: 29–42.

Gramiccia, G. 1988. *The Life of Charles Ledger (1818–1905): Alpacas and Quinine.* London: Macmillan.

Grauds, C. 1997. "Jungle Medicine." *Pharmacy Times* 63: 44–48.

Gray, J. 1737. "Account of the Peruvian or Jesuits Bark, by Mr. John Gray, F. R. S. Now at Cartagena in the Spanish West-Indies; Extracted from Some Papers Given Him by Mr. William Arrot, a Scotch Surgeon, Who Had Gather'd It at the Place Where It Grows in Peru." *Philosophical Transactions of the Royal Society of London* 40: 81–86.

Grenier, L. 1998. *Working with Indigenous Knowledge: A Guide for Researchers.* Ottawa, Canada: International Development Research Centre.

Guèze, M., Luz, A. C., Paneque-Gálvez, J., Macía, M. J., Orta-Martínez, M., Pino, J., and Reyes-García, V. 2014. "Are Ecologically Important Tree Species the Most

Useful? A Case Study from Indigenous People in the Bolivian Amazon." *Economic Botany* 68 (1): 1–15.

Guix, J. 2009. "Amazonian Forests Need Indians and Caboclos." *Orsis* 24: 33–40.

Gunn, G. 2003. *First Globalization: The Eurasian Exchange, 1500–1800.* Lanham, MD: Rowman and Littlefield.

Gyllenhaal, C., and 25 other authors. 2012. "Ethnobotanical Approach versus Random Approach in the Search for New Bioactive Compounds: Support of a Hypothesis." *Pharmaceutical Biology* 50: 30–41.

Hair, P. E. H., Jones, A., and Law, R., eds. 1992. *Barbot on Guinea : The Writings of Jean Barbot on West Africa, 1678–1712.* Vol. 1. London: Hakluyt Society.

Hallam, A., and Reid, J. 2006. "Do Tropical Species Invest More in Anti-herbivore Defence than Temperate Species: A Test in *Eucryphia* (Cunoniaceae) in Eastern Australia." *Journal of Tropical Ecology* 22: 41–51.

Hames, R. 2007. "The Ecologically Noble Savage Debate." *Annual Review of Anthropology* 36: 177–190.

Hanazaki, N., Herbst, D. F., Marques, M. S., and Vandebroek, I. 2013. "Evidence of the Shifting Baseline Syndrome in Ethnobotanical Research." *Journal of Ethnobiology and Ethnomedicine* 9 (1): 1–11.

Hanazaki, N., Tamashiro, J., Leitão-Filho, H., and Begossi, A. 2000. "Diversity of Plant Uses in Two *Caiçara* Communities from the Atlantic Forest Coast, Brazil." *Biodiversity and Conservation* 9: 597–615.

Hanna, W. 1978. *Indonesian Banda: Colonialism and Its Aftermath in the Nutmeg Islands.* Philadelphia: Institute for the Study of Human Issues.

Hardenburg, W. E., and Enock, C. R. 1912. *The Putumayo: The Devil's Paradise.* London: T. F. Unwin.

Hare, J. D. 2011. "Ecological Role of Volatiles Produced by Plants in Response to Damage by Herbivorous Insects." *Annual Review of Entomology* 56: 161–180.

Harper, J. 2005. "The Not-so Rosy Periwinkle: Political Dimensions of Medicinal Plant Research." *Ethnobotany Research & Applications* 3: 295–308.

Harris, S. 2005. "Jesuit Scientific Activity in the Overseas Missions, 1540–1773." *Isis* 96: 71–79.

Harrison, M. 1996. " 'The Tender Frame of Man': Disease, Climate, and Racial Difference in India and the West Indies, 1760–1860." *Bulletin of the History of Medicine* 70: 68–93.

Hartley, T. G., Dunstone, E. A., Fitzgerald, J. S., Johns, S. R., and Lamberton, J. A. 1973. "A Survey of New Guinea Plants for Alkaloids." *Lloydia* 36: 217–319.

Hartlib, S. 1655. *Samuel Hartlib, His Legacy of Husbandry.* London: Richard Wodnothe.

Haynes, D. 2002. "Still the Heart of Darkness: The Ebola Virus and the Meta-Narrative of Disease in *The Hot Zone.*" *Journal of Medical Humanities* 23: 133–145.

Hecht, S. 2013. *The Scramble for the Amazon and the "Lost Paradise" of Euclides da Cunha.* Chicago: University of Chicago Press.

Hecht, S., and Cockburn, A. 2010. *The Fate of the Forest: Developers, Destroyers, and Defenders of the Amazon.* Updated ed. Chicago: University of Chicago Press.

Heinrich, M., and Barerra, N. A. 1993. "Medicinal Plants in a Lowland Mixe Indian Community (Oaxaca, Mexico): Management of Important Resources." *Angewandte Botanik* 67: 141–144.

Helfcrich, G. 2004. *Humboldt's Cosmos: Alexander von Humboldt and the Latin American Journey that Changed the Way We See the World.* New York: Gotham.

Hemming, J. 1978a. *Red Gold: The Conquest of the Brazilian Indians, 1500–1760.* Cambridge, MA: Harvard University Press.

———. 1978b. *The Search for El Dorado.* New York: E. P. Dutton.

Henkel, A. 1904. "Weeds Used in Medicine." *USDA Farmers Bulletin* no. 188. Washington, DC: Government Printing Office, 45 pp.

Hill, K. P. 2015. "Medical Marijuana for Treatment of Chronic Pain and Other Medical and Psychiatric Problems: A Clinical Review." *JAMA* 313 (24): 2474–2483.

Hillary, W. 1759. *Observations on the Changes of the Air and the Concomitant Epidemical Diseases in the Island of Barbados.* London: C. Hitch & L. Hawes.

Hladik, C., and Simmen, B. 1996. "Taste Perception and Feeding Behavior in Nonhuman Primates and Human Populations." *Evolutionary Anthropology* 5: 58–71.

Ho, Ping-ti. 1998. "The Introduction of American Food Plants into China." In *European Intruders and Changes in Behaviour and Customs in Africa, America and Asia before 1800. An Expanding World,* vol. 30, edited by M. J. MacLeod and E. S. Rawski, 283–293. Brookfield, VT: Ashgate.

Hoang, V. S. 2009. *Uses and Conservation of Plant Diversity in Ben En National Park, Vietnam.* PhD diss., National Herbarium of the Netherlands, Leiden University Branch.

Hochschild, A. 1999. *King Leopold's Ghost: A Story of Greed, Terror, and Heroism in Colonial Africa.* Boston: Houghton Mifflin Harcourt.

Hoffman, B. 2013. "Exploring Biocultural Contexts: Comparative Woody Plant Knowledge of an Indigenous and Afro-American Maroon Community in Suriname, South America." In *African Ethnobotany in the Americas,* edited by Robert Voeks and John Rashford, 335–393. New York: Springer.

Hoffmann, R. 2014. *An Environmental History of Medieval Europe.* Cambridge, UK: Cambridge University Press.

Holmsted, B., Wassen, S., and Schultes, R. 1979. "Jaborandi: An Interdisciplinary Appraisal." *Journal of Ethnopharmacology* 1: 3–21.

Honey, M. 2008. *Ecotourism and Sustainable Development: Who Owns Paradise?* 2nd ed. Washington, DC: Island.

Honigsbaum, M. 2002. *The Fever Trail: In Search of the Cure for Malaria.* New York: Farrar, Straus and Giroux.

Hooker, W. D. 1839. *Inaugural Dissertation upon the Cinchonas: Their History, Uses and Effects.* Scotland: Glasgow University Press.

Horsfield, T. 1823. "An Essay on the Oopas, or Poison Tree of Java, Addressed to the Honorable Thomas Stanford Raffles." *The Investigator; or Quarterly Magazine* 7: 76–101.

Howard, P. L., and Nabanoga, G. 2007. "Are There Customary Rights to Plants? An Inquiry among the Baganda (Uganda), with Special Attention to Gender." *World Development* 35: 1542–1563.

Huber, F. K., Ineichen, R., Yang, Y., and Weckerle, C. S. 2010. "Livelihood and Conservation Aspects of Non-wood Forest Product Collection in the Shaxi Valley, Southwest China." *Economic Botany* 64 (3): 189–204.

Hudson, M. M., Neglia, J. P., Woods, W. G., Sandlund, J. T., Pui, C. H., Kun, L. E., and Green, D. M. 2012. "Lessons from the Past: Opportunities to Improve Childhood Cancer Survivor Care through Outcomes Investigations of Historical Therapeutic Approaches for Pediatric Hematological Malignancies." *Pediatric Blood & Cancer* 58 (3): 334–343.

Huffman, M.A. 2001. "Self-Medicative Behavior in the African Great Apes: An Evolutionary Perspective into the Origins of Human Traditional Medicine." *BioScience* 51: 651–661.

Huguet-Termes, T. 2001. "New World Materia Medica in Spanish Renaissance Medicine: From Scholarly Reception to Practical Impact." *Medical History* 45: 359–376.

Hunde, D., Abedeta, C., Birhan, T., and Sharma, M. 2015. "Gendered Division of Labor in Medicinal Plant Cultivation and Management in South West Ethiopia: Implication for Conservation." *Trends in Applied Sciences Research* 10 (2): 77–87.

Huntington, E. 1915. *Civilization and Climate*. New Haven, CT: Yale University Press.

Ichikawa, M. 2001. "The Forest World as a Circulation System: The Impacts of Mbuti Habitation and Subsistence Activities on the Forest Environment." *African Study Monographs* (Suppl.) 26: 157–168.

IUCN (International Union for Conservation of Nature). 2016. The IUCN Red List of Threatened Species. http://www.iucnredlist.org/

Ives, J. D. 1987. "The Theory of Himalayan Environmental Degradation: Its Validity and Application Challenged by Recent Research." *Mountain Research and Development* 7 (3): 189–199.

Jackson, J. 2008. *The Thief at the End of the World: Rubber, Power, and the Seeds of Empire*. New York: Penguin.

Jahoda, G. 1999. *Images of Savages: Ancient Roots of Modern Prejudice in Western Culture*. London: Routledge.

James, P. 1972. *All Possible Worlds: A History of Geographical Ideas*. New York: Odyssey.

Janzen, D. 1988. "Buy Costa Rican Beef." *Oikos* 51: 257–258.

Jarcho, S. 1993. *Quinine's Predecessor. Francesco Torti and the Early History of Cinchona*. Baltimore: Johns Hopkins University Press.

Jarosz, L. 1992. "Constructing the Dark Continent: Metaphor as Geographic Representation of Africa." *Geografiska Annaler: Series B. Human Geography* 74: 105–115.

Jeune, B., Robine, J. M., Young, R., Desjardins, B., Skytthe, A., and Vaupel, J. W. 2010. "Jeanne Calment and Her Successors. Biographical Notes on the Longest Living Humans." In *Supercentenarians*, edited by H. Maier, J. Gampe, B. Jeune, Jean-Marie Robine, and J. W. Vaupel, 285–323. Berlin: Springer. doi:10.1007/978-3-642-11520-2_16

Johns, T. 1990. *The Origins of Human Diet and Medicine*. Tucson: University of Arizona Press.

Jones, A. 1983. *German Sources for West African History, 1599–1669*. Wiesbaden: Franz Steiner Verlag.

Jones, M., Hunt, H., Lightfoot, E., Lister, D., Liu, X., and Motuzaite-Matuzeviciute, G. 2011. "Food Globalization in Prehistory." *World Archaeology* 43 (4): 665–675.

Joseph, B., and Jini, D. 2013. "Antidiabetic Effects of *Momordica charantia* (Bitter Melon) and Its Medicinal Potency." *Asian Pacific Journal of Tropical Disease* 3 (2): 93–102.

Joyce, C. 1992. "Western Medicine Men Return to the Field." *BioScience* 42: 399–403.

Judkins, G., Smith, M., and Keys, E. 2008. "Determinism within Human-Environment Research and the Rediscovery of Environmental Causation." *Geographical Journal* 174: 17–29.

Jungerius, P. D. 1998. "Indigenous Knowledge of Landscape-Ecological Zones among Traditional Herbalists: A Case Study in Keiyo District, Kenya." *GeoJournal* 44: 51–60.

Junqueira, A. B., Shepard Jr., G. H., and Clement, C. R. 2011. "Secondary Forests on Anthropogenic Soils of the Middle Madeira River: Valuation, Local Knowledge, and Landscape Domestication in Brazilian Amazonia." *Economic Botany* 65 (1): 85–99.

Junsongduang, A., Balslev, H., Inta, A., Jampeetong, A., and Wangpakapattana-wong, P. 2013. "Medicinal Plants from Swidden Fallows a___ _ cred Forest of the Karen and the Lawa in Thailand." *Journal of Ethnobiology a d Ethnomedicine* 9 (1): 44.

Junsongduang, A., Balslev, H., Inta, A., Jampeetong, A., and Wangpakapattana-wong, P. 2014. "Karen and Lawa Medicinal Plant Use: Uniformity or Ethnic Divergence?" *Journal of Ethnopharmacology* 151 (1): 517–527.

Kaempfer, E. 1996 [1712]. *Engelbert Kaempfer, Exotic Pleasures, Fascicle III Curious Scientific and Medical Observations*. Translated and with commentary by Robert W. Carrubba. Carbondale: Southern Illinois University Press.

Kamboj, A., and Saluja, A. K. 2009. "*Bryophyllum pinnatum* (Lam.) Kurz.: Phytochemical and Pharmacological Profile: A Review." *Pharmacognosy Reviews* 3 (6): 364–374.

Kaplan, I., Rayko, H., Kessler, A., Sardanello, S., and Denno, R. 2008. "Constitutive and Induced Defenses to Herbivory in Above- and Belowground Plant Tissues." *Ecology* 89 (2): 392–406.

Katz, E., Lopez, C. L., Fleury, M., Miller, R. P., Payê, V., Dias, T., Silva, F., Oliveira, Z., and Moreira, E. 2012. "No Greens in the Forest? Note on the Limited Consumption of Greens in the Amazon." *Acta Societatis Botanicorum Poloniae* 81 (4): 283–293.

Kayser, M., Brauer, S., Weiss, G., Underhill, P., Roewer, L., Schiefenhövel, W., and Stoneking, M. 2000. "Melanesian Origin of Polynesian Y Chromosomes." *Current Biology* 10 (20): 1237–1246.

Keay, J. 2007. *The Spice Route: A History*. Berkeley: University of California Press.

Keen, C. L. 2001. "Chocolate: Food as Medicine/Medicine as Food." *Journal of the American College of Nutrition* 20: 436S–439S.

Kemys, L. 1596. *A Relation of the Second Voyage to Guiana*. London: Thomas Dawson.

Kennedy, J. 2012. "Agricultural Systems in the Tropical Forest: A Critique Framed by Tree Crops of Papua New Guinea." *Quaternary International* 249: 140–150.

Kershaw, E. M. 2000. *A Study of Brunei Dusun Religion: Ethnic Priesthood on a Frontier of Islam*. Monograph Series no. 4. Phillips, ME: Borneo Research Council.

Kevane, M. 2012. "Gendered Production and Consumption in Rural Africa." *Proceedings of the National Academy of Sciences* 109: 12350–12355.

Khuankaew, S., Srithi, K., Tiansawat, P., Jampeetong, A., Inta, A., and Wangpakapattanawong, P. 2014. "Ethnobotanical Study of Medicinal Plants Used by Tai Yai in Northern Thailand." *Journal of Ethnopharmacology* 151 (2): 829–838.

Kiringe, J. 2005. "Ecological and Anthropological Threats to Ethno-medicinal Plant Resources and Their Utilization in Maasai Communal Ranches in the Amboseli Region of Kenya." *Ethnobotanical Research and Applications* 3: 231–241.

Knivet, A. 1625 [1591]. "The Admirable Adventures and Strange Fortunes of Master Antonie Knivet." In *Purchas His Pilgrimes, Contayning a History of the World, in Sea Voyages & Lande-Travells, by Englishmen & Others*, edited by S. Purchas. London: William Stansby.

Knox, J. 1681. *An Historical Relation of the Island Ceylon in the East-Indies*. London: The Royal Society.

Koerner, L. 1996. "Carl Linnaeus in His Time and Place." In *Cultures of Natural History*, edited by N. Jardine, J. Secord, and E. Spary, 145–162. Cambridge, UK: Cambridge University Press.

Kohn, E. O. 1993. "Some Observations on the Use of Medicinal Plants from Primary and Secondary Growth by the Runa of Eastern Lowland Ecuador." *Journal of Ethnobiology* 12: 141–152.

Kosaka, Y. 2013. "Wild Edible Herbs in Paddy Fields and Their Sale in a Mixture in Houaphan Province, the Lao People's Democratic Republic." *Economic Botany* 67 (4): 335–349.

Kothari, B. 2003. "The Invisible Queen in the Plant Kingdom: Gender Perspectives in Medical Ethnobotany." In *Women and Plants: Gender Relations in Biodiversity Management & Conservation*, edited by P. L. Howard, 150–164. London: Zed.

Krech, S. 1999. *The Ecological Indian: Myth and History*. New York: W. W. Norton.

Kueffer, C., and Larson, B. M. 2014. "Responsible Use of Language in Scientific Writing and Science Communication." *BioScience* 64: 1–6.

Kunwar, R. M., Acharya, R. P., Chowdhary, C. L., and Bussmann, R. W. 2015. "Medicinal Plant Dynamics in Indigenous Medicines in Farwest Nepal." *Journal of Ethnopharmacology* 163: 210–219.

Kunwar, R. M., Baral, K., Paudel, P., Acharya, R. P., Thapa-Magar, K. B., Cameron, M., and Bussmann, R. W. 2016. "Land-Use and Socioeconomic Change, Medicinal Plant Selection and Biodiversity Resilience in Far Western Nepal." *PLOS ONE* 11 (12): e0167812.

Kusukawa, S. 2004. "The Medical Renaissance of the Sixteenth Century." In *The Healing Arts: Health, Disease and Society in Europe 1500–1800*, edited by P. Elmer, 58–83. Manchester, UK: The Open University.

Labat, J. B. 1724. *Nouveau Voyage aux Isles de l'Amérique: Contenant l'histoire Naturelle de ces pays, l'origine, les Moeurs, la Religion & le Gouvernement des Habitants Anciens & Modems*. La Haye: Chez R. Alberts, J. van Duren, P. Gosse, P. Husson, T. Johnson, & C. Le Vier.

Landers, J. 2002. "The Central African Presence in Spanish Maroon Communities." In *Central Africans and Cultural Transformations in the American Diaspora*, edited by L. M. Heywood, 227–241. Cambridge, UK: Cambridge University Press.

Lanker, U., Malik, A., Gupta, N., and J. Butola. 2010. "Natural Regeneration Status of the Endangered Medicinal Plant, *Taxus baccata* Hook. F. syn. *T. wallichiana*, in Northwest Himalaya." *International Journal of Biodiversity Science, Ecosystem Services & Management* 6: 20–27.

Larson, B. 2011. *Metaphors for Environmental Sustainability: Redefining Our Relationship with Nature*. New Haven, CT: Yale University Press.

Latorre, J. C., and Latorre, J. C. 2012. "Globalization, Local Communities, and Traditional Forest-Related Knowledge." In *Traditional Forest-Related Knowledge: Sustaining Communities, Ecosystems and Biocultural Diversity*, edited by J. A. Parotta and R. Trosper, 449–490. New York: Springer.

Leblond, J. B. 1813. *Voyage aux Antilles et l'Amerique Méridionale*. Paris: Arthus-Bertrand.

Lee, K. 2010. "Discovery and Development of Natural Product-Derived Chemotherapeutic Agents Based on a Medicinal Chemistry Approach." *Journal of Natural Products* 73: 500–516.

Lee, M. R. 2007. "Solanaceae IV: *Atropa belladonna*, Deadly Nightshade." *Journal of the Royal College of Physicians of Edinburgh* 37: 77–84.

Lee, R., and Balick, M. 2005. "Sweet Wood—Cinnamon and Its Importance as a Spice and Medicine." *Explore: The Journal of Scientific Healing* 1: 61–64.

Leite, S. 1938. *História da Companhia de Jesus no Brasil*. Vol. 1–5. Rio de Janeiro: Civilização Brasileira.

Lemos Coelho, F. 1684. *The Golden Trade, or, A Discovery of River Gambra*. London: Nicholas Oakes.

Lennep, J. R., Schuit, S. C. E., van Bruchem-Visser, R. L., and Özcan, B. 2015. "Unintentional Nutmeg Autointoxication." *Netherlands Journal of Medicine* 73: 46–48.

León Pinelo, A. 1943 [1650]. *El Paraiso en el Nuevo Mundo*. Vol. 1 and 2. Lima: Torres Aguirre.

Leonti, M. 2011. "The Future Is Written: Impact of Scripts on the Cognition, Selection, Knowledge and Transmission of Medicinal Plant Use and Its Implications for Ethnobotany and Ethnopharmacology." *Journal of Ethnopharmacology* 134 (3): 542–555.

Leonti, M., Cabras, S., Eugenia Castellanos, M., Challenger, A., Gertsch, J., and Casu, L. 2013. "Bioprospecting: Evolutionary Implications from a Post-Olmec Pharmacopoeia and the Relevance of Widespread Taxa." *Journal of Ethnopharmacology* 147: 92–107.

Leonti, M., Casu, L., Sanna, F. and L. Bonsignore. 2009. "A Comparison of Medicinal Plant Use in Sardinia and Sicily—De Materia Medica Revisited?" *Journal of Ethnopharmacology* 121 (2): 255–267.

Lestrelin, G., Vigiak, O., Pelletreau, A., Keohavong, B., and Valentin, C. 2012. "Challenging Established Narratives on Soil Erosion and Shifting Cultivation in Laos." *Natural Resources Forum* 36 (2): 63–75.

Levin, D. A. 1976. "Alkaloid-Bearing Plants: An Ecogeographic Perspective." *American Naturalist* 110: 261–284.

Levis, C., Costa, F. R., Bongers, F., Peña-Claros, M., Clement, C. R., Junqueira, A. B., Neves, E. G., Tamanaha, E. K., Figueiredo, F. O., Salomão, R. P., and Castilho, C. V. 2017. "Persistent Effects of Pre-Columbian Plant Domestication on Amazonian Forest Composition." *Science* 355 (6328): 925–931.

Lewis, W. 1791. *An Experimental History of the Materia Medica or of the Natural and Artificial Substances Made Use of in Medicine*. London: J. Johnson.

Lewis, W. H. 2003. "Pharmaceutical Discoveries Based on Ethnomedicinal Plants: 1985 to 2000 and Beyond." *Economic Botany* 57: 126–134.

Lim, J. Y., Fine, P. V., and Mittelbach, G. G. 2015. "Assessing the Latitudinal Gradient in Herbivory." *Global Ecology and Biogeography* 24 (10): 1106–1112.

Livingston, D. 1858. *Missionary Travels and Researches in South Africa: Including a Sketch of Sixteen Years' Residence in the Interior*. New York: Harper & Brothers.

Lloyd, J. 1911. "History of the Vegetable Drugs of the Pharmacopoeia of the United States." *Bulletin of the Lloyd Library* 18: 1–180.

Lody, R. 1992. *Tem dendê, Tem Axé: Etnografia do Dendezeiro*. Rio de Janeiro: Editora Pallas.

Logan, M. H, and Dixon, A. R. 1994. "Agriculture and the Acquisition of Medicinal Plant Knowledge." In *Eating on the Wild Side: The Pharmacologic, Ecologic, and Social Implications of Using Noncultigens*, edited by N. Etkin, 25–45. Tucson: University of Arizona Press.

Long, E. 1774. *History of Jamaica. Or, General Survey of the Ancient and Modern State of that Island: With Reflections on its Situation, Settlements, Inhabitants* . . . London: T. Lownudes.

Lordelo, M., and Marques, J. 2010. "E o Caruru de Iansã se Acabou? Resiliência Cultural de uma Festa de Comida em Salvador e Feira de Santana (BA)." II Congresso Latinamericano de Etnobiologia, Recife, 10 de Novembro.

Lovejoy, A. 1964 [1936]. *The Great Chain of Being: A Study of the History of an Idea.* Cambridge, MA: Harvard University Press.

Lovell, G. 1992. "'Heavy Shadows and Black Night': Disease and Depopulation in Colonial Spanish America." *Annals of the Association of American Geographers* 86: 426–443.

Lucena, R. F. P., and de Albuquerque, U. P. 2005. "Can Apparency Affect the Use of Plants by Local People in Tropical Forests?" *Interciencia: Revista de Ciencia y Tecnología de América* 30 (8): 506–511.

Luoga, E. J., Witkowski, E. T. F., and Balkwill, K. 2000. "Differential Utilization and Ethnobotany of Trees in Kitulanghalo Forest Reserve and Surrounding Communal Lands, Eastern Tanzania." *Economic Botany* 54 (3): 328–343.

Lutzenberger, J. 1982. "The Systematic Demolition of the Tropical Rain Forest in the Amazon." *Ecologist* 12: 248–258.

Luzar, J. B., and Fragoso, J. M. V. 2013. "Shamanism, Christianity and Culture Change in Amazonia." *Human Ecology* 41: 299–311.

Luziatelli, G., Sørensen, M., Theilade, I., and P. Mølgaard. 2010. "Asháninka Medicinal Plants: A Case Study from the Native Community of Bajo Quimiriki, Junín, Peru." *Journal of Ethnobiology and Ethnomedicine* 6: 21. http://www.ethnobiomed.com/content/6/1/21.

Lyon, L. M., and Hardesty, L. H. 2012. "Quantifying Medicinal Plant Knowledge among Non-specialist Antanosy Villagers in Southern Madagascar." *Economic Botany* 66: 1–11.

Ly-Tio-Fane, M. 1958. *Mauritius and the Spice Trade: The Odyssey of Pierre Poivre.* Port Louis, Mauritius: Esclapon Limited.

Maddox, G. 2003. "'Degradation Narratives' and 'Population Time Bombs': Myths and Realities about African Environments." In *South Africa's Environmental History: Cases and Comparisons,* edited by S. Dovers, R. Edgecombe, and B. Guest, 250–258. Athens: Ohio University Press.

Madera, L. M. 2009. "Visions of Christ in the Amazon: The Gospel According to Ayahuasca and Santo Daime." *Journal for the Study of Religion, Nature & Culture* 3 (1): 66–98.

Magalhães, B. 1980. *O Café: Na Historia, No Folclore, e Nas Belas Artes.* São Paulo, Brasil: Companhia Ed. Nacional.

Maia, P. 2012. *Práticas Terapêuticas Jesuíticas no Império Colonial Português: Medicamentos e Boticas no Século XVIII.* PhD diss., Universidade de São Paulo.

Mancall, P. C. 2006. *Travel Narratives from the Age of Discovery: An Anthology.* New York: Oxford University Press.

Manilal, K. S. 1980. "The Implications of Hortus Malabaricus with the Botany and History of Peninsular India." In *Botany and History of Hortus Malabaricus,* edited by K. S. Manilal, 1–5. Rotterdam: A. A. Balkema.

Mann, C. 2011. *1493: Uncovering the New World Columbus Created.* New York: Alfred A. Knopf.

Mann, J. 2000. *Murder, Magic and Medicine.* Oxford, UK: Oxford University Press.

Manniche, L. 1989. *An Ancient Egyptian Herbal.* Austin: University of Texas Press.

Markham, C. 1880. *Peruvian Bark: A Popular Account of the Introduction of Cinchona Cultivation into British India, 1860–1880.* London: John Murray.

———. 1894. *The Letters of Amerigo Vespucci. Letter to Lorenzo Pietro Francesco Medici (1503).* London: Hakluyt Society.

Marquis, R., Ricklefs, R., and Abdala-Roberts, L. 2011. "Testing the Low Latitude/High Defense Hypothesis for Broad-Leaved Tree Species." *Oecologia* 169: 811–820.

Marshall, C. A., and Hawthorne, W. D. 2012. "Regeneration Ecology of the Useful Flora of the Putu Range Rainforest, Liberia." *Economic Botany* 66 (4): 398–412.

Martin, P. S. 1973. "The Discovery of America." *Science* 179: 969–974.

Masi, S., Gustafsson, E., Saint Jalme, M., Narat, V., Todd, A., Bomsel, M. C., and Krief, S. 2012. "Unusual Feeding Behavior in Wild Great Apes, a Window to Understand Origins of Self-Medication in Humans: Role of Sociality and Physiology on Learning Process." *Physiology & Behavior* 105 (2): 337–349.

Mathez-Stiefel, S., and Vandebroek, I. 2012. "Distribution and Transmission of Medicinal Plant Knowledge in the Andean Highlands: A Case Study from Peru and Bolivia." *Evidence-Based Complementary and Alternative Medicine*, Article ID 959285. doi:10.1155/2012/959285

Maverick, L. A. 1941. "Pierre Poivre: Eighteenth Century Explorer of Southeast Asia." *Pacific Historical Review* 10: 165–177.

McCann, J. 1997. "The Plow and the Forest: Narratives of Deforestation in Ethiopia, 1840–1992." *Environmental History* 2 (2): 138–159.

———. 2005. *Maize and Grace: Africa's Encounter with a New World Crop, 1500–2000.* Cambridge, MA: Harvard University Press.

McCarter, J., and Gavin, M. C. C. 2014. "Local Perceptions of Changes in Traditional Ecological Knowledge: A Case Study from Malekula Island, Vanuatu." *Ambio* 43 (3): 288–296.

———. 2015. "Assessing Variation and Diversity of Ethnomedical Knowledge: A Case Study from Malekula Island, Vanuatu." *Economic Botany* 69 (3): 251–261.

McClintock, A. 1995. *Imperial Leather: Race, Gender, and Sexuality in the Colonial Contest.* New York: Routledge.

McDade, T. W., Reyes-García, V., Blackinton, P., Tanner, S., Huanca, T., and Leonard, W. R. 2007. "Ethnobotanical Knowledge Is Associated with Indices of Child Health in the Bolivian Amazon." *Proceedings of the National Academy of Sciences* 104 (15): 6134–6139.

Mechoulam, R. 2012. "Cannabis—A Valuable Drug that Deserves Better Treatment." *Mayo Clinic Proceedings* 87 (2): 107–109.

Meek, J. C. K. 1931. *Sudanese Kingdom: An Ethnographical Study of the Jukun-Speaking Peoples of Nigeria.* London: Kegan Paul, Trench, Trubneer.

Mendelsohn, R., and Balick, M. 1995. "The Value of Undiscovered Pharmaceuticals in Tropical Forests." *Economic Botany* 49: 223–228.

Mercader, J. 2003. "Introduction: The Paleolithic Settlement of Rain Forests." In *Under the Canopy: The Archaeology of Tropical Rain Forests*, edited by J. Mercader, 1–31. New Brunswick, NJ: Rutgers University Press.

Metcalf, A. C. 2005. *Go-Betweens and the Colonization of Brazil—1500–1600*. Austin: University of Texas Press.

Metcalf, P. A. 1976. "Who Are the Berawan? Ethnic Classification and the Distribution of Secondary Treatment of the Dead in Central North Borneo." *Oceania* 47 (2): 85–105.

Meyer, R., Bamshad, M., Fuller, D., and Litt, A. 2014. "Comparing Medicinal Uses of Eggplant and Related Solanaceae in China, India, and the Philippines Suggests Independent Development of Uses, Cultural Diffusion, and Recent Species Substitutions." *Economic Botany* 68 (2): 137–152.

Miller, J. I. 1969. *The Spice Trade of the Roman Empire 29 B.C. to A.D. 641*. London: Clarendon.

Miller, J. S. 2011. "The Discovery of Medicines from Plants: A Current Biological Perspective." *Economic Botany* 65 (4): 396–407.

Milton, G. 1999. *Nathaniel's Nutmeg, Or, The True and Incredible Adventures of the Spice Trader Who Changed the Course of History*. New York: Macmillan.

Miranda, T., Hanazaki, N., Govone, J., and Alves, D. 2011. "Existe Utilização Efetiva dos Recursos Vegetais Conhecidos em Comunidades Caiçaras da Ilha do Cardoso, Estado de São Paulo, Brasil?" *Rodriguesia* 62 (1): 153–169. http://rodriguesia.jbrj.gov.br

Moerman, D. E., Pemberton, R. W., Kiefer, D., and Berlin, B. 1999. "A Comparative Analysis of Five Medicinal Floras." *Journal of Ethnobiology* 19 (1): 49–70.

Molares, S., and Ladio, A. 2009. "Ethnobotanical Review of the Mapuche Medicinal Flora: Use Patterns on a Regional Scale." *Journal of Ethnopharmacology* 122 (2): 251–260.

Moles, A., Bonser, S., Poore, A., Wallis, I., and Foley, W. 2011. "Assessing the Evidence for Latitudinal Gradients in Plant Defense and Herbivory." *Functional Ecology* 25: 380–388.

Momsen, J. 2004. *Gender and Development*. New York: Routledge.

Monardes, N. 1580. *Joyful Newes out of the New Found World, Wherein are Declared the Rare and Singular Vertues of Divers and Sundrie Herbs, Trees, Dyes, Plants, and Stones*. Translated by John Frampton. London: William Norton.

Montúfar, R., Anthelme, F., Pintaud, J., and Balslev, H. 2011. "Disturbance and Resilience in Tropical American Palm Populations and Communities." *Botanical Review* 77: 426–461.

Mora, C., Tittensor, D., Adl, S., Simpson, A., and Worm, B. 2011. "How Many Species Are There on Earth and in the Ocean?" *PLoS Biology* 9: 1–8.

Moraes, P. L. R. 2012. "Linnaeus's Plantae Surinamenses Revisited." *Phytotaxa* 41 (1): 1–86.

Moreau, J. P. 1994 [1618–1620]. *Un Flibustier Français dans la Mer des Antilles (1618–1620)*. Paris: Payot & Rivages.

Morens, D. 1999. "Death of a President." *New England Journal of Medicine* 341 (24): 1845–1848.

Moret, E. S. 2013a. "Hegemony, Identity, and Trans-Atlantic Modernity: Afro-Cuban Religion (Re)politicization and (De)legitimization in the Post-Soviet Era." *Journal for the Study of Religion, Nature & Culture* 6 (4): 421–446.

———. 2013b. "Trans-Atlantic Diaspora Ethnobotany: Legacies of West African and Iberian Mediterranean Migration in Central Cuba." In *African Ethnobotany in the Americas*, edited by R. Voeks and J. Rashford, 217–245. New York: Springer.

Morgan, J. L. 1997. "'Some Could Suckle over Their Shoulder': Male Travelers, Female Bodies, and the Gendering of Racial Ideology, 1500–1770." *William & Mary Quarterly* 54 (1): 167–192.

Mori, S., Boom, B. M., Carvalho, A. M., and Santos, T. S. 1983. "Southern Bahian Moist Forests." *Botanical Review* 49: 155–232.

Morison, S. E., ed. 1963. *Journals and Other Documents on the Life and Voyages of Christopher Columbus*. New York: Heritage Press.

Morison, S. E., and Obregón, M. 1964. *The Caribbean as Columbus Saw It*. Boston: Little, Brown.

Mort, M. E., Soltis, D. E., Soltis, P. S., Francisco-Ortega, J., and Santos-Guerra, A. 2001. "Phylogenetic Relationships and Evolution of Crassulaceae Inferred from matK Sequence Data." *American Journal of Botany* 88 (1): 76–91.

Müller, J. G., Boubacar, R., and Guimbo, I. D. 2014. "The 'How' and 'Why' of Including Gender and Age in Ethnobotanical Research and Community-Based Resource Management." *Ambio* 44: 67–78.

Müller-Schwartz, N. 2006. "*Antes* and *Hoy Dia*: Plant Knowledge and Categorization as Adaptations to Life in Panama in the Twenty-First Century." *Economic Botany* 60: 321–334.

Mutheeswaran, S., Pandikumar, P., Chellappandian, M., and Ignacimuthu, S. 2011. "Documentation and Quantitative Analysis of the Local Knowledge on Medicinal Plants among Traditional Siddha Healers in Virudhunagar District of Tamil Nadu, India." *Journal of Ethnopharmacology* 137 (1): 523–533.

Muzik, T. J. 1952. "Citrus in West Africa, with Special Reference to Liberia." *Economic Botany* 6 (3): 246–251.

Myers, N. 1980. "The Present Status and Future Prospects of Tropical Moist Forests. *Environmental Conservation* 7 (2): 101–114.

———. 1981. "The Hamburger Connection: How Central America's Forests Become North America's Hamburgers." *Ambio* 10 (1): 3–8.

———. 1984. *The Primary Source: Tropical Forests and Our Future*. New York: W. W. Norton.

———. 1997. "Biodiversity's Genetic Library." In *Nature's Services*, edited by G. Daily, 255–273. Washington, DC: Island.

Nail, T. 2015. "Migrant Cosmopolitanism." *Public Affairs Quarterly* 29: 187–199.

Newman, D. J., and Cragg, G. M. 2016. "Natural Products as Sources of New Drugs from 1981 to 2014." *Journal of Natural Products* 79 (3): 629–661.

Newson, L. 2006. "Medical Practice in Early Colonial Spanish America: A Prospectus." *Bulletin of Latin American Research* 25: 367–391.

Niane, D. T. 1984. "Relationships and Exchanges among the Different Regions." In *General History of Africa*. Vol. 4, *Africa from the Twelfth to the Sixteenth Century*, edited by D. T. Niane, 615–634. Berkeley: University of California Press.

Nixon, R. 2011. *Slow Violence and the Environmentalism of the Poor*. Cambridge, MA: Harvard University Press.

Noble, L. 2011. *Medicinal Cannibalism in Early Modern English Literature and Culture*. New York: Palgrave Macmillan.

Norton, M. 2008. *Sacred Gifts, Profane Pleasures: A History of Tobacco and Chocolate in the Atlantic World*. Ithaca, NY: Cornell University Press.

Nygren, A. 2006. "Representations of Tropical Forests and Tropic Forest-Dwellers in Travel Accounts of *National Geographic*." *Environmental Values* 15: 505–525.

Oliver, B. 1984. In: Global Nonviolent Action Database. http://nvdatabase.swarthmore.edu/content/us-activists-stop-burger-king-importing-rainforest-beef-1984-1987

Oliver, J. E., and Wood, T. 2014. "Medical Conspiracy Theories and Health Behaviors in the United States." *JAMA Internal Medicine* 174 (5): 817–818.

Orellana, S. 1977. "Part Five: Aboriginal Medicine in Highland Guatemala." *Medical Anthropology* 1: 113–156.

Orta, G. de. 1913 [1563]. *Colloquies on the Simples and Drugs of India*. Translated by Sir Clements Markham. London: Henry Sotheran.

Osseo-Asare, A. 2008. "Bioprospecting and Resistance: Transforming Poisoned Arrows into *Strophantin* Pills in Colonial Gold Coast, 1885–1922." *Social History of Medicine* 21: 269–290.

———. 2014. *Bitter Roots: The Search for Healing Plants in Africa*. Chicago: University of Chicago Press.

Padoch, C., Coffey, K., Mertz, O., Leisz, S., Fox, J., and Wadley, R. L. 2007. "The Demise of Swidden in Southeast Asia? Local Realities and Regional Ambiguities." *Danish Journal of Geography* 107: 29–41.

Pagden, A. 1993. *European Encounters with the New World: From Renaissance to Romanticism*. New Haven, CN: Yale University Press.

Palmer, C. T. 2004. "The Inclusion of Recently Introduced Plants in the Hawaiian Ethnopharmacopoeia." *Economic Botany* 58 (1): S280–S293.

Panghal, M., Arya, V., Yadav, S., Kumar, S., and Yadav, J. P. 2010. "Indigenous Knowledge of Medicinal Plants Used by Saperas Community of Khetawas, Jhajjar District, Haryana, India." *Journal of Ethnobiology and Ethnomedicine* 6 (4). http://www.ethnobiomed.com/content/6/1/4

Parsons, J. J. 1976. "Forest to Pasture: Development or Destruction?" *Revista de Biología Tropical* 24: 121–138.

Patchett, A. 2011. *A State of Wonder*. New York: Harper Collins.

Peachi, J. 1694a. *Some Observations Made upon the Banellas, Imported from the Indies*. London: NP.

———. 1695. *Some Observations made upon the Mexico Seeds, Imported from the Indies.* London: NP.

Pearson, M. N. 1996. "First Contacts between Indian and European Medical Systems: Goa in the Sixteenth Century," In *Warm Climates and Western Medicine*, edited by D. Arnold, 20–41. Amsterdam: P. V. Rodopi.

———. 2011. "Medical Connections and Exchanges in the Early Modern World." Health and Borders across Time and Cultures: China, India and the Indian Ocean Region Special Issue, edited by B. C. García and D. Ghosh. *PORTAL Journal of Multidisciplinary International Studies* 8 (2): 1–15.

Peluso, N. L. 1996. "Fruit Trees and Family Trees in an Anthropogenic Forest: Ethics of Access, Property Zones, and Environmental Change in Indonesia." *Comparative Studies in Society and History* 38: 510–548.

Pereira, N. A., Jaccoud, R. J., and Mors, W. B. 1996. "Triaga Brasilica: Renewed Interest in a Seventeenth-Century Panacea." *Toxicon* 34 (5): 511–516.

Peters, C. M. 2000. "Precolumbian Silviculture and Indigenous Management of Neotropical Forests." In *Imperfect Balance: Landscape Transformations in the Precolumbian Americas*, edited by D. Lentz, 203–223. New York: Columbia University Press.

Petersen, J., and Crock, J. 2007. "Handsome Death: The Taking, Veneration and Consumption of Human Remains in the Insular Caribbean and Greater Amazonia." In *The Taking and Displaying of Human Body Parts as Trophies by Amerindians*, edited by R. Chacon and D. Dye, 547–574. New York: Springer.

Petiver, J. 1697 (September 22). "Letter Sent to Dr. Hans Sloane, A Catalogue of some Guinea Plants, with their Native Names and Virtues." *Philosophical Transactions* 19: 677–686.

Pfeiffer, J., and Butz, R. 2005. "Assessing Cultural and Ecological Variation in Ethnobiological Research: The Importance of Gender." *Journal of Ethnobiology* 25: 240–278.

Pfeiffer, J., and Voeks, R. 2008. "Biological Invasions and Biocultural Diversity: Linking Ecological and Cultural Systems." *Environmental Conservation* 35 (4): 281–293.

Phillips, O., and Gentry, A. 1993. "The Useful Plants of Tambopata, Peru: II. Additional Hypothesis Testing in Quantitative Ethnobotany." *Economic Botany* 47 (1): 33–43.

Pimm, S. L., and Joppa, L. N. 2015. "How Many Plant Species Are There, Where Are They, and at What Rate Are They Going Extinct?" *Annals of the Missouri Botanical Garden* 100 (3): 170–176.

Pinheiro, C. 1997. "Jaborandi (*Pilocarpus* sp. Rutaceae): A Wild Species and Its Rapid Transformation into a Crop." *Economic Botany* 51: 49–58.

Pinto, E. P. 2012. "Viagem de Descrobimento ao Rio Madeira e Suas Vertentes por Francisco de Melo Palheta." *Revista Veredas Amazônicas* 2 (2): 69–86.

Pires, M. 1998. "Mutual Influences between Portugal and China." *Review of Culture* 6: 76–83.

Pires, T. 1967 [1512–1515]. "Suma Oriental." In *The Suma Oriental of Tomé Pires*, edited by A. Cortesao, 1–289. Nendeln, Liechtenstein: Hakluyt Society.

Piso, W. 1948 [1648]. *História Natural do Brasil Ilustrada*. Translated by A. Taunay. São Paulo: Companhia Editora Nacional.

Piu, C. 2009. "Toward a Total Cure for Acute Lymphoblastic Leukemia." *Journal of Clinical Oncology* 27: 5121–5123.

Pliny the Elder. 1890 [77–79 AD]. *The Natural History of Pliny the Elder*. Vol. 2. Translated by J. Bostock and H. T. Riley. London: George Bell & Sons.

Plotkin, M. 1993. *Tales of a Shaman's Apprentice: An Ethnobotanist Searches for New Medicines in the Amazon Rainforest*. New York: Viking.

Poderoso, R. A., Hanazaki, N., and Junior, A. D. 2012. "How Is Local Knowledge about Plants Distributed among Residents near a Protected Area?" *Ethnobiology and Conservation* 1: 1–26. http://ethnobioconservation.com

Poivre, P. 1770. *Travels of a Philosopher: Or Observations on the Manners and Arts of Various Nations in Africa and Asia*. Glasgow: R. Urie.

Polari de Alverga, A. 1999. *Forest of Visions: Ayahuasca, Amazonian Spirituality, and the Santo Daime Tradition*. South Paris, ME: Park Street.

Pollio, A., De Natale, A., Appetiti, E., Aliotta, G., and Touwaide, A. 2008. "Continuity and Change in the Mediterranean Medical Tradition: *Ruta* spp. (Rutaceae) in Hippocratic Medicine and Present Practices." *Journal of Ethnopharmacology* 116 (3): 469–482.

Pope, A. 1828. *Pope's Essay on Man, and other Poems*. Vol. 2. London: John Sharpe.

Posey, D. A. 1985. "Indigenous Management of Tropical Forest Ecosystems: The Case of the Kayapó Indians of the Brazilian Amazon." *Agroforestry Systems* 3 (2): 139–158.

Pouliot, M. 2011. "Relying on Nature's Pharmacy in Rural Burkina Faso: Empirical Evidence of the Determinants of Traditional Medicine Consumption." *Social Science & Medicine* 73 (10): 1498–1507.

Prandi, R. 2002. "Candomblé and Time: Concepts of Time, Knowing and Authority from Africa to African-Brazilian Religions." *Brazilian Review of Social Sciences* 2: 7–22.

Prayag, G., Mura, P., Hall, M., and Fontaine, J. 2015. "Drug or Spirituality Seekers? Consuming Ayahuasca." *Annals of Tourism Research* 52: 175–177.

Prest, J. 1981. *The Garden of Eden: The Botanic Garden and the Re-creation of Paradise*. New Haven, CN: Yale University Press.

Price, R. 1979. "Kwasimukamba's Gambit." *Bijdragen tot de Taal-, Land- en Volkenkunde* 135: 151–169.

———. 1996. "Introduction." In *Maroon Societies: Rebel Slave Communities in the Americas*, edited by R. Price, 1–32. Baltimore: Johns Hopkins University Press.

———. 2007. "Africans Discover America: The Ritualization of Gardens, Landscapes, and Seascapes by Suriname Maroons." In *Sacred Gardens and Landscapes: Ritual and Agency*, edited by M. Conan, 221–236. Washington, DC: Dumbarton Oaks Research Library and Collection.

————. 2010. "Uneasy Neighbors: Maroons and Indians in Suriname." *Tipití: Journal of the Society for the Anthropology of Lowland South America* 8 (2): 1–15.

Principe, P. P. 1991. "Valuing the Biodiversity of Medicinal Plants." In *Conservation of Medicinal Plants*, edited by O. Akerele, V. Heywood, and H. Synge, 79–124. Cambridge, UK: Cambridge University Press.

————. 1996. "Monetizing the Pharmacological Benefits of Plants." In *Medicinal Resources of the Tropical Forest: Biodiversity and Its Importance to Human Health*, edited by M. J. Balick, E. Elisabetsky, and S. Laird, 191–218. New York: Columbia University Press.

Purchas, S. 1613. *Purchas His Pilgrimage, Or Relations of the World and the Religions Observed in All Ages*. London: William Stansby.

————. 1625. *Purchas His Pilgrimes, Contayning a History of the World, in Sea Voyages & Lande-Travells, by Englishmen & Others*. London: William Stansby.

Puri, R. K. 2005. "Post-abandonment Ecology of Penan Fruit Camps: Anthropological and Ethnobiological Approaches to the History of a Rain Forested Valley in East Kalimantan." In *Conserving Nature in Culture: Case Studies from Southeast Asia*, edited by M. R. Dove, P. E. Sajise, and A. Doolittle, 25–82. New Haven, CN: Yale University Press.

Purseglove, J. 1974. *Tropical Crops: Dicotyledons*. London: Longman.

Querino, M. 1955. *A Raca Africana e os Seus Costumes*. Salvador, Brasil: Progresso.

Quinlan, M. B., and Quinlan, R. J. 2007. "Modernization and Medicinal Plant Knowledge in a Caribbean Horticultural Village." *Medical Anthropology Quarterly* 21 (2): 169–192.

Quinlan, M. B., Quinlan, R. J., Council, S. K., and Roulette, J. W. 2016. "Children's Acquisition of Ethnobotanical Knowledge in a Caribbean Horticultural Village." *Journal of Ethnobiology* 36 (2): 433–456.

Quinn, D. B. 1990. *Explorers and Colonies: 1500–1625*. London: Hambledon.

Raffauf, R. 1970. "Some Notes on the Distribution of Alkaloids in the Plant Kingdom." *Economic Botany* 24: 34–38.

Raffles, H. 2002. *In Amazonia: A Natural History*. Princeton, NJ: Princeton University Press.

Rahi, D. K., and Malik, D. 2016. "Diversity of Mushrooms and Their Metabolites of Nutraceutical and Therapeutic Significance." *Journal of Mycology* Article ID 7654123. http://dx.doi.org/10.1155/2016/7654123

Ralegh, Sir W. 1596. *The Discoverie of the Large, Rich, and Beautiful Empire of Guiana*. London: Robert Robinson.

Ram, H. Y. 2005. "On the English Edition of Van Rheede's Hortus Malabaricus by K. S. Manilal, 2003." *Current Science* 89: 1632–1680.

Rambo, A. T. 1985. "Primitive Polluters: Semang Impact on the Malaysian Tropical Rain Forest Ecosystem." *Anthropological Papers 76*. Ann Arbor: University of Michigan Museum of Anthropology.

Ramirez, C. R. 2007. "Ethnobotany and the Loss of Traditional Knowledge in the 21st Century." *Ethnobotany Research & Applications* 5: 245–247.

Rangan, H., Carney, J., and Denham, T. 2012. "Environmental History of Botanical Exchanges in the Indian Ocean World." *Environment and History* 18 (3): 311–342.

Raven, C. E. 1950. *John Ray, Naturalist: His Life and Works*. Cambridge, UK: Cambridge University Press.

Ravindran, P. N., and Babu, K. N. 2004. "Introduction." In *Ginger: The Genus Zingiber*, edited by P. N. Ravindran and K. N. Babu, 1–14. New York: CRC.

Ray, J. 1692. *The Wisdom of God Manifested in the Works of the Creation*. London: Samuel Smith.

Reichel, E. 1999. "Cosmology, Worldview and Gender-Based Knowledge Systems among the Tanimuka and Yukuna (Northwest Amazon)." *Worldviews: Environment, Culture, Religion* 3: 213–242.

Reichel-Dolmatoff, G. 1972. *Amazonian Cosmos: The Sexual and Religious Symbolism of the Tukano Indians*. Chicago: University of Chicago Press.

Renner, S. 2004. "Plant Dispersal across the Tropical Atlantic by Wind and Sea Currents." *International Journal of Plant Sciences* 165 (S4): S23–S33.

Reuters News Service. 2015. "South African Park Rangers Have Reportedly Killed 500 Poachers in the Past Five Years." September 21.

Reyes-García, V., Guèze, M., Luz, A. C., Paneque-Gálvez, J., Macía, M. J., Orta-Martínez, M., Pino, J., and Rubio-Campillo, X. 2013a. "Evidence of Traditional Knowledge Loss among a Contemporary Indigenous Society." *Evolution and Human Behavior* 34 (4): 249–257.

Reyes-García, V., Luz, A. C., Gueze, M., Paneque-Gálvez, J., Macía, M. J., Orta-Martínez, M., and Pino, J. 2013b. "Secular Trends on Traditional Ecological Knowledge: An Analysis of Changes in Different Domains of Knowledge among Tsimane' Men." *Learning and Individual Differences* 27: 206–212.

Richards, P. W. 1973. "Africa, the 'Odd Man Out.'" In *Tropical Forest Ecosystems in Africa and South America: A Review*, edited by B. J. Meggers, E. S. Ayensu, and W. D. Duckworth, 21–26. Washington, DC: Smithsonian Institution.

Riddle, J. M. 1985. "Ancient and Medieval Chemotherapy for Cancer." *Isis* 76: 319–330.

Rival, L. M. 2002. *Trekking through History: The Huaorani of Amazonian Ecuador*. New York: Columbia University Press.

Rochefort, C. de. 1681. *Histoire Naturelle et Morale des iles Antilles de l'Amerique*. Rotterdam: Reinier Lerrs.

Rocheleau, D., and Edmunds, D. 1997. "Women, Men and Trees: Gender, Power and Property in Forest and Agrarian Landscapes" *World Development* 25: 1351–1371.

Rodenburg, J., Both, J., Heitkönig, I. M., Van Koppen, C. S. A., Sinsin, B., Van Mele, P. and P. Kiepe. 2012. "Land Use and Biodiversity in Unprotected Landscapes: The Case of Noncultivated Plant Use and Management by Rural Communities in Benin and Togo." *Society & Natural Resources* 25 (12): 1221–1240.

Rodríguez, I. 2004. *Transatlantic Topographies*. Minneapolis: University of Minnesota Press.

Rolander, D. 2008 [1755]. "The Suriname Journal: Composed during an Exotic Journey, 1754–1756." In *The Linnaeus Apostles: Global Science & Adventure,* edited by L. Hansen, 1215–1576. London: IK Foundation.

Roosevelt, A. C. 2013. "The Amazon and the Anthropocene: 13,000 Years of Human Influence in a Tropical Rainforest." *Anthropocene* 4: 69–87.

Roosevelt, T. 1914. *Through the Brazilian Wilderness.* New York: C. Scribner's Sons.

Ross, N. 2011. "Modern Tree Species Composition Reflect Ancient Maya 'Forest Gardens' in Northwest Belize." *Ecological Applications* 21: 75–84.

Rozin, P., and Schiller, D. 1980. "The Nature and Acquisition of a Preference for Chili Pepper by Humans." *Motivation and Emotion* 4: 77–101.

Rull, V., Cañellas-Boltà, N., Margalef, O., Pla-Rabes, S., Sáez, A., and Giralt, S. 2016. "Three Millennia of Climatic, Ecological, and Cultural Change on Easter Island: An Integrative Overview." *Frontiers in Ecology and Evolution* 4: 1–4. http://dx.doi.org/10.3389/fevo.2016.00029

Rutten, A. M. G. 2000. *Dutch Transatlantic Medicine Trade in the Eighteenth Century under the Cover of the West India Company.* 2nd ed. Rotterdam: Erasmus.

Ryu, S. H., Werth, L., Nelson, S., Scheerens, J. C., and Pratt, R. C. 2013. "Variation of Kernel Anthocyanin and Carotenoid Pigment Content in USA/Mexico Borderland Land Races of Maize." *Economic Botany* 67 (2): 98–109.

Salazar, S., and Marquis, R. 2012. "Herbivore Pressure Increases toward the Equator." *Proceedings of the National Academy of Sciences* 109: 12616–12620.

Sale, K. 1990. *The Conquest of Paradise: Christopher Columbus and the Columbian Legacy.* New York: Knopf.

Salmon, T. 1746. *Modern History, or the Present State of All Nations.* Vol. 3. 3rd ed. London: Longman and Shewell.

Sander, L., and Vandebroek, I. 2016. "Small-Scale Farmers as Stewards of Useful Plant Diversity: A Case Study in Portland Parish, Jamaica." *Economic Botany* 70 (3): 303–319.

Sanín, M. J., Anthelme, F., Pintaud, J. C., Galeano, G., and Bernal, R. 2013. "Juvenile Resilience and Adult Longevity Explain Residual Populations of the Andean Wax Palm *Ceroxylon quindiuense* after Deforestation." *PLOS ONE* 8 (10): e74139. doi:10.1371/journal.pone.0074139

Santana, B. F., Voeks, R. A., and Funch, L. S. 2016. "Ethnomedicinal Survey of a Maroon Community in Brazil's Atlantic Tropical Forest." *Journal of Ethnopharmacology* 181: 37–49.

Santos, F. 2009. *As Plantas Brasileiras, os Jesuítas e os Indígenas do Brasil: História e Ciência na Triaga Brasílica (séc. XVII–XVIII).* São Paulo, Brasil: Novo Autor Editora.

Santos, J. dos. 1609. *Ethiopia Oriental, e Varia Historia de Cousas Notaveis de Oriente.* NP: Impressa no Conuento de S. Domongos.

Santos, L. L., do Nascimento, A. L. B., Vieira, F. J., da Silva, V. A., Voeks, R. A., and Albuquerque, U. P. 2014. "The Cultural Value of Invasive Species: A Case Study from Semi-arid Northeastern Brazil." *Economic Botany* 68 (3): 283–300.

Saslis-Lagoudakis, C., Klitgaard, B., Forest, F., Francis, L., Savolainen, V., William-son, E., and Hawkins, J. 2011. "The Use of Phylogeny to Interpret Cross-Cultural Patterns in Plant Use and Guide Medicinal Plant Discovery: An Example from *Pterocarpus* (Leguminosae)." *PLOS ONE* 6 (7): e22275.

Sauer, C. O. 1963. "Man in the Ecology of Tropical America." In *Land and Life: A Selection from the Writings of Carl Ortwin Sauer*, edited by J. Leighly, 182–193. Berkeley: University of California Press.

Scarpa, A., and Guerci, A. 1982. "Various Uses of the Castor Oil Plant (*Ricinus communis* L.): A Review." *Journal of Ethnopharmacology* 5: 117–137.

Schaan, D., Pärssinen, M., Saunaluoma, S., Ranzi, A., Bueno, M., and Barbosa, A. 2012. "New Radiometric Dates for Precolumbian (2000–700 b.p.) Earthworks in Western Amazonia, Brazil." *Journal of Field Archaeology* 37 (2): 132–142.

Schedel, H. 1493. *Liber Chronicarum* (or *Nuremberg Chronicle*). Fol. 12. Nuremburg: Anton Koberger.

Schemske, D., Mittelbach, G., Cornell, H., Sobel, J., and Roy, K. 2009. "Is There a Latitudinal Gradient in the Importance of Biotic Interactions?" *Annual Review of Ecology, Evolution, and Systematics* 40: 245–269.

Schep, L.J., Temple, W. A., Butt, G. A. and Beasley, M. D. 2009. "Ricin as a Weapon of Mass Terror—Separating Fact from Fiction." *Environment International* 35 (8): 1267–1271.

Schiebinger, L. 2004. *Plants and Empire: Colonial Bioprospecting in the Atlantic World*. Cambridge, MA: Harvard University Press.

Schrepfer, S. R. 2005. *Nature's Altars: Mountains, Gender, and American Environmentalism*. Lawrence: University of Kansas Press.

Schroeder, R. 1993. "Shady Practice: Gender and the Political Ecology Resource Stabilization in Gambian Garden/Orchards." *Economic Geography* 69: 349–365.

Schultes, R. E. 1960. "Tapping Our Heritage of Ethnobotanical Lore." *Economic Botany* 14 (4): 257–262.

Schultes, R. E., and Raffauf, R. F. 1990. *The Healing Forest*. Portland, OR: Dioscorides Press.

Schwartz, S. B. 1985. *Sugar Plantations in the Formation of Brazilian Society: Bahia, 1550–1835*. Cambridge, UK: Cambridge University Press.

———. 1992. *Slaves, Peasants, and Rebels: Reconsidering Brazilian Slavery*. Urbana: University of Illinois Press.

Scott, H. V. 2010. "Paradise in the New World: An Iberian Vision of Tropicality." *Cultural Geographies* 17: 77–101.

Semple, E. 1911. "The Influence of Climate." In *The Geographic Environment*, edited by E. Semple, 607–635. New York: Henry Holt.

Shackleton, C., Pandey, A., and Ticktin, T., eds. 2015. *Ecological Sustainability for Non-timber Forest Products: Dynamics and Case Studies of Harvesting*. London: Routledge.

Shanley, P., and Rosa, N. A. 2004. "Eroding Knowledge: An Ethnobotanical Inventory in Eastern Amazonia's Logging Frontier." *Economic Botany* 58 (2): 135–160.

Shaw, E. 1992. *Plants of the New World: The First 150 Years.* Cambridge, MA: Harvard College Library.

Shenton, J., Ross, N., Kohut, M., and Waxman, S. 2011. "Maya Folk Botany and Knowledge Devolution: Modernization and Intra community Variability in the Acquisition of Folkbotanical Knowledge." *Ethos* 39 (3): 349–367.

Shepard, G., and Ramirez, H. 2011. " 'Made in Brazil': Human Dispersal of the Brazil Nut (*Bertholletia excelsa*, Lecythidaceae) in Ancient Amazonia." *Economic Botany* 65 (1): 44–65.

Sherman, I. 2011. *Magic Bullets to Conquer Malaria: From Quinine to Quinghaosu.* Washington, DC: ASM.

Sherman, P. W., and Billing, J. 1999. "Darwinian Gastronomy: Why We Use Spices." *BioScience* 49 (6): 453–463.

Simoons, F. J. 1998. *Plants of Life: Plants of Death.* Madison: University of Wisconsin Press.

Slater, C. 2002. *Entangled Edens: Visions of the Amazon.* Berkeley: University of California Press.

Sloane, H. 1707 and 1725. *A Voyage to the Islands Madera, Barbados, Nieves, S. Christophers and Jamaica with the Natural History of the Herbs and Trees, Four-footed Beasts, Fishes, Birds, Insects, Reptiles, & C. of that Place, with Some Relations Concerning the Neighbouring Continent, and Islands of America.* London: Printed for BM by the author.

Sluyter, A. 1999. "The Making of the Myth in Postcolonial Development: Material-Conceptual Landscape Transformation in Sixteenth-century Veracruz." *Annals of the Association of American Geographers* 89 (3): 377–401.

Smith, A. 2002. *Peanuts: The Illustrious History of the Goober Pea.* Urbana: University of Illinois Press.

Smith, B. 1985. *European Visions and the South Pacific.* 2nd ed. New Haven, CT: Yale University Press.

Smith, D. L., and Panaitiu, I. 2015. "Aping the Human Essence: Simianization as Dehumanization." In *Simianization: Apes, Gender, Class, and Race*, vol. 6, edited by W. D. Hund, C. W. Mills, and S. Sebastiani, 77–104. Zurich: LIT Verlag.

Smith, N. J. 1980. "Anthrosols and Human Carrying Capacity in Amazonia." *Annals of the Association of American Geographers* 70 (4): 553–566.

Smith, W. 1774. *A New Voyage to Guinea, Describing the Customs, Manners, Soils, etc.* London: John Nourse.

Soejarto, D. D., and Farnsworth, N. R. 1989. "Tropical Rain Forests: Potential Source of New Drugs?" *Perspectives in Biology and Medicine* 32: 244–256.

Sogbohossou, O. E., Achigan-Dako, E. G., Komlan, F. A., and Ahanchede, A. 2015. "Diversity and Differential Utilization of *Amaranthus* spp. along the Urban-Rural Continuum of Southern Benin." *Economic Botany* 69 (1): 9–25.

Sousa, G. S. de.1971 [1587]. *Tratado Descritivo do Brasil em 1587*. 4th ed. São Paulo, Brasil: Editora Nacional.

Souto, T., and Ticktin, T. 2012. "Understanding Interrelationships among Predictors (Age, Gender, and Origin) of Local Ecological Knowledge." *Economic Botany* 66 (2): 149–164.

Spary, E. C. 2000. *Utopia's Garden: French Natural History from Old Regime to Revolution*. Chicago: University of Chicago Press.

Srithi, K., Trisonthi, C., Wangpakapattanawong, P., Srisanga, P., and Balslev, H. 2012. "Plant Diversity in Hmong and Mien Homegardens in Northern Thailand." *Economic Botany* 66 (2): 192–206.

Staden, H. 1928 [1557]. *Hans Staden: The True History of his Captivity, 1557*. Translated by Malcolm Letts. London: George Routledge & Sons.

Stagegaard, J., Sorensen, M., and Kvist, L. 2002. "Estimations of the Importance of Plant Resources Extracted by Inhabitants of the Peruvian Amazon Flood Plain Forests." *Perspectives in Plant Ecology, Evolution, and Systematics* 5: 103–122.

Stanley, D., Voeks, R., and Short, L. 2012. "Is Non-timber Forest Product Harvest Sustainable in the Less Developed World? A Meta-analysis of the Recent Economic and Ecological Literature." *Ethnobiology and Conservation* 1: 1–39.

Stannard, J. 1964. "A Fifteenth-Century Botanical Glossary." *Isis* 55: 353–367.

———. 1999. "Dioscorides and Renaissance Materia Medica." In *Herbs and Herbalism in the Middle Ages and Renaissance*, edited by J. Stannard and R. Kay, 1–21. Brookfield, VT: Ashgate.

Stearn, W. T. 1988. "Carl Linnaeus's Acquaintance with Tropical Plants." *Taxon* 37: 776–781.

Stedman, J. G. 1988 [1790]. *Narrative of a Five Years Expedition Against the Revolted Negroes of Surinam*. Edited by R. Price and S. Price. Baltimore, MD: Johns Hopkins University Press.

Steinberg, M. 2002. "The Second Conquest: Religious Conversion and the Erosion of the Cultural Ecological Core among the Mopan Maya." *Journal of Cultural Geography* 20: 91–105.

Stepan, N. 2001. *Picturing Tropical Nature*. Ithaca, NY: Cornell University Press.

Stepp, J. R. 2004. "The Role of Weeds as Sources of Pharmaceuticals." *Journal of Ethnopharmacology* 92 (2): 163–166.

Stepp, J. R., and Moerman, D. E. 2001. "The Importance of Weeds in Ethnopharmacology." *Journal of Ethnopharmacology* 75 (1): 19–23.

Stewart, K. M. 2003. "The African Cherry (*Prunus africana*): Can Lessons Be Learned from an Over-exploited Medicinal Tree?" *Journal of Ethnopharmacology* 89: 3–13.

Stolarczyk, J., and Janick, J. 2011. "History-Carrot: History and Iconography." *Chronica Horticulturae* 51 (2): 13. http://www.ishs.org/chronica-horticulturae/vol51nr2

Stopp, K. 1963. "Medicinal Plants of the Mt. Hagen People (Mbowamb) in New Guinea." *Economic Botany* 17 (1): 16–22.

Sugiyama, M. S. 2001. "Food, Foragers, and Folklore: The Role of Narrative in Human Subsistence." *Evolution and Human Behavior* 22 (4): 221–240.

Sumner, J. 2000. *The Natural History of Medicinal Plants*. Portland, OR: Timber.

Svetaz, L., Zuljan, F., Derita, M., Petenatti, E., Tamayo, G., Cáceres, A., Cechinel Filho, V., Giménez, A., Pinzón, R., Zacchino, S.A., and Gupta, M. 2010. "Value of the Ethnomedical Information for the Discovery of Plants with Antifungal Properties. A Survey among Seven Latin American Countries." *Journal of Ethnopharmacology* 127: 37–158.

Swanson, T. D. 2009. "Singing to Estranged Lovers: Runa Relations to Plants in the Ecuadorian Amazon." *Journal for the Study of Religion, Nature & Culture* 3 (1): 36–65.

Sylvester, O., and Avalos, G. 2009. "Illegal Palm Heart (*Geonoma edulis*) Harvest in Costa Rican National Parks: Patterns of Consumption and Extraction." *Economic Botany* 63 (2): 179–189.

Talbor, R. 1672. *A Rational Account of the Cause & Cure of Agues, With their Signes Diagnostick & Prognostick*. London: R. Robinson.

Taussig, M. 1987. *Shamanism, Colonialism, and the Wild Man: A Study in Terror and Healing*. Chicago: University of Chicago Press.

Tavernier, J. 2001 [1676]. *Travels in India: 1640-1676*. Translated by V. Ball. Edited by W. Crooke. 2 vols. New Delhi: Asian Educational Services.

Teklehaymanot, T. 2009. "Ethnobotanical Study of Knowledge and Medicinal Plants Use by the People in Dek Island in Ethiopia." *Journal of Ethnopharmacology* 124 (1): 69–78.

Terborgh, J., Davenport, L. C., Niangadouma, R., Dimoto, E., Mouandza, J. C., Schultz, O., and Jaen, M. R. 2016. "The African Rainforest: Odd Man Out or Megafaunal Landscape? African and Amazonian Forests Compared." *Ecography* 39 (2): 187–193.

Ter Steege, H., and 119 other authors. 2013. "Hyperdominance in the Amazonian Tree Flora." *Science* 342 (6156): 1243092.

Thevet, A. 1625. Cited in: *Purchas His Pilgrimes, Contayning a History of the World, in Sea Voyages & Lande-Travells, by Englishmen & Others*, edited by S. Purchas. London: William Stansby.

Thoden van Velzen, H., and van Wetering, W. 2004. *In the Shadow of the Oracle: Religion as Politics in a Suriname Maroon Society*. Long Grove, IL: Waveland.

Thomas, E., Semo, L., Morales, M., Noza, Z., Nuñez, H., Cayuba, A., and Van Damme, P. 2011. "Ethnomedicinal Practices and Medicinal Plant Knowledge of the Yuracarés and Trinitarios from Indigenous Territory and National Park Isiboro-Sécure, Bolivian Amazon." *Journal of Ethnopharmacology* 133 (1): 153–163.

Thomas, E., and Van Damme, P. 2010. "Plant Use and Management in Homegardens and Swiddens: Evidence from the Bolivian Amazon." *Agroforestry Systems* 80 (1): 131–152.

Thomas, E., Vandebroek, I., Van Damme, P., Goetghebeur, P., Douterlungne, D., Sanca, S., and Arrazola, S. 2009. "The Relation between Accessibility, Diversity and Indigenous Valuation of Vegetation in the Bolivian Andes." *Journal of Arid Environments* 73: 854–861.

Thompson, R. C. 1924. "Assyrian Medical Texts." *Proceedings of the Royal Society of Medicine* 17 (Sect Hist. Med): 1–34.

Thornton, J. 1992. *Africa and Africans in the Making of the Atlantic World, 1400– 1680.* Cambridge, UK: Cambridge University Press.

Titanji, V. P., Zofou, D., and Ngemenya, M. N. 2008. "The Antimalarial Potential of Medicinal Plants Used for the Treatment of Malaria in Cameroonian Folk Medicine." *African Journal of Traditional, Complementary, and Alternative Medicines* 5 (3): 302–321.

Toledo, V. M., Batis, A. I., Becerra, R., Esteban, M., and Ramos, C. H. 1992. "Products from the Tropical Rain Forests of Mexico: An Ethnoecological Approach." In *Sustainable Harvest and Marketing of Rain Forest Products*, edited by M. Plotkin and L. Famolare, 99–109. Washington, DC: Island.

Tomášková, S. 2013. *Wayward Shamans: The Prehistory of an Idea.* Berkeley: University of California Press.

Torres-Avilez, W., Medeiros, P. M. D., and Albuquerque, U. P. 2016. "Effect of Gender on the Knowledge of Medicinal Plants: Systematic Review and Meta-Analysis." *Evidence-Based Complementary and Alternative Medicine*, Article ID 6592363. doi:10.1155/2016/6592363

Touwaide, A., and Appetiti, E. 2013. "Knowledge of Eastern Materia Medica (Indian and Chinese) in Pre-modern Mediterranean Medical Traditions. A Study in Comparative Historical Ethnopharmacology." *Journal of Ethnopharmacology* 148: 361–378.

———. 2015. "Food and Medicines in the Mediterranean Tradition. A Systematic Analysis of the Earliest Extant Body of Textual Evidence." *Journal of Ethnopharmacology* 167: 11–29.

Towns, A. M., Ruysschaert, S., van Vliet, E., and van Andel, T. 2014. "Evidence in Support of the Role of Disturbance Vegetation for Women's Health and Childcare in Western Africa." *Journal of Ethnobiology and Ethnomedicine* 10 (1): 42.

Tragett, C. 2012. "From Romanticized Women to Politicized Gender: How Has the Literature Addressed Gender and Natural Resource Use?" PhD diss., Imperial College London.

Trapham, T. 1679. *A Discourse of the State of Health in the Island of Jamaica.* London: R. Boulter.

———. 1694. *Some Observations Made upon the Bermudas Berries Imported from the Indies Shewing their Admirable Virtues in Curing the Green-Sickness.* London: NP.

Turnbull, C. 1972. *The Mountain People.* New York: Simon and Schuster.

Váczy, C. 1980. "Hortus Indicus Malabaricus and Its Importance for the Botanical Nomenclature." In *Botany and History of Hortus Malabaricus*, edited by K. S. Manilal, 25–34. Rotterdam: A. A. Balkema.

Van Andel, T. 2010. "African Rice (*Oryza glaberrima* Steud.): Lost Crop of the Enslaved Africans Discovered in Suriname." *Economic Botany* 64: 1–10.

————. 2015. "The Reinvention of Household Medicine by Enslaved Africans in Suriname." *Social History of Medicine* doi:10.1093/shm/hkv014

Van Andel, T., and Carvalheiro, L. G. 2013. "Why Urban Citizens in Developing Countries Use Traditional Medicines: The Case of Suriname." *Evidence-Based Complementary and Alternative Medicine* 127: 694–701.

Van Andel, T., Croft, S., van Loon, E. E., Quiroz, D., Towns, A. M., and Raes, N. 2015. "Prioritizing West African Medicinal Plants for Conservation and Sustainable Extraction Studies Based on Market Surveys and Species Distribution Models." *Biological Conservation* 181: 173–181.

Van Andel, T., de Korte, S., Koopmansb, D., Behari-Ramdas, J., and Ruysschaert, S. 2008. "Dry Sex in Suriname." *Journal of Ethnopharmacology* 116: 84–88.

Van Andel, T., and Fundiko, M. 2016. "The Trade in African Medicinal Plants in Matonge-Ixelles, Brussels (Belgium)." *Economic Botany* 70 (4): 405–415.

Van Andel, T., Maas, P., and Dobreff, J. 2012. "Ethnobotanical Notes from Daniel Rolander's *Diarium Surinamicum* (1754–1756): Are These Plants Still Used in Suriname Today?" *Taxon* 61: 852–863.

Van Andel, T., Meyer, R. S., Aflitos, S. A., Carney, J. A., Veltman, M. A., Copetti, D., Flowers, J. M., Havinga, R. M., Maat, H., Purugganan, M. D., and Wing, R. A. 2016. "Tracing Ancestor Rice of Suriname Maroons Back to Its African Origin." *Nature Plants* 2: 16149.

Van Andel, T., Ruysschaert, S., Van de Putte, K., and Groenendijk, S. 2013. "What Makes a Plant Magical? Symbolism and Sacred Herbs in Afro-Surinamese Winti Rituals." In *African Ethnobotany in the Americas*, edited by R. Voeks and J. Rashford, 247–284. New York: Springer.

Van Andel, T., van der Velden, A., and Reijers, M. 2015. "The 'Botanical Gardens of the Dispossessed' Revisited: Richness and Significance of Old World Crops Grown by Suriname Maroons." *Genetic Resources and Crop Evolution* 63: 695–710.

Van Andel, T., van't Klooster, C. I., Quiroz, D., Towns, A. M., Ruysschaert, S., and van den Berg, M. 2014. "Local Plant Names Reveal that Enslaved Africans Recognized Substantial Parts of the New World Flora." *Proceedings of the National Academy of Sciences* 111 (50): E5346–E5353.

Van Andel, T., and Westers, P. 2010. "Why Surinamese Migrants in the Netherlands Continue to Use Medicinal Herbs from Their Home Country." *Journal of Ethnopharmacology* 127: 694–701.

Van der Heijden, R., Jacobs, D. I., Snoeijer, W., Hallard, D., and Verpoorte, R. 2004. "The *Catharanthus* Alkaloids: Pharmacognosy and Biotechnology." *Current Medicinal Chemistry* 1 (1): 51.

Van Gemerden, B. S., Olff, H., Parren, M. P., and Bongers, F. 2003. "The Pristine Rain Forest? Remnants of Historical Human Impacts on Current Tree Species Composition and Diversity." *Journal of Biogeography* 30 (9): 1381–1390.

Vandebroek, I. 2010. "The Dual Intracultural and Intercultural Relationship between Medicinal Plant Knowledge and Consensus." *Economic Botany* 64 (4): 303–317.

Vandebroek, I., and Balick, M. 2012. "Globalization and Loss of Plant Knowledge: Challenging the Paradigm." *PLOS ONE* 7 (5): e37643. d 0.1371/journal .pone.0037643

———. 2014. "Lime for Chest Congestion, Bitter Orange for Diabeɩes: Foods as Medicines in the Dominican Community in New York City." *Economic Botany* 68 (2): 177–189.

Vandebroek, I., Calewaert, J., De Jonckheere, S., Sanca, S., Semo, L., Van Damme, P., Van Puyvelde, L., and De Kimpe, N. 2004a. "Use of Medicinal Plants and Pharmaceuticals by Indigenous Communities in the Bolivian Andes and Amazon." *Bulletin of the World Health Organization* 82 (4): 243–250.

Vandebroek, I., Van Damme, P., Van Puyvelde, L., Arrazola, S. and De Kimpe, N. 2004b. "A Comparison of Traditional Healers' Medicinal Plant Knowledge in the Bolivian Andes and Amazon." *Social Science & Medicine* 59 (4): 837–849.

Vasconcellos, S. de. 1668. *Notícias Curiosas, e Necessárias das Coisas do Brasil*. Lisbon: João da Costa.

Verger, Pierre. 1952. "Cartas de um Brasileiro Estabelicido no Século XIX na Costa dos Escravos." *Anhembi* 6: 212–253.

———. 1967. *Awon Ewe Osanyin: Yoruba Medicinal Leaves*. Ife, Nigeria: University of Ife.

———. 1976–77. *Use of Plants in Traditional Medicine and Its Linguistic Approach*. Seminar series no. 1, part 1, 242–297. Ife, Nigeria: University of Ife.

———. 1981. *Orixás*. São Paulo: Corrupio.

———. 1995. *Ewé: O Uso das Plantas na Sociedade Iorubá*. São Paulo: Editora Schwarcz.

Vespucci, A. 1992 [1504]. *Letters from a New World: Americo Vespucci's Discovery of America*. Translated by David Jacobson, edited by L. Formisano. New York: Marsilio.

Vibrans, H. 2016. "Ethnobotany of Mexican Weeds." In *Ethnobotany of Mexico: Interaction of People and Plants in Mesoamerica*, edited by R. Lira, A. Casas, and J. Blancas, 287–317. New York: Springer.

Vicente do Salvador. 1931 [1627?]. *História do Brasil: 1500–1627*. 3rd ed. São Paulo, Brasil: Companhia Melhoramentos.

Virey, J. J. 1837. *Natural History of the Negro Race*. Translated by J. H. Guenebault. Charleston, SC: D. J. Dowling.

Voeks, R. A. 1988. "The Brazilian Fiber Belt: Management and Harvest of the Piassava Fiber Palm (*Attalea funifera*)." *Advances in Economic Botany* 6: 262–275.

———. 1993. "African Medicine and Magic in the Americas." *Geographical Review* 83: 66–78.

———. 1995. "Candomblé Ethnobotany: African Medicinal Plant Classification in Brazil." *Journal of Ethnobiology* 15: 257–280.

———. 1996. "Tropical Forest Healers and Habitat Preference." *Economic Botany* 50 (3): 354–373.

———. 1997. *Sacred Leaves of Candomblé: African Magic, Medicine, and Religion in Brazil*. Austin: University of Texas Press.

———. 2004. "Disturbance Pharmacopoeias: Medicine and Myth from the Humid Tropics." *Annals of the Association of American Geographers* 94 (4): 868–888.

———. 2007a. "Are Women Reservoirs of Traditional Plant Knowledge? Gender, Ethnobotany and Globalization in Northeastern Brazil." *Singapore Journal of Tropical Geography* 28 (1): 7–20.

———. 2007b. "Ethnobotanical Knowledge and Mode of Subsistence: Between Foraging and Farming in Northern Borneo." In: *Beyond the Green Myth: Borneo's Hunter-Gatherers in the 21st Century*, edited by P. Sercombe and B. Sellato, 333–352. Copenhagen: Nordic Institute of Asian Studies.

———. 2009. "Traditions in Transition: African Diaspora Ethnobotany in Lowland South America." In *Mobility and Migration in Indigenous Amazonia: Contemporary Ethnoecological Perspectives*, edited by M. Alexiades, 275–294. London: Berghahn.

———. 2010. "Ecotourism and Ethnobotanical Erosion: A Possible Rescue Effect in Brazil's Chapada Diamantina?" In *Recent Developments and Case Studies in Ethnobotany*, edited by U. P. Albuquerque and N. Hanazaki, 228–245. Recife, Brazil: NPPEEA.

———. 2013. "Ethnobotany of Brazil's African Diaspora: The Role of the Columbian Exchange." In *African Ethnobotany in the Americas*, edited by R. Voeks and J. Rashford, 395–416. New York: Springer.

Voeks, R. A., and Leony, A. 2004. "Forgetting the Forest: Assessing Medicinal Plant Erosion in Eastern Brazil." *Economic Botany* 58 (supplement): 294–306.

Voeks, R. A., and Nyawa, S. 2001. "Healing Flora of the Brunei Dusun." *Borneo Research Bulletin* 32: 178–195.

Voeks, R. A., and Sercombe, P. 2000. "The Scope of Hunter-Gatherer Ethnomedicine." *Social Science & Medicine* 51: 679–690.

Von Humboldt, A. 1821. "Account of the Cinchona Forests of South America; Drawn up during Five Years Residence and Travels on the South American Continent." In *An Illustration of the Genus Cinchona; Comprising Descriptions of all the Officinal Peruvian Barks Including Several New Species*, edited by A. B. Lambert, 19–59. London: J. Searle.

———. 1850 [1807]. *Views on Nature, or Contemplations on the Sublime Phenomena of Creation.* Translated by E. C. Otté and Henry Bohn. London: Harrison and Son.

Von Humboldt, A., and Bonpland, A. 1818. *Personal Narrative of Travels to the Equinoctial Regions of the New Continent during the Years 1799–1804.* Vol. 3. Translated by Helen Maria Williams. London: Longman, Hurst, Rees, Orme and Brown.

———. 1827. *Personal Narrative of Travels to the Equinoctial Regions of the New Continent during the Years 1799–1804.* Vol. 5, part. 1. Translated by Helen Maria Williams. London: Longman, Hurst, Rees, Orme and Brown.

———. 1852. *Personal Narrative of Travels to the Equinoctial Regions of the New Continent during the Years 1799–1804.* Vol. 2. Translated by Thomasina Ross. London: Henry G. Bohn.

Von Hutten, U. 1536. *Of the VVood called Guaiacum, that Healeth the Frenche Pockes, and also Healeth the Goute in the Feete, the Stoone, the Palsey, Lepree, Dropsy,*

Fallynge Euyll, and other Diseases. Translated by Thomas Paynell. London: Aedibus Tho. Bertheleti.

Von Martius, K. F. P. 1939. *Natureza, Doenças, Medicina e Remédios dos Indios Brasileiros (1844).* São Paulo, Brasil: Companhia Editora Nacional.

Von Spix, J. B., and Von Martius, C. F. P. 1928 [1824]. *Atraves da Bahia: Excertos da "Obra Reise in Brasilien.* Translated by P. da Silva and P. Wolf. Bahia, Brasil: Imprensa Official.

Wadstrom, C. B. 1795. *An Essay on Colonization, Particularly Applied to the West Coast of Africa, with some Free Thoughts on Cultivation and Commerce.* London: Darton & Harvey.

Waitz, T. 1863. *Introduction to Anthropology.* London: Longman, Green, Longman & Roberts.

Walker, T. D. 2013. "The Medicines Trade in the Portuguese Atlantic World: Acquisition and Dissemination of Healing Knowledge from Brazil (c. 1580–1800). *Social History of Medicine* 26 (3): 403–431.

Wallace, A. R. 1862. "On the Trade of the Eastern Archipelago with New Guinea and its Islands." *Journal of the Royal Geographical Society* 32: 127–137.

Wallis, P. 2012. "Exotic Drugs and English Medicine: England's Drug Trade, c. 1550–c. 1800." *Social History of Medicine* 25: 20–46.

Walsh, V., and Goodman, J. 1999. "Cancer Chemotherapy, Biodiversity, Public and Private Property: The Case of the Anti-cancer Drug Taxol." *Social Science & Medicine* 49: 1215–1225.

Wang, G., Hub, W., Huanga, B., and Qin, L. 2011. "*Illicium verum*: A Review on its Botany, Traditional Use, Chemistry and Pharmacology." *Journal of Ethnopharmacology* 136: 10–20.

Waterman, P. G., and McKey, D. 1989. "Herbivory and Secondary Compounds in Rain-Forest." In *Ecosystems of the World: Tropical Rain Forest Ecosystems*, edited by H. Lieth and M. J. A. Werger, 513–536. Amsterdam: Elsevier.

Watkins, C., and Voeks, R. 2016. "A Mata Transatlântica: Afrodescendentes e Transformação Socio-ambiental no Litoral Baiano, 1500–1888." In *Metamorphoses Florestais: Culturas, Ecologias, e as Transformações Históricas da Mata Atlântica Brasileira*, edited by D. Cabral and A. Bustamante, 150–174. Rio de Janeiro: FAPESP.

Watling, J., Iriarte, J., Mayle, F. E., Schaan, D., Pessenda, L. C., Loader, N.J., Street-Perrott, F. A., Dickau, R. E., Damasceno, A., and Ranzi, A. 2017. "Impact of Pre-Columbian 'Geoglyph' Builders on Amazonian Forests." *Proceedings of the National Academy of Sciences* 114 (8): 1868–1873.

Wayland, C., and Walker, L. S. 2014. "Length of Residence, Age and Patterns of Medicinal Plant Knowledge and Use among Women in the Urban Amazon." *Journal of Ethnobiology and Ethnomedicine* 10 (1): 25.

Wear, A. 2004. "Medicine and Health in the Age of European Colonization." In *The Healing Arts, Health, Disease and Society in Europe, 1500–1800*, edited by P. Elmer, 315–341. Manchester, UK: Manchester University Press.

Weaver, K. K. 2002. "The Enslaved Healers of Eighteenth-Century Saint Domingue." *Bulletin of the History of Medicine* 76: 429–461.

Weese, T. L., and Bohs, L. 2010. "Eggplant Origins: Out of Africa, into the Orient." *Taxon* 59 (1): 49–56.

Weil, A. 1965. "Nutmeg as a Narcotic." *Economic Botany* 19 (3): 194–217.

Weiss, E. A. 1979. "Some Indigenous Plants Used Domestically by East African Coastal Fishermen." *Economic Botany* 33 (1): 35–51.

Welwitsch, F. 1955 [1862]. "Amostras de Drogas Medicinais de Plantas Filamentosas e Tecidas e de Várious Outros objectos Marmente Etnograficos, Coligidos em Angola." In *Colectônia de Escritos Doutrinarios, Floristocos e Fitogeograficos de Frederico Welwitsch*. Compiled by Ascensão Mendonça. 295–333. Lisboa: Agência Geral das Colónias.

Wezel, A., and Bender, S. 2003. "Plant Species Diversity of Homegardens of Cuba and Its Significance for Household Food Supply." *Agroforestry Systems* 57: 39–49.

Whistler, W. A. 2009. *Plants of the Canoe People: An Ethnobotanical Voyage through Polynesia*. Kalaheo, HI: National Tropical Botanical Garden.

Whitehead, N. L. 2000. "Hans Staden and the Cultural Politics of Cannibalism." *Hispanic American Historical Review* 80: 721–751.

Wiessner, P. W. 2014. "Embers of Society: Firelight Talk among the Ju/'hoansi Bushmen." *Proceedings of the National Academy of Sciences* 111 (39): 14027–14035.

Williams, V. L., Victor, J. E., and Crouch, N. R. 2013. "Red Listed Medicinal Plants of South Africa: Status, Trends, and Assessment Challenges." *South African Journal of Botany* 86: 23–35.

Wilson, D., and Wilson, A. 2013. "Figs as a Global Spiritual and Material Resource for Humans." *Human Ecology* 41: 459–464.

Wink, M. 2008. "Ecological Role of Alkaloids: Structure, Isolation, Synthesis and Biology." In *Modern Alkaloids*, edited by E. Fattorusso and O. Taglialatela-Scafati, 3–24. Weinheim, Germany: Wiley-VCH Verlag.

Wink, M., and Schimmer, O. 1999. "Modes of Action of Defensive Secondary Metabolites." In *Functions of Plant Secondary Metabolites and Their Exploitation in Biotechnology*, edited by M. Wink, 17–133. Sheffield, UK: Academic.

Winkelman, M. 2010. "The Shamanic Paradigm: Evidence from Ethnology, Neuropsychology and Ethology." *Time and Mind* 3 (2): 159–181.

WinklerPrins, A. M., and de Souza, P. S. 2005. "Surviving the City: Urban Home Gardens and the Economy of Affection in the Brazilian Amazon." *Journal of Latin American Geography* 4 (1): 107–126.

Winterbottom, A. E. 2014. "Of the China Root: A Case Study of the Early Modern Circulation of *Materia Medica*." *Social History of Medicine* 28: 22–48.

Wiseman, F. M. 1978. "Agricultural and Historical Ecology of the Maya Lowlands." In *Pre-Hispanic Maya Agriculture*, edited by P. D. Harrison and B. L. Turner, 63–116. Albuquerque: University of New Mexico Press,

Withering, W. 1785. *An Account of the Foxglove and Some of its Medical Uses*. London: G. G. J. and G. Robinson.

Wolff, P., and Medin, D. L. 2001. "Measuring the Evolution and Devolution of Folk-Biological Knowledge." In *On Biological Diversity: Linking Language,*

Knowledge and the Environment, edited by L. Maffi, 212–262. Washington, DC: Smithsonian Institution Press.

Woodward, J. 1696. *Brief Instructions for Making Observations in All Parts of the World, as Also for Collecting, Preserving, and Sending Over Natural Things.* London: Richard Wilkin.

Woodward, M. 1927. *Gerard's Herball. The Essence thereof Distilled by Marcus Woodward from the Edition of Th. Johnson, 1636.* London: Spring.

Worth, S. E. 2008. "Storytelling and Narrative Knowledge: An Examination of the Epistemic Benefits of Well Told Stories." *Journal of Aesthetic Education* 42: 42–56.

Zent, S. 2001 "Acculturation and Ethnobotanical Knowledge Loss among the Piaroa of Venezuela: Demonstration of a Quantitative Method for the Empirical Study of Traditional Ecological Knowledge Change." In *On Biocultural Diversity: Linking Language, Knowledge and the Environment,* edited by L. Maffi, 190–211. Washington, DC: Smithsonian Institution Press.

Zimmerer, K. 2006. "Humboldt's Nodes and Modes of Interdisciplinary Environmental Science in the Andean World." *Geographical Review* 96: 335–360.

Zohary, M. 1982. *Plants of the Bible: A Complete Handbook to All Plants.* Cambridge, MA: Cambridge University Press.

Zumbroich, T. 2005. "The Introduction of Nutmeg (*Myristica fragrans* Houtt.) and Cinnamon (*Cinnamomum verum* J. Presl) to America." *Acta Botánica Venezuelica* 28: 1–5.

Zurita-Benavides, M. G., Jarrín, P., and Rios, M. 2016. "Oral History Reveals Landscape Ecology in Ecuadorian Amazonia: Time Categories and Ethnobotany among Waorani People." *Economic Botany* 70 (1) 1–14.

Index

Page numbers in italics refer to illustrations.